MATHEMATICS + PHYSICS

Lectures on Recent Results

Volume 2

Editor: L Streit

World Scientific

Published by

World Scientific Publishing Co Pte Ltd.
P. O. Box 128, Farrer Road, Singapore 9128
242, Cherry Street, Philadelphia PA 19106-1906, USA

Library of Congress Cataloging-in-Publication Data
(Revised for vol. 2)

Mathematics + physics.

   Collection of lectures presented at the Center for Interdisciplinary Research, University of Bielefeld.
   Includes bibliographies.
   1. Mathematical physics. 2. Mathematics.
I. Streit, Ludwig, 1938-   II. Universität Bielefeld. Zentrum für Interdisziplinäre, Forschung.
QC20.5.M36   1985   530.1'5   85-9249
ISBN 9971-966-63-8 (v. 1)
ISBN 9971-966-64-6 (pbk. : v. 1)
ISBN 9971-978-40-7 (v. 2)
ISBN 9971-978-60-1 (pbk. : v. 2)

Copyright © 1986 by World Scientific Publishing Co Pte Ltd.

All rights reserved. This book, or parts thereof, may not be reproduced in any form or by any means, electronic or mechanical, including photocopying, recording or any information storage and retrieval system now known or to be invented, without written permission from the Publisher.

Printed in Singapore by Kim Hup Lee Printing Co Pte Ltd.

74,06
80E

MATHEMATICS + PHYSICS
Lectures on Recent Results

PREFACE
(Preface to Volume 1)

In the past decade the Center for Interdisciplinary Research (ZiF) of Bielefeld University has established its name as an international scientific forum in many fields of academic research. It has done so by hosting hundreds of interdisciplinary workshops and symposia over the years and, most importantly, through Research Years which are devoted in each case to a prescribed interdisciplinary field and which bring together a large number of scientists for extended periods of residence and joint work in Bielefeld.

For mathematics and physics in particular Bielefeld has thus become an established meeting place for joint work and for the exchange of new results and ideas. It need hardly be emphasized how much of their intrinsic growth these two disciplines owe to the dialogue with each other. This has been so since the earliest times and indications are that the dialogue will continue to be necessary and potentially fruitful in the foreseeable future. ZiF has been contributing to this dialogue through a long series of symposia, the "Bielefeld Encounters in Mathematics and Physics" [1]. These meetings have "interpolated" between two Research Years in mathematics and physics, the first one in 1975-76, and "Project No. 2" in the academic year 1983-84.

Project No. 2 has seen about one hundred mathematicians and physicists of various specializations from 25 countries for extended periods of residence, as well as a comparable number of short-term visitors. ZiF Research Years serve the double purpose of furthering scientific production as well as interdisciplinary dialogue. The production aspect of Project No. 2 is documented by a bibliography of some 130 scientific articles written at ZiF under its auspices. Scientific dialogue was served by a large number of mathematics and physics seminars from resident and guest speakers and by a number of workshops and symposia within the framework of Project No. 2 [2]. Apart from these more technical presentations, a number of resident mathematicians and physicists have offered "Lecture Series" of a more expository nature, with the goal of acquainting a larger interdisciplinary audience with their particular field of research. The present book and its companion volume present a collection of these lectures.

Looking at the scope of methods and applications gathered in these two volumes it is instructive to think back for comparison to the first ZiF Research Year in 1975-76. Those years were the harvest time of the constructive quantum field theory program which had developed like an avalanche, but then stopped short of its great goal, the construction of a fundamental theory of matter. At present the interdisciplinary field of mathematics and physics brings in a harvest of a different kind, yet of the same origin: while the first impetus of constructive field theory did not suffice to break a path all the way to a description of particle phenomena, it generated a vast amount of insight and renewed confidence in mathematical methods, from established ones such as analysis, operator and group theory to refined and well-founded numerical computations, with probabilistic methods in a central role. This second wave is one of a large breadth of physical applications and of mathematical research, as is exemplified by the contributions to these volumes. Advances on so many fronts encourage one to think that relativistic quantum field theory may yet, before reaching the age of one hundred, learn to describe the world.

Thanks are due to my friends and colleagues for their advice, and to the staff of ZiF. Everybody at ZiF had adopted Project No. 2 as their own with unlimited amounts of patience and enthusiasm, and all have worked for its success far beyond the call of duty. Preparation of the manuscript has profited from the skills and the help of Ms. A. Roggen and Dr. G. Rupp. It could not have been done so well and so quickly without the clerical and psychological expertise of Ms. L. Jegerlehner whose hard work and gentle insistence are acknowledged with warm gratitude.

[1] "Quantum Dynamics: Models and Mathematics" (L. Streit, ed.), Springer, Vienna (1976)

"Quantum Fields – Algebras, Processes" (L. Streit, ed.), Springer, Vienna (1980)

"Dynamics and Processes" (Ph. Blanchard, L. Streit, eds.), Springer, L. N. M. **1031**, Berlin (1983).

"Trends and Developments in the Eighties" (S. Albeverio, Ph. Blanchard, eds.), World Scientific (1985)

[2] "Resonances – Models and Phenomena" (S. Albeverio, L. S. Ferreira, L. Streit, eds.), Springer L. N. P. **211**, Berlin (1984)

CONTENTS

Preface v

S. ALBEVERIO
Non-Standard Analysis: Applications to Probability Theory and
Mathematical Physics 1

L. ALVAREZ-GAUMÉ
Topology and Anomalies 50

D. BOLLÉ
Sum Rules in Scattering Theory and Applications to Statistical Mechanics 84

J. GINIBRE, G. VELO
Non Linear Evolution Equations. Cauchy Problem and Scattering Theory 154

H. GROSSE
The Inverse Method in Quantum Mechanics and Field Theory 187

G. JONA-LASINIO, G. MARTINELLI, E. SCOPPOLA
Tunneling in One Dimension: General Theory, Instabilities, Rules of
Calculation, Applications 227

D. KASTLER
Geometric Aspects of BRS and Anomalies 261

C. B. LANG
Computer Quantum Field Theory 271

NON-STANDARD ANALYSIS: APPLICATIONS TO PROBABILITY THEORY AND MATHEMATICAL PHYSICS

Sergio ALBEVERIO

Institut für Mathematik
Ruhr-Universität Bochum
Bochum, FRG and

Research Center Bielefeld-Bochum-Stochastics,
Bielefeld University, D-4800 Bielefeld 1, FRG

Abstract

We expose in a rather short way the basic framework of non-standard analysis as needed for elementary and more advanced applications. We indicate some applications in the areas of differential equations, probability theory and mathematical physics.

0. Introduction

Non-standard analysis is an extension of analysis and related domains (functional analysis, measure theory, probability theory, differential geometry, ...) based on the use of a larger number system $^*\mathbb{R}$ (resp. $^*\mathbb{C}$) than $\mathbb{R}$ (resp. $\mathbb{C}$), containing infinitesimal and infinite numbers. It can be looked upon as a justification (and vindication) of "infinitesimal methods". Those methods have a long historical tradition e.g. Euclid, Eudoxus, Archimedes, ..., Nicolaus Cusanus, ..., Cavalieri (who incidentically tried to "win to the cause" Galileo Galilei) up to the fathers of "calculus", Leibniz, Newton (and Seki Kowa in Japan)). Since the advent of the modern use of infinitesimals by "non-standard analysis" (in the late fifties) the interest in the history of "infinitesimal methods" has greatly increased, so we can refer the interested reader to numerous publications discussing history (e.g. [1], [2], [3], [4] and references therein). It suffices here perhaps to recall Euler's extensive use of infinitesimal and infinite numbers (like e.g.

$\sum_{n=0}^{\omega} x^n = \frac{x^{\omega+1} - 1}{x-1}$, $2\sinh x = (1+\frac{x}{\omega})^{\omega} - (1-\frac{x}{\omega})^{\omega}$, for ω infinite) and perhaps Cauchy's position, sort of intermediary between full use of infinitesimals and avoidance of them. Of course with Weierstrass' "ε-δ method" "officially" infinitesimals disappear from the main stream of analysis, but survive very well e.g. in differential geometry, in the study of formal series, in physics, and in heuristic considerations. The breakthrough in a rigorous foundation of infinitesimal methods is mainly due to A. Robinson (see [1]), who used methods of mathematical logic (non-standard models of arithmetic have been introduced already by Skolem in '34, the field $^*\mathbb{R}$ itself was known well before Robinson's work and even a tool of model theory, yielding the "transfer principle", basic for non-standard analysis, was known before Robinson (Łos '55). However, it was Robinson who put all tools together to not only found non-standard analysis but also to pave the way to important concrete applications of it (e.g. Bernstein-Robinson solution of the invariant subspace problem for operator powers which are compact, see [1], [5], [12])).

It should be reminded, however, that a partial solution of the problem of finding a correct frame for infinitesimal methods and a good one for many uses, was found shortly before Robinson by Schmieden and Laugwitz, see [2].

The first applications of non-standard tools have been in algebra (see e.g. [6]), classical analysis (new proofs and new insights into classical results), algebra/topology, functional analysis, see e.g. [1] - [3], [5] - [20], [23], [24], [86], [114], [126], [132].

In 1975 an important tool in measure theory was found ("Loeb measure"), since then applications of non-standard analytic methods in measure theory, probability theory, stochastic analysis, partial differential equations, mathematical physics have been flourishing (we shall give below some references). Also the work by Nelson [26], introducing a new set-theoretic framework, has been most stimulating.

In these lectures (based on and conceived as an invitation to [5]), we shall attempt to give a rather short introduction to the basic tools and methods of non-standard analysis (Sects. I-III), which contains the basic material essentially sufficient for reading advanced papers involving applications of non-standard analysis. In Sect. IV, we shall describe shortly some applications to differential equations and operators as well as measure theory, potential theory, probability theory and mathematical physics. For a more thorough discussion and further applications we refer to [5].

I. The Number Field $^*\mathbb{R}$ of Non-Standard Analysis

We shall see how one can solve, in a simple way, the following problem: Construct an enlargement of $\mathbb{R}$, with the same algebraic and order properties as $\mathbb{R}$, containing infinitesimal and infinite numbers.

Let m be a given 0-1-valued, finitely additive normalized measure defined on all subsets of $\mathbb{N}$ and vanishing on all finite subsets of $\mathbb{N}$, i.e.

$$0 \leq m(A) \leq 1 \quad \forall A \subset \mathbb{N}; \quad m(\emptyset) = 0, \quad m(\mathbb{N}) = 1;$$

$$m\left(\bigcup_{i=1}^{n} A_i\right) = \sum_{i=1}^{n} m(A_i) \quad \text{if} \quad A_i \cap A_j = \emptyset \quad i \neq j, \quad A_i \subset \mathbb{N};$$

$$m(A) = 0 \quad \text{if} \quad A \quad \text{is finite.}$$

Remark: The existence of such m is actually equivalent with the so-called "ultrafilter theorem", which follows from Zorn's lemma (but is actually strictly weaker than Zorn's lemma or, equivalently, Zermelo's axiom of choice: see e.g. [21]).

Definition: Sets $A \subset \mathbb{N}$ such that $m(A) = 0$ will be called small, sets $A \subset \mathbb{N}$ such that $m(A) = 1$ will be called big.

Proposition 1:

a) Finite sets are small

b) **Complements** of finite sets (the so-called <u>cofinite sets</u>) are big.

   c) There are infinite subsets of $\mathbb{N}$ which are small.

<u>Proof</u>:
   a) is part of the definition of m.

   b) follows from $m(A) = m(\mathbb{N}-C) = 0$ if $A = \mathbb{N}-C$, A finite, which together with $m(\mathbb{N}) = 1$ yields $m(C) = 1$.

   c) E.g. either $A = \{2n-1, n \in \mathbb{N}\}$ or $B = \{2n, n \in \mathbb{N}\}$ is small, because if both were big, we would have $m(A) = m(B) = 1$ and then $m(A \cup B) = m(A) + m(B) = 2$, on the other hand, $A \cup B = \mathbb{N}$ hence $m(\mathbb{N}) = 1$, a contradiction.

<u>Remark</u>: Clearly whether $m(A) = 0$ or $m(B) = 0$ in c) depends on m. Call U the family of big subsets of $\mathbb{N}$, i.e. $U = \{A \subset \mathbb{N} \mid m(A) = 1\}$. Then b), c) above can be expressed by $U \supset \text{Cof } \mathbb{N}$, where $\text{Cof } \mathbb{N}$ is the family of all cofinite subsets of $\mathbb{N}$, i.e. $\text{Cof } \mathbb{N} \equiv \{B \subset \mathbb{N} \mid \mathbb{N}-B \text{ is finite}\}$.

<u>Proposition 2</u>: U is an ultrafilter, i.e.

U is a filter: F1) $\phi \notin U$

F2) $\mathbb{N} \in U$

F3) $A \in U, B \supset A \Rightarrow B \in U$

F4) $A, B \in U \Rightarrow A \cap B \in U$

and: UF) U has the maximality property with respect to inclusion or, equivalently, has the property that for <u>any</u> $A \subset \mathbb{N}$ we have either $A \in U$ or $\mathbb{N} - A \in U$ (but not both!).

<u>Proof</u>: The filter properties F1) - F4) follow easily from the properties of m. The specific ultrafilter property UF) follows from the fact that $A \subset \mathbb{N} \Rightarrow$ either $m(A) = 1$ or $m(A) = 0$, since m is 0-1-valued. In the first case we have $A \in U$, by definition of U, in the second case we have $m(\mathbb{N}-A) = 1$, hence $\mathbb{N}-A \in U$.

<u>Proposition 3</u>: There is a 1-1 correspondence between ultrafilters U extending $\text{Cof } \mathbb{N}$ and measures of type m, given by $m(A) = 1 \Leftrightarrow A \in U$.

Proof: The previous Proposition 2 shows that, given m, U defined as $\{A|m(A) = 1\}$ is an ultrafilter. Conversely, if V is an ultrafilter and we define m by $m(A) = 1$ iff $A \in V$, $m(A) = 0$ iff $A \notin V$, then we have $m(\phi) = 0$, $m(\mathbb{N}) = 1$ and $m(A \cup B) = 1$ if $A \in V$ since $A \cup B \supset A$ hence $A \cup B \in V$, hence $m(A \cup B) = m(A) + m(B)$, for $A \cap B = \phi$ (since $A \in V \rightarrow B \notin V$, hence $m(B) = 0$, when $A \cap B = \phi$, otherwise we would have a contradiction to F1), F4)).

Remark: Any 0-1-valued normalized additive measure m vanishing on finite subsets of $\mathbb{N}$ is not σ-additive.

Proof: We have $m(\bigcup_{n=1}^{\infty} \{1,...,n\}) = m(\mathbb{N}) = 1$, but $m(\{1,...,n\}) = 0$ $\forall n \in \mathbb{N}$.

Remark: It is sometimes useful to think of m as a (finitely additive) probability measure and use the corresponding terminology: m-almost surely, m-almost everywhere, with m-probability one ...

As mentioned in the Introduction, we would like to distinguish sequences in $\mathbb{R}$ converging with different speed to the same number, like $\frac{1}{n}$ and $\frac{1}{n^2}$ as $n \to \infty$. The trick is to use equivalence classes modulo an ultrafilter U (like the one associated to the 0-1-measure m we started from).

Let $\mathbb{R}^{\mathbb{N}}$ be the set of all sequences with values in $\mathbb{R}$, i.e. $f \in \mathbb{R}^{\mathbb{N}} \leftrightarrow f$ maps $\mathbb{N}$ into $\mathbb{R}$, i.e. $f(i) \in \mathbb{R}$ $\forall i \in \mathbb{N}$. We introduce the following equivalence $\sim$ between sequences f,g: $f \sim g \leftrightarrow \{i \in \mathbb{N} | f(i) = g(i)\} \in U$,

i.e. $f \sim g$ iff the set of components i s.t. $f(i) = g(i)$ is big, i.e. $f \sim g$ iff $f = g$ (m-)almost surely.

That $\sim$ is an equivalence relation is a consequence of the filter properties (and is very clear for the "almost sure interpretation")!

Let $<f>$ be the equivalence class to the sequence f with respect to $\sim$, i.e. $<f> = <g>$ whenever $f \sim g$. We shall write $<f> = \{f(i), i \in \mathbb{N}\}/U$ (since the equivalence $\sim$ is determined by U). Let $^*\mathbb{R} \equiv \mathbb{R}^{\mathbb{N}}/U$ be the set of all equivalence classes $<f>$. We define of course equality in $^*\mathbb{R}$ by $<f> = <g> \leftrightarrow f \sim g$, and inequality in $^*\mathbb{R}$ by $\langle f \rangle \neq \langle g \rangle \Leftrightarrow f \not\sim g \Leftrightarrow f \neq g$ a.s. (here the ultrafilter property of U is used).

There exists a <u>natural embedding</u> $*$ of $\mathbb{R}$ into $^*\mathbb{R}$:

$$r \in \mathbb{R} \Rightarrow \langle f \rangle = \{f(i) = r \ \forall i \in \mathbb{N}\}/U$$
$$= \{r,r,\ldots,r,\ldots\}/U \ .$$

We shall also denote this sequence $\langle f \rangle$, i.e. $\{r,r,\ldots,r,\ldots\}/U$ modulo U, by $^*r$ or $\langle r \rangle$.

We shall call $^\sigma\mathbb{R}$ the set of all $^*r$ with $r \in \mathbb{R}$.

We have $^\sigma\mathbb{R} \subset {^*\mathbb{R}}$ (we shall see below $^\sigma\mathbb{R} \subsetneq {^*\mathbb{R}}$).

$\mathbb{R}$ is a field with operations $+,\cdot$, and neutral elements 0 resp. 1. We want $^*\mathbb{R}$ to be a field, too. Therefore, we must introduce in $^*\mathbb{R}$ the operations addition and multiplication and corresponding neutral elements.

It is practical to denote addition in $^*\mathbb{R}$ by $+$ again and multiplication by $\cdot$, as in $\mathbb{R}$. We define $+$ in $^*\mathbb{R}$ by

$\langle h \rangle = \langle f \rangle + \langle g \rangle \Leftrightarrow h = f + g$ a.s., and define $\cdot$ in $^*\mathbb{R}$ by

$\langle h \rangle = \langle f \rangle \cdot \langle g \rangle \Leftrightarrow h = f \cdot g$ a.s. (often later on $\cdot$ will be dropped, i.e. $\langle f \rangle\langle g \rangle$ stands then for $\langle f \rangle \cdot \langle g \rangle$).

By the filter property, these definitions are independent of the chosen representatives.

The neutral element for the sum in $^*\mathbb{R}$ is ^{*0}, the neutral element for the multiplication in $^*\mathbb{R}$ is ^{*1}, as easily verified.

There are no zero-divisors, i.e.

$$\langle f \rangle\langle g \rangle = {^*0} \Leftrightarrow \langle f \rangle = {^*0} \text{ or } \langle g \rangle = {^*0} \ .$$

In fact, if $\langle f \rangle \neq {^*0}$, $\langle g \rangle \neq {^*0}$ then $A \equiv \{i | f(i) \neq 0\}$ and $B \equiv \{i | g(i) \neq 0\}$ are big, so $C \equiv \{i | f(i) g(i) \neq 0\} \supset A \cap B$ is big (by the filter properties F3), F4)), hence defining $\langle h \rangle$ by $h(i) = f(i) g(i)$ for $i \in C$, $h(i) = 0$ for $i \in \mathbb{N} - C$, we have $\langle h \rangle = \langle f \rangle\langle g \rangle \neq {^*0}$. Thus $\langle f \rangle\langle g \rangle = {^*0} \Rightarrow \langle f \rangle = {^*0}$ or $\langle g \rangle = {^*0}$.

Conversely, if $\langle f \rangle = {^*0}$ then if $\langle h \rangle \equiv \langle f \rangle \cdot \langle g \rangle$ we have for a big set A of i's: $h(i) = f(i)g(i)$. But then, with $B \equiv \{i | f(i) = 0\}$

(a big set), we have $A \cap B$ big, and $h(i) = 0$ for i in this big set, hence $<f><g> = {}^*0$. Similarly for $<g> = {}^*0$.

Example: $<f> = \{1,0,3,0,\ldots,0,2n+1,0,\ldots\}/U$

$<g> = \{0,2,0,4,\ 0,2n,0,\ldots\}/U$.

Then $<f> \cdot <g> = {}^*0$, and indeed either $<f> = {}^*0$ or $<g> = {}^*0$, which corresponds to the cases $A = \{2n-1,\ n \in \mathbb{N}\} \notin U$ resp. $\mathbb{N} - A = \{2n, n \in \mathbb{N}\} \notin U$, since in the first case $f(i) \ne 0$ only on A, hence $<f> = {}^*0$, A being small, in the second case $g(i) \ne 0$ only on $\mathbb{N} - A$, hence $<g> = {}^*0$, $\mathbb{N} - A$ being small. Incidentally, this example illustrates well the role of the basic ultrafilter in excluding zero divisors.

How does the inverse $<f>^{-1}$ of $<f> \ne {}^*0$ in ${}^*\mathbb{R}$ look like? We can define $<f>^{-1}$ as $<g>$, with $g(i) = 1/f(i)$ for those i for which $f(i) \ne 0$ (there is a big set of such i's), and $g(i) = 1$ (e.g.) for those i for which $f(i) = 0$.
Then we are sure that $f(i)\,g(i) = 1$ on a big set of i's, hence indeed $<f><g> = <g><f> = {}^*1$, thus $<g> = <f>^{-1}$.
In this way then we see that ${}^*\mathbb{R}$ is made into a field of numbers. What about a total order in ${}^*\mathbb{R}$ (recall that $\mathbb{R}$ is totally ordered)?

We <u>define</u> for any $<f>, <g> \in {}^*\mathbb{R}$:

$<f><<g>$ iff $f(i) < g(i)$ for a.e. i

$<f>><g>$ iff $f(i) > g(i)$ for a.e. i.

We then have the

<u>Proposition 4</u>: ${}^*\mathbb{R}$ is totally ordered with respect to $<$ (i.e. for any given $<f> \ne <g>$ either $<f><<g>$ or $<g><<f>$ and $<f><<g>, <g><<h>$ implies $<f><<h>$).

Proof: By the ultrafilter property of U we have that either is the set $\{i | f(i) < g(i)\}$ in U or its complement in $\mathbb{N}$, $\{i | f(i) \geq g(i)\}$, is in U. Hence either $<f><<g>$ or $<f> \geq <g>$.

Example: Is $(-1,2,-1,2,\ldots)/U < (-3,3,-3,3,\ldots)/U$?

Answer: Yes if $\{2,4,6,\ldots\} \in U$
no if $\{1,3,5,\ldots\} \in U$

(exactly one of the two cases occurs, by the ultrafilter property of U).

Remark: The embedding $*$ of $\mathbb{R}$ in $*\mathbb{R}$ (through $^\sigma\mathbb{R} \subset {}^*\mathbb{R}$) preserves the order, i.e. $r < s$ in $\mathbb{R}$ implies $*r < *s$. (Since $*r = \{r,r,\ldots r\ldots\}/U$ and $r < s$ implies all components of $*r$ are less than those of $*s$).

What have we gained in going from $\mathbb{R}$ to $*\mathbb{R}$?

Theorem 5: $*\mathbb{R}$ is a proper extension of $^\sigma\mathbb{R}$ (hence of $\mathbb{R}$ as embedded in $*\mathbb{R}$), i.e. $*\mathbb{R} \supsetneq {}^\sigma\mathbb{R}$. In fact, $*\mathbb{R}$ contains numbers which are strictly larger than any number in $^\sigma\mathbb{R}$ (i.e. than any real number).

Proof: Let us consider $<f> \equiv \{1,2,\ldots n,\ldots\}/U$.

We have by definition $<f> \in {}^*\mathbb{R}$, moreover $<f> > {}^*r \; \forall r \in \mathbb{R}$, since $\{i \in \mathbb{N} \,|\, f(i) > r\} = \{i \in \mathbb{N} \,|\, i > r\}$ is big (being a cofinite set, containing all numbers $r+1, r+2, \ldots$).

Thus $<f> \notin {}^\sigma\mathbb{R} = \{*r, r \in \mathbb{R}\}$: qed.

It will be convenient, from now on, to use a simpler notation for elements in $*\mathbb{R}$, namely to denote elements of $*\mathbb{R}$ by letters. Let ω be the element $<f> = \{1,2,\ldots,n\ldots\}/U$ occurring in the above proof. We have seen above that $\omega > {}^*r \; \forall r \in \mathbb{R}$. Using the algebra and order in $*\mathbb{R}$ we can then produce many other elements in $*\mathbb{R}$ which are not in $^\sigma\mathbb{R}$, e.g. $\omega^2 > {}^*r$, in fact $\omega^2 > \omega$. Moreover, $n\omega > \omega, \forall n \in \mathbb{N}, n > 1$, $\omega + k > \omega \; \forall k \in \mathbb{R}_+$, etc. Moreover, we can produce easily elements in $*\mathbb{R}$ which are smaller than any $*r, r > 0, r \in \mathbb{R}$: e.g. $\omega^{-1} = \{1, \frac{1}{2}, \ldots, \frac{1}{n}, \ldots\}/U$.

We shall need some simple definitions to classify these numbers. For $s \in {}^*\mathbb{R}$ we set $|s| = s$ if $s \geq {}^*0$, $|s| = -s$ if $s \leq {}^*0$. $s \in {}^*\mathbb{R}$ is called <u>positive</u> (or nonnegative) if $s \geq {}^*0$, <u>negative</u> (or nonpositive) if $s \leq {}^*0$.

__Definition__ $s \in {}^*\mathbb{R}$ is called __infinite__ iff $|s| > {}^*n \,\forall n \in \mathbb{N}$, __finite__ iff $|s| < {}^*n$ for some $n \in \mathbb{N}$. $s \in {}^*\mathbb{R}$ is called __infinitesimal__ iff either $s = {}^*0$ or $s = \omega^{-1}$ for some infinite ω.

__Remark:__ We have s infinitesimal iff $|s| < {}^*r \,\forall r > 0, r \in \mathbb{R}$. Thus for $\omega = \{1,2,\ldots,n,\ldots\}/U$ we have ω^{-1} infinitesimal, but also e.g. $(\omega^{-2}) < \omega^{-1}$ infinitesimal, $r\omega^{-1}$ infinitesimal for all $r \in {}^\sigma\mathbb{R}$, $r \neq 0$.

__Definition:__ $s \in {}^*\mathbb{R}$ is called __standard__ (real) iff $s \in {}^\sigma\mathbb{R}$ (i.e. if s is a real number as embedded in ${}^*\mathbb{R}$; i.e. the components of s in a big set are all identical), $s \in {}^*\mathbb{R}$ is called __non-standard__ iff $s \in {}^*\mathbb{R} - {}^\sigma\mathbb{R}$.

All the numbers in ${}^*\mathbb{R}$ (standard and non-standard) are called __hyperreal__ (and ${}^*\mathbb{R}$ is the field of hyperreal numbers).

__Notation:__ It is often convenient in the following to drop the distinction between $\mathbb{R}$ and ${}^\sigma\mathbb{R}$ i.e. to think $\mathbb{R}$ as embedded in ${}^*\mathbb{R}$. In the same spirit one writes then r instead of ${}^*r$ (in particular 0 instead of ^{*0}, 1 instead of ^{*1}).

We shall now exploit infinitesimals to introduce a "nearness" concept (a substitute of topology!) in ${}^*\mathbb{R}$.

__Definition:__ For any given number $s \in {}^*\mathbb{R}$ one defines __the monad__ $\mu(s)$ of s as follows:

$$\mu(s) \equiv \{t \in {}^*\mathbb{R} \mid \exists \text{ infinitesimal } \varepsilon \text{ such that } t = s + \varepsilon\}.$$

I.e. $\mu(s)$ consists of s and all numbers in ${}^*\mathbb{R}$ which differ from s only by a (positive or negative) infinitesimal (intuitively we can think of $\mu(s)$ as an "infinitesimal cloud" around s).

E.g. the monad $\mu({}^*0)$ of ^{*0} is the set of all infinitesimals.

__Definition:__ For any two numbers $s_1, s_2 \in {}^*\mathbb{R}$ we say that s_1 is "(infinitely) close to s_2" or "(infinitely) near to s_2" iff $s_1 - s_2$ is an infinitesimal, i.e. the monads of s_1 and s_2 coincide. We then write $s_1 \approx s_2$.

E.g. for ε infinitesimal, we have $\varepsilon \approx {}^*0$. The monad $\mu(s)$ of s

can be written as $\mu(s) = \{t \in {}^*\mathbb{R} \mid t \approx s\}$.

A little less frequent, but also pittoresque, is the use of the concept of <u>galaxy</u> Gal(s) of a number $s \in {}^*\mathbb{R}$: $\text{Gal}(s) \equiv \{t \in {}^*\mathbb{R} \mid t-s \text{ finite}\}$. E.g. if ε is a positive infinitesimal, $\varepsilon \neq 0$: $\text{Gal}(\varepsilon^{-1}) = \{\varepsilon^{-1} + s\}, \forall s \in {}^*\mathbb{R}$, s finite.

Intuitively, the finite numbers in ${}^*\mathbb{R}$ should be near to standard numbers, i.e. we should be able to extract from them a "standard part". This is the content of the following theorem, in which we denote by F the finite numbers of ${}^*\mathbb{R}$.

<u>Theorem 6</u>: a) There exists an uniquely defined map st (also denoted by °) from F into $\mathbb{R}$ such that for any $x \in F$ we have x-stx is infinitesimal (i.e. $x \approx \text{st} x$ i.e. $x \in \mu(\text{st } x)$). st is a surjective homomorphism (relative to the field operations), in particular, st $(x+y)$ = st x + st y, st (xy) = st x st y. st x is the least upper bound in $\mathbb{R}$ of $\{r \in \mathbb{R} \mid {}^*r < x\}$.

b) ${}^\sigma\mathbb{R}$ is isomorphic (hence can be identified) (as an ordered field) with the quotient F/I of F modulo the ideal I of infinitesimal numbers in F. (In short: real numbers are finite numbers modulo infinitesimals).

<u>Proof</u>: a) Uniqueness is clear. suppose $r, s \in \mathbb{R}$ s.t. $r \approx x$, $s \approx x$, then $r-s \approx 0$, but r, s being real this implies $r = s$.
As to the existence, let $x \in F$ and set $S_x \equiv \{r \in \mathbb{R} \mid {}^*r < x\}$. Then $S_x \neq \emptyset$, since $|x| < {}^*s$ for some $s \in \mathbb{R}$, x being finite, hence $-s \in S_x$. s is an upper bound of S_x, and by the Dedekind completeness of $\mathbb{R}$, S_x has then a least upper bound t in $\mathbb{R}$. Then $x \leq t + r$ $\forall$ real $r > 0$, thus $x - t \leq r$. On the other hand, $t - r \leq x$ (otherwise $t-r$ would be an upper bound to S_x, smaller than t!). Hence $-r \leq x - t \leq r \ \forall r \in \mathbb{R}, r > 0$ hence $x - t \approx 0$.
This proves a).

b) is proven using that I is a maximal ideal in F and F/I is a totally ordered, archimedean field, thus isomorphic to a subfield of $\mathbb{R}$. Since it contains however ${}^\sigma\mathbb{R}$ it must be isomorphic to $\mathbb{R}$ (for details see [8], [12], [22]).

Remark: Clearly, we have at disposal an "algebraic" way of expressing "near", rather than a topological one. $^*\mathbb{R}$ can be made into a normal Hausdorff topological space in an interval topology, but it is disconnected and the embedding of $\mathbb{R}$ into $^*\mathbb{R}$ is not topological (in fact, the topology induced on $\mathbb{R}$ by the interval topology in $^*\mathbb{R}$ is the discrete one). It is better altogether, at least at the beginning, not to think at all topologically when handling with $^*\mathbb{R}$ (those interested in these aspects might look at [20], [22], e.g.).

II. Basic Structures for Analysis over $^*\mathbb{R}$

For analysis we need functions, spaces of functions, etc., not only a field. The first question we might like to answer is: Let F be a function from $\mathbb{R}$ into $\mathbb{R}$. How can we extend it to a function $^*F$ from $^*\mathbb{R}$ into $^*\mathbb{R}$?

The following definition is quite natural:

Definition: $^*F(<f>) \equiv <g> \Leftrightarrow$
$\qquad F(f) = g$ a.s.

i.e. F is extended to $^*F$ "componentwise" in the sense that F evaluated at the sequence $<f>$ is the sequence $<g>$ (modulo the ultrafilter U, of course) iff $F(f(i)) = g(i)$ for a big set of components i.

E.g. $^*\sin x = y \Leftrightarrow \sin x_i = y_i$ for a big set of i, where $x = \{x_i, i \in \mathbb{N}\}/U$, $y = \{y_i, i \in \mathbb{N}\}/U$.

Of course functions $^*F$ of the above form are very special among the set of all functions from $^*\mathbb{R}$ into $^*\mathbb{R}$. Below we shall discuss more general ones. Let us first remark that in the same way as we extended unary functions $F(x)$, $x \in \mathbb{R}$ to $^*F(x)$, $x \in {}^*\mathbb{R}$ we can extend n-ary functions $F(x_1,\ldots,x_n)$ (and relations).

We also can extend sets, and this will be important in the following.

Definition: To any subset E of $\mathbb{R}$ we associate a subset $^*E$ of

$^*\mathbb{R}$ defined as $^*E = \{<f> \in {}^*\mathbb{R} \mid f(i) \in E$ for a big set of $i\}$. Thus $^*E$ consists of numbers for which a big set of components are in E.

Remark: We immediately realize that $^*\mathbb{R}$ as defined in Section I is indeed the * of the set $\mathbb{R}$, since all numbers in $^*\mathbb{R}$ have components in $\mathbb{R}$.

For any $E \subset \mathbb{R}$ let us set $^\sigma E = \{{}^*r \mid r \in E\}$. Then $^\sigma E$ coincides, for $E = \mathbb{R}$, with $^\sigma\mathbb{R}$ as defined in Section I. Similarly as we proved that $^\sigma\mathbb{R} \subsetneq {}^*\mathbb{R}$ we can show $^\sigma E \subset {}^*E$ and, see e.g. [5], [12], [16], [18], that $^\sigma E = {}^*E$ iff E is finite.

E.g. for $E = [0,1]$ we have $^\sigma E = $ standard numbers in $[0,1]$, whereas $^*E = \{x \in {}^*\mathbb{R} \mid 0 \leq x \leq 1\}$ contains all positive infinitesimals, the monads of all real in $(0,1)$ and the part of the monad of ^{*1} consisting of numbers less or equal ^{*1}. In fact, by the definition of $^*E$ we have $x \in {}^*E \Leftrightarrow x(i) \in E$ for a big set of i, and this by the definition of E means $0 \leq x(i) \leq 1$ for a big set of i. ε positive infinitesimal means that its i-th component $\varepsilon(i)$ is, for a.e. i, $< \frac{1}{n}$ for each given $n \in \mathbb{N}$, hence $\varepsilon \in {}^*E$, etc.

It is easy to verify that $*$ is a homomorphism of Boolean algebras (i.e. $^*(A \cap B) = {}^*A \cap {}^*B$, etc.) and is injective (i.e. $^*A = {}^*B$ iff $A = B$). It might help the intuition to realize that the following are true and easily verified (cfr. [5]):

$E \subset \mathbb{R}$ open $\Leftrightarrow$ all monads of numbers in E are in $^*E$

$E \subset \mathbb{R}$ closed $\Leftrightarrow$ all finite x in $^*E$ have standard parts in E

$E \subset \mathbb{R}$ compact $\Leftrightarrow$ all x in $^*E$ are finite and their standard parts are in E.

Particular sets of the form $^*E$ are the "hyperfinite integers" $^*\mathbb{N}$. The numbers in $^*\mathbb{N} - \mathbb{N}$ are infinite and are called infinite (hyperfinite) integers.

E.g. the number $\omega = \{1,2,\ldots n,\ldots\}/U$ introduced in Section 1 is an infinite hyperfinite integer. Also $\omega - 1 \in {}^*\mathbb{N} - \mathbb{N}$, in fact $\omega - k \in {}^*\mathbb{N} - \mathbb{N} \;\forall\; k \in \mathbb{N}$.

We summarize what we have achieved up to now, in going from $\mathbb{R}$ to $^*\mathbb{R}$ and extending sets and functions by $*$ in the following "transfer principle", stated informally:

"Any statement φ involving $+,\cdot,0,1,=,<,$ real functions, real subsets and real numbers, and the logical and set theoretical operations "=", "$\in$", "and", "or", "not", "$\Rightarrow$", $\forall x$, $\exists y$ for x,y real variables is true in $\mathbb{R}$ iff φ taken with $*$ everywhere is true in $^*\mathbb{R}$."

We have given this statement informally, it is however easy to provide a precise formulation using a bit of terminology from elementary logic, see e.g. [5], [12], [14].

Remark: Since we have been using a model for $^*\mathbb{R}$ the transfer principle is here an (easy) theorem. In an axiomatic approach it is a postulate, see e.g. [16].

Examples: $\sin^2 x + \cos^2 x = 1$ is true for all $x \in \mathbb{R}$, hence is true "by transfer" for all $x \in {}^*\mathbb{R}$ (with $\sin^2$ replaced by $^*\sin^2$, $\cos^2$ by $^*\cos^2$, 1 by ^{*1}, and $=$ by the corresponding operations in $^*\mathbb{R}$).

As an application of the principle we might discuss continuity properties of functions.

Proposition 7: Let f be a real function on $\mathbb{R}$. f is continuous at $c \in \mathbb{R} \Leftrightarrow {}^*f(x) \approx {}^*f({}^*c)$ for all $x \in {}^*\mathbb{R}$ s.t. $x \approx {}^*c$.

Proof: $\Rightarrow$: From the assumption we have that for each $n \in \mathbb{N}$ there exists $\delta_n > 0$ such that for all $x \in \mathbb{R}$, $|x-c| < \delta_n \Rightarrow |f(x)-f(c)|<1/n$. Then by transfer for each $n \in {}^*\mathbb{N}$ $\exists \delta_n > {}^*0$ such that for all $x \in {}^*\mathbb{R}$, $|x - {}^*c| < {}^*\delta_n \Rightarrow |{}^*f(x) - {}^*f({}^*c)| < {}^*1/n$. If now $x \approx {}^*c$ for $x \in {}^*\mathbb{R}$ then $|x - {}^*c| < {}^*\delta_n$, hence $|{}^*f(x) - {}^*f({}^*c)| < {}^*1/n$, thus $^*f(x) \approx {}^*f({}^*c)$.

$\Leftarrow$: By assumption there exist $\varepsilon, \delta > 0$ infinitesimal such that whenever $|x - {}^*c| < \delta$ then $|{}^*f(x) - {}^*f({}^*c)| < \varepsilon$. By transfer for given $\varepsilon > 0$, $\varepsilon \in \mathbb{R}$ there exists $\delta > 0$, $\delta \in \mathbb{R}$ such that $|x-c| < \delta \Rightarrow |f(x) - f(c)| < \varepsilon$, hence f is continuous. ∎

What is then the characterization of uniform continuity? (a quite important problem within the old "infinitesimal methods": recall the historical difficulties, the "interactions" between Abel and Cauchy, see e.g. [4]).

Proposition 8: f is uniformly continuous on an interval I of $\mathbb{R} \Leftrightarrow {}^*f(x) \approx {}^*f(y)$ for all $x,y \in {}^*I$ such that $x \approx y$.

Proof: An immediate consequence of transfer.

Application: $f(x) = x^{-1}$ in $(0,1)$ is <u>not</u> uniformly continuous, since for $x = \omega^{-1} \in {}^*\mathbb{N} - \mathbb{N}$ we have $x^2 \approx x \approx 0$ but ${}^*f(x) = \omega \not\approx {}^*f(x^2) = \omega^2$!

As another exercise in transfer we might mention the following nonstandard expression of convergence for sequences:

s_n a real sequence, r a real number:

$$s_n \to r \text{ as } n \to \infty \Leftrightarrow \text{st}\,{}^*s_\omega = {}^*r \; \forall \, \omega \in {}^*\mathbb{N} - \mathbb{N},$$

where ${}^*s_\omega$ is the extension of the function (sequence) $s: \mathbb{N} \to \mathbb{R}$ to ${}^*\mathbb{N} \to {}^*\mathbb{R}$.

Remark: One often uses the convention to drop $*$ on functions ${}^*F$ which are extensions of functions F on $\mathbb{R}$, e.g. to use the symbol exp for the real-valued exponential function on $\mathbb{R}$ as well as its extension from ${}^*\mathbb{R}$ to ${}^*\mathbb{R}$.

Remark: In the above framework much of classical analysis can be rewritten and reinterpreted using non-standard analysis, see e.g. [14], [15],[24][1]) regaining in particular the original intuitive insight which was at the basis of the historical "method of infinitesimals". However, for more advanced applications, like in the theory of stochastic processes and applications to quantum (field) theory also spaces of functions, measures, etc. should be extended. This will be done in the next section.

III. Superstructure, Internal Quantities

We first need a (standard) set-theoretic concept. Let X be any non-void set (e.g. $X = \mathbb{R}$).

<u>Definition</u>: The superstructure $V(X)$ over X is by definition $V(X) = \bigcup_n V_n(X)$ with $V_n(X)$ defined recursively by $V_0(X) \equiv X$, $V_{n+1}(X) \equiv V_n(X) \cup P(V_n(X))$, where P of a set S means the power set of S, i.e. the family of all subsets of S.

<u>Remark</u>: For those not used to set theoretical terminology this construction might appear not so intuitive. However, it's easy to convince ourselves that in the case $X = \mathbb{R}$, $V(X)$ contains "all is needed for analysis (and functional analysis, measure theory, probability ...)." In fact, functions on $\mathbb{R}$ with values in $\mathbb{R}$ can be looked upon as graphs, i.e. as subsets of $\mathbb{R}^2$. $\mathbb{R}^2$ itself can be looked upon as a set of ordered pairs and an ordered pair $<x,y>$ can be looked upon as a subset of $P(\mathbb{R})$, namely the one with elements $\{x\}$ and $\{x,y\}$ (since $<x_1,y_1> = <x_2,y_2> \Leftrightarrow x_1 = x_2, y_1 = y_2 \Leftrightarrow \{x_1\} = \{x_2\}$ and $\{x_1,y_1\} = \{x_2,y_2\}$).

Thus a function can be looked as a subset of $V_2(\mathbb{R})$, i.e. as an element of $V_3(\mathbb{R})$. A space of functions like $C_0^\infty(\mathbb{R})$ can thus be looked upon as a subset of $V_3(\mathbb{R})$, i.e. as an element of $V_4(\mathbb{R})$.

In a similar way it is not difficult to identify the space of distributions $\mathcal{D}'(\mathbb{R})$ as an element in $V(\mathbb{R})$, etc.

So we are going to use $V(\mathbb{R})$ as the framework for "all analysis". According to the general definition, we might also consider the superstructure $V(^*\mathbb{R})$. Do we have set theoretic notions like "$=$" and "$\in$" in $V(^*\mathbb{R})$ so that statements true in $V(\mathbb{R})$ go over into statements true in $V(^*\mathbb{R})$?

Some caution is obviously needed, e.g. every set $A \subset \mathbb{R}$ with an upper bound in $\mathbb{R}$ has a last upper bound in $\mathbb{R}$, but that is <u>not</u> true for all subsets of $^*\mathbb{R}$, indeed the set of all infinitesimals has an upper bound but no least upper bound.

What is going on here? For which elements in $V(^*\mathbb{R})$ go properties over, i.e. do we have transfer? We shall see below that this holds only for the so-called <u>internal</u> elements, which are elements of a suitably defined $^*V(\mathbb{R})$.

Technically we have to introduce *-entities (somewhat similarly as we already did for subsets E of $\mathbb{R}$, when we defined $^*E$, in particular $^*\mathbb{R}$), in such a way that they and their elements (like $^*E$ and their elements, which are hyperreal numbers) behave (with respect to equality, belonging and "logics") "like quantities in the standard superstructure $V(\mathbb{R})$".
Of course the *-entities we already defined, like $^*r$ for $r \in \mathbb{R}$, $^*E$ for $E \subset \mathbb{R}$, $^*F$ for F an n-ary real-valued function, should be special cases of the newly defined *-entities.

The general construction of *-entities is somewhat technical and not so intuitive at first glance (except for those familiar with methods of model theory). The main justification is that it works and suffices for all applications. In analogy with the construction of $^*\mathbb{R}$ as $\mathbb{R}^{\mathbb{N}}/U$ we define first $V(\mathbb{R})^{\mathbb{N}}$ as the set of all sequences with values in $V(\mathbb{R})$ (this is the analogue of $\mathbb{R}^{\mathbb{N}}$) and introduce equivalence classes modulo U (the analogue of $\mathbb{R}^{\mathbb{N}}/U$): for $f,g \in V(\mathbb{R})^{\mathbb{N}}$ define $f \sim g \Leftrightarrow f(i) = g(i)$ for a big set of i, where "=" is the (set theoretical) equality in $V(\mathbb{R})$. By the ultrafilter properties "$\sim$" is an equivalence relation and we denote by $<f>$ the equivalence class with representative f.

Let us call $V(\mathbb{R})^{\mathbb{N}}/U$ the family of all such equivalence classes.

When we were dealing with $\mathbb{R}$ instead of $V(\mathbb{R})$ we saw that $\mathbb{R}^{\mathbb{N}}/U$ was $^*\mathbb{R}$. It is natural to try to relate now $V(\mathbb{R})^{\mathbb{N}}/U$ with $V(^*\mathbb{R})$ (the superstructure built about $^*\mathbb{R}$).

Moreover, we would like to embed $V(\mathbb{R})$ into $V(^*\mathbb{R})$, as we did before with $\mathbb{R}$ in $^*\mathbb{R}$.

Due to the complications of superstructures, which include both individual (like $r \in \mathbb{R}$, which are <u>not</u> sets) and sets, and in which e.g.

both $V_n(\mathbb{R}) \subset V_{n+1}(\mathbb{R})$ and $V_n(\mathbb{R}) \in V_{n+1}(\mathbb{R})$ hold, we have to take care in defining the quantities in $V({}^*\mathbb{R})$ "coming from $V(\mathbb{R})^{\mathbb{N}}/U$" by specifying their elements.

It turns out that this can be done by considering only the part $\pi_U V(\mathbb{R}) \equiv \bigcup_n \pi_U V_n(\mathbb{R})$ of $V(\mathbb{R})^{\mathbb{N}}/U$, where $\pi_U V_n(\mathbb{R}) = \{<f> \in V(\mathbb{R})^{\mathbb{N}}/U | f(i) \in V_n(\mathbb{R}) \text{ for a large set of } i\}$. $\pi_U V(\mathbb{R})$ is the so-called "<u>bounded ultrapower over $V(\mathbb{R})$</u>".

We can define equality $=_U$ (modulo U) and belonging $\in_U$ in $\pi_U V(\mathbb{R})$ componentwise:

$<f> =_U <g> \leftrightarrow \{i | f(i) = g(i)\}$ is big

$<f> \in_U <g> \leftrightarrow \{i | f(i) \in g(i)\}$ is big.

These definitions are independent of representatives, by the ultrafilter properties.

We have $\pi_U V_0(\mathbb{R}) = {}^*\mathbb{R}$.

Similarly as for the embedding of $\mathbb{R}$ into ${}^*\mathbb{R} = \mathbb{R}^{\mathbb{N}}/U$ in Section I there is a natural embedding i of $V(\mathbb{R})$ in $\pi_U V(\mathbb{R}) \subset V(\mathbb{R})^{\mathbb{N}}/U$, by:

$i(x) \equiv \{x,x,\ldots,x,\ldots\}/U$ for all $x \in V(\mathbb{R})$.

We remark that for $x = r \in \mathbb{R}$ we have $i(r) = {}^*r$, hence $i = *$ on $V_0(\mathbb{R}) = \mathbb{R}$. In particular, we already see that i restricted to $\mathbb{R}$ is not surjective, in fact, its image is ${}^\sigma\mathbb{R} \subsetneq {}^*\mathbb{R}$ (it is easy to convince oneself that also $i \upharpoonright V_n(\mathbb{R})$ is not surjective).

For any $E \subset \mathbb{R}$, we have $r = \{r_1,\ldots,r_j,\ldots\}/U \in_U (E,\ldots,E,\ldots)/U = i(E)$ (by definition of i), whenever $r_j \in E$ for a large set of j. On the other hand, in Section I we defined ${}^*E = \{<f> | f(j) \in E$ for a bit set of $j\}$. Thus $i(E) = {}^*E$. As remarked before, ${}^*E \neq \{{}^*r, r \in E\} = \{i(r), r \in E\}$, thus $i(E) \neq \{i(r), r \in E\}$, as sets (note, however, that $=_U$ is not defined on $\{i(r), r \in E\}$, which is not in $V(\mathbb{R})^{\mathbb{N}}/U$). We see here an instance that one has to be careful about specifying which elements belong to a given set.

We have now to indicate which quantities of $V(^*\mathbb{R})$ "come from $V(\mathbb{R})^{\mathbb{N}}/U$." To do this, one introduces the so-called "Mostowski collapse function" j mapping $\pi_U V(\mathbb{R})$ injectively into $V(^*\mathbb{R})$ and defined inductively and "gradewise" as follows:

1) j is the identity on $\pi_U V_0(\mathbb{R}) = {}^*\mathbb{R}$ (it is easily verified that $^*\mathbb{R}$ has the correct elements in the sense that $<f> \in_U {}^*\mathbb{R} \Leftrightarrow f(i) = r \in \mathbb{R}$ for a big set of $i \Leftrightarrow <f> = \{r,r,\ldots,r\ldots\}/U \in {}^*\mathbb{R}$);

2) if j is already defined on $\pi_U V_n(\mathbb{R})$ for $n \geq 0$, then on $\pi_U V_{n+1}(\mathbb{R}) - \pi_U V_n(\mathbb{R})$ j shall be defined by:

$$j(<f>) = \{j(<g>) \mid <g> \in_U <f>\}$$

(in this way by construction the elements of $j(<f>)$ will be precisely the j-images of the $\in_U$-elements of $<f>$).

Remark: $<g> \in_U <f>$ is equivalent with $\{i \mid g(i) \in f(i)\}$ big. $<f> \in \pi_U V_{n+1}(\mathbb{R})$ and $<g> \in_U <f>$ imply $<g> \in \pi_U V_n(\mathbb{R})$, by the fact that $V_{n+1}(\mathbb{R}) = V_n(\mathbb{R}) \cup P(V_n(\mathbb{R}))$.

It is not difficult to verify, using the properties of the ultrafilter U, that j is well-defined by 1), 2) and j maps $\pi_U V_n(\mathbb{R})$ into $V_n(^*\mathbb{R})$ (the latter can be proven by induction).

We are now at the end of our construction of "good sets" in $V(^*\mathbb{R})$.

We shall define $\alpha \equiv j \circ i$. Then α is an embedding (injective) of $V(\mathbb{R})$ into $V(^*\mathbb{R})$.

Remark: a) For $r \in \mathbb{R}$ we have $\alpha(r) = (j \circ i) = j(i(r)) = j(^*r) = {}^*r$, the last equality coming from 1).

b) For $E \subset \mathbb{R}$ i.e. $E \in V_1(\mathbb{R})$ we have

$$\alpha(E) = j(i(E)) = \{j(<b>) \mid <b> \in_U \{E,E,\ldots,E,\ldots\}/U,$$
$$<b> \in {}^*\mathbb{R}\} = \{<b> \mid b_i \in E \text{ for a big set of } i\} = {}^*E.$$

c) On any n-ary function F we have

$$\alpha(F) = {}^*F .$$

It is therefore natural to denote α by * (extending our previous use of *).

Proposition: $\text{Im} \, {}^* \subsetneq \text{Im} \, j \subsetneq V({}^*\mathbb{R})$, all inclusions being proper embeddings (injective).

Proof: The inclusions are clear by construction. That they are proper can be seen on the model of $\text{Im}({}^*\restriction \mathbb{R}) = {}^\sigma\mathbb{R} \subsetneq {}^*\mathbb{R} = V_0({}^*\mathbb{R})$, and of ${}^\sigma\mathbb{N} \notin \text{Im} \, j$ (for the latter see [18]).

Definition: Entities of $V({}^*\mathbb{R})$ in $\text{Im}\,{}^*$ are called "*-standard".[2] Elements of $V({}^*\mathbb{R})$ in $\text{Im} \, j$ are called "internal".

It follows immediately from the above proposition that *-standard $\Rightarrow$ internal. (We shall see below that the converse is not true.)

Elements of $V({}^*\mathbb{R})$ which are not internal are called <u>external</u>.

Elements of $V(\mathbb{R})$ are called <u>real</u>. They can be looked upon as elements of $V(\mathbb{R}^*)$ by identifying $A \in V(\mathbb{R}) - \mathbb{R}$ with ${}^\sigma A = \{{}^*a | a \in A\}$ and $r \in \mathbb{R}$ with ${}^*r$. Note that ${}^\sigma\mathbb{R}$ coincides with the previously defined one. Often one drops the distinction between A and ${}^\sigma A$ and looks upon A as already lying in $V({}^*\mathbb{R})$.

Examples of external sets are: $\mathbb{R}, \mathbb{N}, I, F, {}^*\mathbb{R} - \mathbb{R}, \ldots$. ($I \equiv$ infinitesimals, $F \equiv$ finite hyperreal numbers).

An example of an internal set which is not *-standard is $[0,\omega] = \{x \in {}^*\mathbb{R} \mid 0 \leq x \leq \omega\}$, with $\omega \in {}^*\mathbb{N} - \mathbb{N}$ (see below, Application).

Verification of these properties becomes easier once a little more is found out in general about the structure of standard and internal sets, which we are going to do shortly.

The following theorem is very instrumental for this:

Theorem: The following are equivalent:
a) A is an internal entity of $V({}^*\mathbb{R})$

b) A is an element of a *-standard set i.e. $A \in {}^*B$ for some $B \in V(\mathbb{R})$

c) $A \in {}^*V_n(\mathbb{R})$ for some n

d) $A = j(\{A_1, A_2, \ldots\}/U)$ with $A_i \in V_n(\mathbb{R})$ for a big set of i.

<u>Proof</u>: a) ⇒ c): By definition of internal entity we have $A \in \text{Im } j$, i.e. $A = j(<f>)$ for some $<f> \in \pi_U V_n(\mathbb{R})$, some n.
Then $f \in_U (V_n(\mathbb{R}), \ldots, V_n(\mathbb{R}), \ldots)$ and
${}^*V_n(\mathbb{R}) = j\{(V_N(\mathbb{R}), \ldots, V_N(\mathbb{R}), \ldots)/U\} \ni j(<f>) = A$, by construction of j (since j of X has as elements the j-maps of the elements of X).

c) ⇒ b): trivial by $V_n(\mathbb{R}) \subset V(\mathbb{R})$.

b) ⇒ a): trivial since $\text{Im }^* \subset \text{Im } j$ and $A \in \text{Im }^*$ by assumption.

d) ⇒ a): trivial, as in b) ⇒ a).

a) ⇒ d): A internal means $A = j(<f>)$ for some $<f> \in \pi_U V_n(\mathbb{R})$. Set $A_i = f(i)$.

<u>Remark</u>: *-standard sets are characterized by

$$A = j(\{A_0, A_0, \ldots\}/U),$$

for some $A_0 \in V_n(\mathbb{R})$, some n.

Hence they involve one and the same element A_0 of $V_n(\mathbb{R})$ in all components, whereas by the above Theorem (d) internal sets involve different elements in each component. In particular,
$\{r_1, r_2, \ldots, r_n, \ldots\}/U \in {}^*\mathbb{R}$, $r_i \in \mathbb{R}$ is an internal number,
$\{r, r, \ldots, r, \ldots\}/U$, $r \in \mathbb{R}$ is a standard number.

<u>Remark</u>: The above theorem specialized to $n = 1$ yields the following characterization of internal subsets of ${}^*\mathbb{R}$: A is an internal subset of ${}^*\mathbb{R} \Leftrightarrow A = \{<f>, f(i) \in A_i\}$ for a big set of i's, for some subsets A_i of $\mathbb{R}$.

<u>Application</u>: Let $\omega = \{1, 2, \ldots, n, \ldots\}/U$ (so that ω is an infinite hyperinteger). Define $[0, \omega] = \{x \in {}^*\mathbb{R} \mid 0 \leq x \leq \omega\}$.

It is easily verified that $[0,\omega] = j(A_1, A_2, \ldots /U)$, with $A_n = [0,n]$. By the above characterization we see that $[0,\omega]$ is internal (but non-*-standard).

As an exercise we might ask to prove that the function $\sin\omega x$, $x \in {}^*\mathbb{R}$, ω infinite, is internal but not *-standard. In addition, we might ask to prove that the function st (standard part) is external (but not real).

Remark: Since $V_n(\mathbb{R}) \in V(\mathbb{R})$ we have by the construction of j: ${}^*V_n(\mathbb{R}) \in {}^*V(\mathbb{R})$. ${}^*V(\mathbb{R})$ is the largest internal set and contains as elements all internal sets. Clearly ${}^*V(\mathbb{R}) \subsetneq V({}^*\mathbb{R})$, since as we mentioned above sets like ${}^\sigma\mathbb{R}, {}^\sigma\mathbb{N}$ are contained in $V({}^*\mathbb{R})$ and external.

Analogously as we did in Section II, we can summarize the result involved in the construction of the *-embedding of $V(\mathbb{R})$ in $V({}^*\mathbb{R})$ and the identification of internal sets in $V({}^*\mathbb{R})$ in the following

"Łos and Transfer Theorem", which we state again informally[3]: Let $\varphi(x_1, \ldots, x_n)$ be a statement with variables x_i in $V(\mathbb{R})$ involving $=, \in$ and logical operations (or, not, $\Rightarrow$), operations in $\mathbb{R}$ we discussed in the previous transfer theorem, quantifiers in the form $\forall x \in y$, $\exists x \in y$, with $y \in V(\mathbb{R})$.

Let $<f^\ell> \in \pi_U V(\mathbb{R})$, $\ell = 1, \ldots, n$. Then ${}^*\varphi(j(<f^1>), \ldots, j(<f^n>))$ holds iff $\varphi(f^1(i), \ldots, f^n(i))$ holds for a big set of i. In particular, if $f^\ell \in V(\mathbb{R})$ then ${}^*\varphi({}^*f^1, \ldots, {}^*f^n)$ holds iff $\varphi(f^1, \ldots, f^n)$ holds.

This is actually a special case of a general model theoretical result of Łos, its practical value here is that we can "go back and forth between standard world (with φ, $V(\mathbb{R})$ etc.) and non-standard world" (with ${}^*\varphi$, ${}^*V(\mathbb{R})$ etc.), provided "we replace all elements of $V(\mathbb{R})$ ("constants") with the correspondent elements of ${}^*V(\mathbb{R})$."

This is actually what is used in practice (rather than going back all the time to the very definition of *, j, etc.)

Examples

1) The statement φ:

   $\exists n \in \mathbb{N}_0, n \leq k \ \forall k \in \mathbb{N}$ is true (take $n = 0$), hence $^*\varphi: \exists n \in {}^*\mathbb{N}$, $n \leq k \ \forall k \in {}^*\mathbb{N}_0$ is true (take $n = {}^*0$)

   But the statement $\exists n \in {}^*\mathbb{N} - \mathbb{N}, n \leq k \ \forall k \in {}^*\mathbb{N} - \mathbb{N}$ is wrong (it is not $*$ of a true statement φ involving only elements of $V(\mathbb{R})$).

2) $\forall n \in \mathbb{N} \Rightarrow n - 1 \in \mathbb{N}$ or $n = 1$ is true, hence $\forall n \in {}^*\mathbb{N} \Rightarrow n - {}^*1 \in {}^*\mathbb{N}$ or $n = {}^*1$ is true.

3) The statement: all non-void upper bounded subsets of $\mathbb{R}$ have a least upper bound is <u>not</u> true, when one replaces $\mathbb{R}$ by $^*\mathbb{R}$ (e.g. the infinitesimals are upper bounded but do not have a least upper bound): it is however true, that <u>internal</u> subsets of $^*\mathbb{R}$ have this property as seen (we leave it as an exercise) by formalizing the statement. This can be used incidentally to prove that $\mathbb{R}$ and $\mathbb{N}$, $^*\mathbb{R} - \mathbb{R}$, $^*\mathbb{N} - \mathbb{N}$, the set I of infinitesimal, the set F of finite numbers are external sets.

Since internal sets are those "for which the properties of real sets can be carried over", it is useful to get further intuition about them and we list here some properties:

a) $A \subset {}^*\mathbb{N}, A \neq \emptyset$, A internal $\Rightarrow$ A has a least element.
   E.g. $\{\omega, \omega + 1, \ldots, \omega + n\}$, with $\omega \in {}^*\mathbb{N} - \mathbb{N}, n \in \mathbb{N}$ is internal (this is proven by transfer).

b) A internal, $A \neq \emptyset$, A has upper bound $\Rightarrow$ A has a least upper bound (proof as in a)).

c) $A \subset {}^*\mathbb{N}$, A internal, $A \supset \mathbb{N} \Rightarrow \exists \omega \in {}^*\mathbb{N} - \mathbb{N}, \omega \in A$ (this is a particular case of a general principle called "overflow").

d) A internal, $A \subset \mathbb{R} \Leftrightarrow A$ is finite ($\Leftarrow$ is clear; for $\Rightarrow$ see e.g. [17], p. 44)) (of course $A \subset \mathbb{R}$ is looked upon as $^\sigma A \in V(^*\mathbb{R})$).

e) A internal, $A \supset {}^*\mathbb{N} - \mathbb{N} \Rightarrow \exists n \in \mathbb{N}, n \in A$ (this is a particular case of a principle called "underflow").

f) A internal, $A \ni I_+ \equiv$ positive infinitesimal $\Rightarrow \exists r \in \mathbb{R}_+$, $r \in A$ ("infinitesimal overflow") (this is proven ad absurdum, by reduction to b) ; similarly we prove c), d), e)).

g) A internal, $A \supset \mathbb{R}_+$, $\Rightarrow \exists \varepsilon \in I_+$, $\varepsilon \in A$ ("infinitesimal underflow").

h) Let $^*\varphi$ be a formula as in the transfer principle. Then $A = \{b \in {}^*V(\mathbb{R}) \mid {}^*\varphi(b) \text{ is true}\}$ is internal (this is called "internal definition principle" or "internality theorem"). E.g. $\{x \in {}^*\mathbb{R} \mid 0 \leq x \leq \omega\}$ for some $\omega \in {}^*\mathbb{R}$ is internal.

i) A,B internal $\Rightarrow A \cap B$, $A \cup B$ and $A \times B$ are internal (this follows from h)).

j) $A \in B$, B internal $\Rightarrow A$ internal.

l) The set of all internal subsets of an internal set is internal.

Remark: c), d), e), f) are examples of "prolongation theorems".

m) A function is internal iff its graph is internal.

IV. Some Applications

We have described the basic framework[4,5] of non-standard analysis (including nonstandard functional analysis, measure theory, etc.!), the reader who has followed us up to now is certainly very eager to see applications. Let us differentiate between several types of applications.

The first type consists of getting a new look at results or problems of classical analysis in the broad sense. E.g. all of differential and integral calculus can be rewritten in non-standard analysis terms. This brings us back to calculus, to some extent at least, the basic intuitive reasoning based on infinitesimal and infinities which permeated analysis until the advent of "ε,δ"'s (and even beyond that, as we mentioned already in the introduction). This kind of application has been well-described in books and several expository papers (e.g. [14], [15], [24],[1] so we shall not go into this here.

IV.1 Differential equations, dynamical systems, Schrödinger operators

We should, however, spend a few words to mention shortly an application to the study (qualitative but also quantitative) of (systems of) differential equations. Roughly speaking, by allowing infinitesimal or infinite coefficients new phenomena can be discovered (and have been discovered!) which would be difficult to unravel by limiting procedures. Such studies have been the work mainly of the "Strasbourg-Mulhouse-Oran group" (around G. Reeb), see e.g. [33] - [35], [48], [56], [64] - [67], [87], [90], [112], [113].

Just to get started, let us look at the simple differential equation $\frac{d}{dt} x = 3x^{2/3}$ in the interval $[0,1]$. The initial condition $x(0) = 0$ does not fix the solution, in fact $x_a(t) = 0$ $t \in [0,a]$, $x_a(t) = (t-a)^3$, $t \in [a,1]$ are solutions, different for all $0 \leq a \leq 1$. Discretization and limiting procedure yields only the solution $x_1(t) = 0 \quad \forall t \in [0,1]$.

If we look at the solution v_δ of the discretized equation with "time slicing" $1/\omega, 2/\omega, \ldots \omega/\omega = 1$, $\omega \in {}^*\mathbb{N} - \mathbb{N}$ (we use the convention to denote $*r$ by r for $r \in \mathbb{R}$) with initial condition $v_\delta(0) = \delta$, δ a positive infinitesimal, then we can show that every solution $x_a(t)$ of the above type can be obtained as the standard part of some solution v_δ for some suitable δ (for $\delta = 0$ we have x_1). In fact, the solution of the equation $\frac{d}{dt} x = 3x^{2/3}$ with initial value δ real positive is unique, hence by transfer v_δ is unique. The internal set $\{\delta \in {}^*\mathbb{R} \mid \delta > 0 \text{ and } v_\delta(1) > 1\}$ contains every standard real $r > 0$, hence by infinitesimal underflow it contains some positive infinitesimal $\delta_o > 0$. Then $v_o(1) = 0, v_{\delta_o}(1) > 1$. By transfer, $\delta \to v_\delta(1)$ is internal and continuous in the sense of $V({}^*\mathbb{R})$, hence by continuity for any $r \in {}^*[0,1]$ there exists $\delta \in [0,\delta_o]$, $\delta \in {}^*\mathbb{R}$ s.t. $v_\delta(1) = r$. But the standard part u_δ of v_δ satisfies $u_\delta(0) = 0$, $u_\delta(1) = \text{str}$, hence it is x_a, with $(1-a)^3 = \text{str}$. Thus in a sense infinitesimals give a "resolution" of "non-uniqueness" associated with above equation.

For more general results of this type see [128]. [5].

We shall now come to the type of new phenomena studied by the above mentioned group. Let us look at the van der Pol equation

$$\varepsilon \frac{d^2}{dt^2} x + (x^2 - 1) + x - a = 0,$$

with ε positive infinitesimal, $a \in {}^*\mathbb{R}_+$.

By transfer, the only stationary or equilibrium point is $(a,F(a))$, with $F(x) = \frac{x^3}{3} - x$. By Liénard's substitution $x \to u = F(x) + \varepsilon x$ this study is reduced to the one of the vector field

$$\begin{cases} \varepsilon \frac{dx}{dt} = u - F(x) \\ \frac{du}{dt} = a - x. \end{cases}$$

By transfer from the situation ε real positive we have, also for ε infinitesimal positive, that $(a,F(a))$ is the only stationary or equilibrium point, stable for $a > 1$, unstable for $a < 1$, with a stable limit cycle around it. But for infinitesimal ε more can indeed be said. For $a = 1 - \frac{\varepsilon}{8} - \varepsilon\eta$, for some infinitesimal η, a new type of limit cycle is observed (sta = 1, thus the corrections in ε can only be obtained by standard means indirectly, through limit procedures).

A suitable analysis of what happens for such values of a needs "rescaling" by an infinite amount, for details see [35], [5]. Clearly much more should be said about this type of analysis, see the excellent surveys [48],[87] e.g. it would be very interesting to extend further this type of analysis to partial differential equations and to stochastic perturbations of classical dynamical systems, in particular with the hope of getting new insights into problems of stability.

Let us now look at some examples from the theory of Schrödinger operators with singular coefficients, of the type arising e.g. in nuclear physics, solid state physics, the theory of electric antennas, polymer physics, see e.g. [91] – [93], and references therein.

Just "pour fixer les idées" let us consider a formal operator of the type $-\Delta + \lambda \delta(x)$, $x \in \mathbb{R}^d$ in $L^2(\mathbb{R}^d, dx)$. To define properly an "operator" like this one has, in non-standard analysis, basically two procedures, either by "discretization" (taking a hyperfinite lattice $\varepsilon \mathbb{Z}^d$ "realization", ε infinitesimal, of $\mathbb{R}^d$ and replacing consequently Δ and $\delta(x)$ by their hyperdiscrete analogues) or by "a smooth non-standard realization" of the singular coefficient, in this case of $\delta(x)$ (realized e.g. as $(\frac{4}{3}\pi\varepsilon^3)^{-1} \chi(|x|/\varepsilon)$, with χ the characteristic function of the unit ball $\{|x| \le 1\}$ in $\mathbb{R}^3$). Let us give a few details of both procedures.

First, we describe shortly the latter method, for $d = 3$. The operator we want to realize acts thus in $L^2(\mathbb{R}^3)$. Let ε be a positive infinitesimal and let us consider the function (non-standard "realization" of Dirac's δ-function) $\delta_\varepsilon(x) \equiv (\frac{4}{3}\pi\varepsilon^3)^{-1} \chi(|x|/\varepsilon)$. We want to use δ_ε as a "perturbation of the Laplacian". Let $^*L^2(\mathbb{R}^3)$ be the non-standard Hilbert space in which operators are going to be defined, let Δ be the self-adjoint operator obtained by transfer to $^*L^2(\mathbb{R}^3)$ from the self-adjoint Laplacian in $L^2(\mathbb{R}^3)$. For any finite $\lambda_\varepsilon \in {^*\mathbb{R}}$, $H_\varepsilon = -\Delta + \lambda_\delta \delta_\varepsilon(x)$ is a well-defined self-adjoint operator in $^*L^2(\mathbb{R}^3)$ (by transfer, $\lambda_\varepsilon \delta_\varepsilon$ being *-bounded).

By transfer, we can split, in the same way as with $L^2(\mathbb{R}^3)$, $^*L^2(\mathbb{R}^3)$ into a rotationally symmetric part $^*H_s$ and its orthogonal complement $^*H_s^\perp$. H_ε acts as $-\Delta$ in $^*H_s^\perp$ and in fact its standard part (defined e.g. through its resolvent) is $-\Delta$ in $^*H_s^\perp$. The restriction of H_ε to $^*H_s$ is unitary equivalent (in the sense of $^*L^2(\mathbb{R}^3)$) with $A_\varepsilon = -\frac{d^2}{dr^2} + \lambda_\varepsilon \delta_\varepsilon(r)$ ($|x| = r$), acting in $^*L^2(\mathbb{R}_+, dr)$, with Dirichlet boundary conditions at $r = 0$.

For $\operatorname{Im} k^2 \ne 0$, $\operatorname{Re}\sqrt{k^2} > 0$ we have for any finite $x, y \in {^*\mathbb{R}_+}$, by transfer:

$$(A_\varepsilon - k^2)^{-1}(x,y) = (2ak)^{-1} \begin{cases} v_1(x)\, v_2(y) & x \le y \\ v_1(y)\, v_2(x) & y \le x \end{cases} ,$$

with

$$v_1(r) \equiv \begin{cases} \sin(\sqrt{k^2-\lambda}\; r), & 0 \le r \le \varepsilon \\ a\, e^{kr} + b\, e^{-kr}, & \varepsilon \le r \end{cases}$$

$$a \equiv \tfrac{1}{2} e^{-k\varepsilon}[\sin(\sqrt{k^2-\lambda}\,\varepsilon) + \tfrac{\sqrt{k^2-\lambda}}{2} \cos(\sqrt{k^2-\lambda}\;\varepsilon)],$$

$$b \equiv \tfrac{1}{2} e^{k\varepsilon}[\sin(\sqrt{k^2-\lambda}\,\varepsilon) - \tfrac{\sqrt{k^2-\lambda}}{2} \cos(\sqrt{k^2-\lambda}\;\varepsilon)]$$

v_2 is given by a similar formula, see [5].

This Green's function is found by transfer and Sturm-Liouville theory for the situation with ε standard, the coefficients of the trigonometric functions being fixed by continuity conditions at $r = \varepsilon$. From these formulae we see easily that for $\lambda_\varepsilon / (\tfrac{4}{3}\pi\varepsilon^3)$ finite the standard part of $(A_\varepsilon - k^2)^{-1}(x,y)$, $x \ne y$ is $(-\tfrac{d^2}{dr^2} - k^2)^{-1}(\text{sty},\text{sty})$, with $-\tfrac{d^2}{dr^2}$ acting in $L^2(\mathbb{R}_+, dr)$ with Dirichlet boundary conditions at 0. Thus in this case we do not get any interaction. For $\lambda_\varepsilon / (\tfrac{4}{3}\pi\varepsilon^3)$ infinite then the standard part of $(A_\varepsilon - k^2)^{-1}(x,y)$ exists only when a,b are finite and a is non-infinitesimal, which then requires $\sin(\sqrt{k^2-\lambda}\;\varepsilon)$ and $\sqrt{k^2-\lambda}\cos(\sqrt{k^2-\lambda}\;\varepsilon)$ both near standard. But then $\cos(\sqrt{k^2-\lambda}\;\varepsilon)$ must be infinitesimal, which yields $\sqrt{k^2-\lambda}\;\varepsilon = (\gamma+\tfrac{1}{2})\pi+\eta$ for some infinitesimal η and any $\gamma \in \mathbb{R}$ such that $\sqrt{k^2-\lambda}\cos(\sqrt{k^2-\lambda}\;\varepsilon)(-1)^\gamma$ is near standard. An elementary calculation then shows that this is only possible when

$$\lambda_\varepsilon = \tfrac{4}{3}\pi[-(\gamma+\tfrac{1}{2})\pi^2\varepsilon + 4\pi\alpha\varepsilon^2 + \beta\varepsilon^3]$$

for arbitrary $\beta \in \mathbb{R}$. In this case the standard part of $(A_\varepsilon - k^2)^{-1}(x,y)$ is independent of β,γ and H_ε defines, by taking standard parts in the resolvent, a self-adjoint, lower bounded operator H^α, depending on the parameter $\alpha \in \mathbb{R}$. H^α is thus a one-parameter family of realizations of "$-\Delta + \lambda\delta(x)$". It can be shown, of course, that H^α coincides with the limit, in the norm resolvent sense, of $-\Delta + \lambda_\varepsilon \delta_\varepsilon(x)$ for $\varepsilon > 0$ real, as $\varepsilon \downarrow 0$, when λ_ε is chosen as above. In a sense, however, to think of H^α as $-\Delta$ perturbed by a potential with infinitesimal support is closer to the physical intuition.

For more details about this model and various extensions using non-standard analysis see [122], [5] (see also [91] - [93] for further information about the study of models of this type).

Below we shall look at the approach to singular perturbations using "hyperfinite discretization". Before doing so, however, we would like to introduce a little bit of non-standard measure and integration theory, which is also useful for our later discussion of probabilistic applications.

IV.2 A basic instrument from non-standard measure and integration theory: the Loeb measure

The natural measures in non-standard analysis are additive internal, not σ-additive, measures. Loeb [129] realized in '75 that to any such measure one can associate a standard σ-additive measure (on a non-standard space): this was a breakthrough, which made it possible to apply non-standard, even hyperfinite, tools to measure theory and related domains, like the theory of stochastic processes (but also, the study of certain non-linear partial differential equations through the concept of "Loeb solution" [94], [95], [5] as well as to statistical mechanics [5], [85], [96], [97], [110], [116], [117], [124].

Let X be some internal set, A an algebra of internal subsets of X, ν a finitely additive measure on (X,A) with values in $^*[0,1]$, with $\nu(X) = 1$. (X,A,ν) is called an internal measure space. Can one associate to ν a σ-additive [0,1]-valued measure? The answer is yes, for a simple reason. It is namely possible to show in our model of non-standard analysis discussed in Sects. I-III that the following "saturation property" holds:

Theorem: If $A_n \neq \emptyset$ are internal for all $n \in \mathbb{N}$ then $\bigcap_{n \in \mathbb{N}} A_n \neq \emptyset$

Proof: We can assume $A_n \supset A_{n+1} \forall n$. Since A_n are internal, we can write $A_n = j(A_n')$, for some $A_n' \in \pi_U V(\mathbb{R})$.

Let k be such that $A'_1 \in \pi_u V_k(\mathbb{R})$ and take $A'_o \in \pi V_{k+1}(\mathbb{R})$ s.t. $A'_o(i) = V_k(\mathbb{R})$ for all $i \in \mathbb{N}$. We can assume $A'_n \supset_U A'_{n+1}$, $A'_n \neq_U \emptyset$ (with $\supset_U$ defined in the natural sense that each element of A'_{n+1} belongs in the $\in_U$-sense to A'_n).

Let $I_o \equiv \mathbb{N}$, $I_k = \{i \in \mathbb{N} | i > k\}$. Then $I_o \supset I_1 \supset \ldots \supset I_n \supset \ldots$, $\cap I_n = \emptyset$
Let $I'_n \equiv I_n \cap \{i \in \mathbb{N}\ A'_o(i) \supset \ldots \supset A'_n(i) \neq \emptyset\}$, $n \in \mathbb{N}$. Then $I'_o = \mathbb{N}$, $I'_n \supset I'_{n+1}$, $I'_n \in U$ (since $I_n \in U$, being cofinite, and $\{i \in \mathbb{N} | A'_o(i) \supset \ldots \supset A'_n(i) \neq \emptyset\}$ being in U, since $A'_o \supset_U A'_1 \supset_U \ldots$).
Define $n(i) \equiv \{\max n | I'_n \ni i\}$ (which is finite, using $I_o = \mathbb{N}$ and $\cap I_n = \emptyset$). Since $A'_n(i) \neq \emptyset$ for a large set of i, we can take an element $a(i)$ in $A'_{n(i)}(i)$ and consider $B \equiv \{i \in \mathbb{N} | a(i) \in A'_m(i)\}$. This contains I'_m (since $n(i) \geq m$ from the definition of $n(i)$, hence by the isotony of the I'_n: $A'_{n(i)}(i) \subset A'_m(i)$, thus $a(i) \in A'_{n(i)}(i) \subset A'_m(i)$, hence $a(i) \in A'_m(i)$).

Hence $B \in U$, thus $a \in_U A'_m \forall m \in \mathbb{N}$, thus $a \in \bigcap_{m \in \mathbb{N}} A'_m$.
By the fact that $A_m = j(A'_m)$ and the definition of j this implies $\cap_m A_m = j(\cap_m A'_m) \ni j(a)$, hence that $\cap_m A_m \neq \emptyset$ ∎

The above saturation theorem shows intuitively that internal sets are "rich". E.g. $\cap_{n \in \mathbb{N}} (0,n) = \emptyset$ ($(0,n)$ are external, since $(0,n) = \{x \in \mathbb{R} | 0 < x < n\}$, but $\mathbb{R}$ is external), whereas $\cap\ ^*(0,n) = $ (positive infinitesimals different from zero) $\neq \emptyset$ ($^*(0,n)^{n \in \mathbb{N}}$ are internal, even *-standard, being of the form $\{x \in\ ^*\mathbb{R} | 0 < x < n\}$ with $^*\mathbb{R}$ *-standard). In fact, as an immediate corollary of the above we have:

<u>Corollary 1</u>: Internal elements of $^*V(\mathbb{R})$ are either finite or uncountable.

Proof: Assume A is internal and A is countable, i.e. $A = \{a_n, n \in \mathbb{N}\}$, $a_i \in\ ^*V(\mathbb{R})$. Define $A_o \equiv A$, $A_n \equiv A - \{a_1, \ldots, a_n\}$. Then the A_i are internal and $\cap A_n = \emptyset$, on the other hand, by saturation $\cap A_n \neq \emptyset$ if $A_n \neq \emptyset$. Hence to avoid contradiction we must have $A_{n_0} = \emptyset$ for some n_0 (and then for all $n \geq n_0$), but then A is finite. Or we have to reject the "ad absurdum" assumption, i.e. A is uncountable. ∎

<u>Corollary 2</u>: ("countable comprehensiveness")

If A_n, $n \in \mathbb{N}$ are internal and in $*V_k(\mathbb{R})$ for some k then the sequence $\{A_n, n \in \mathbb{N}\}$ can be prolonged to an internal sequence $\{A_n, n \in *\mathbb{N}\}$.

<u>Proof</u>: Let for any $n \in \mathbb{N}$, $B_n \equiv \{f \in *V_k(\mathbb{R})^{*\mathbb{N}} \mid f(i) = A_i \, \forall i \leq n\}$. Then B_n is internal and $\neq \emptyset$ (since e.g. f such that $f(i) = A_i$, $i \leq n$, $f(i) = *V_k(\mathbb{R})$, $i > n$, $i \in *\mathbb{N}$ is in B_n), thus by saturation $\bigcap_{n \in \mathbb{N}} B_n \neq \emptyset$. But then take $f \in \bigcap_{n \in \mathbb{N}} B_n$, then f is internal, as an element of the internal set B_n. We then have, by the definition of the B_n's, $f(i) = A_i \, \forall i \in \mathbb{N}$, hence the required prolonged internal sequence is $\{A_n, n \in *\mathbb{N}\}$, with $A_n = f(n)$ for $\forall n \in *\mathbb{N} - \mathbb{N}$. ∎

Now, let (X, A, ν) be an "internal measure space". We shall see, using the saturation theorem, that ν can be used to obtain a standard, σ-additive measure, "near ν". Let $°\nu$ be the standard additive, $[0,1]$-valued normalized measure on the algebra A defined by $°\nu(A) = st(\nu(A)) \, \forall A \in A$.

<u>Theorem (Loeb)</u>: $°\nu$ has a Caratheodory extension to a σ-additive probability measure $L(\nu)$ on the σ-algebra generated by A. $(X, \sigma(A), L(\nu))$ is a probability space. $L(\nu)$ is "near ν" in the sense that, to any $A \in A, \exists B \in \sigma(A)$ such that $L(\nu)(A \Delta B) = 0$, where Δ means symmetric difference of sets.

<u>Proof</u>: Having the saturation theorem, the proof is immediate from Caratheodory's criterium. In fact, from $A_n \in A$, $A_n \downarrow \emptyset$ we have by saturation that there exists an $n_0 \in \mathbb{N}$ such that $A_n = \emptyset$ for all $n \geq n_0$ (otherwise $\bigcap_{n \in \mathbb{N}} A_n \neq \emptyset$, the A_n being internal!). But then of course $°\nu(A_n) = 0 \, \forall n \geq n_0$, hence $°\nu(A_n) \to 0$ as $n \to \infty$, thus Caratheodory's theorem condition is satisfied and $°\nu$ has a σ-additive extension to a probability measure on $\sigma(A)$. ∎

The measure $L(\nu)$ (or often its completion) is called <u>Loeb measure</u> associated with ν. Correspondingly $(X,\sigma(A),L(\nu))$ is called Loeb measure space. The following Corollary is easily shown, see [5]:

Corollary: For any $L(\nu)$-integrable real-valued function f on X there exists an internal function F from X into $^*\mathbb{R}$ s.t. st $F(x) = f(x)$ $L(\nu)$-a.e. $x \in X$, and viceversa. F is S-integrable in the sense that $\int |F| \, d\nu$ is finite and $\int_A F d\nu \approx 0$ if $\nu(A) \approx 0$. One has $\int F d\nu \approx \int f dL(\nu)$. F is called "<u>lifting</u>" of f.

Remark: As mentioned above, Loeb's construction is also possible when ν is not $^*[0,1]$-valued, see [5], [46], [89].

The usefulness of Loeb measures is that they are in the above sense entirely approximable by internal additive ones, which are easy to operate with, as we shall now see in the particularly important case of the one yielding a realization of Lebesgue measure (as a "uniform measure") on $[0,1]$ (say). Such (special) Loeb measures are associated with an important class of internal measure spaces, (X,A,ν), the so-called <u>hyperfinite</u> ones. An internal set X is called hyperfinite if there exists an internal bijective map of a "hyperfinite segment" of $^*\mathbb{N}$, $\{n \in {}^*\mathbb{N} \mid n \leq \omega$ for some $\omega \in {}^*\mathbb{N}\}$, onto X (ω is then called the "internal cardinality" of X).

Hyperfinite sets behave in a sense as finite sets, although they can be used to "modellize a continuum". E.g. a hyperfinite model of the interval $[0,1]$ is the hyperfinite set X described as follows: Let ε^{-1} be some hyperfinite integer, i.e. $\varepsilon^{-1} \in {}^*\mathbb{N} - \mathbb{N}$ and let $X = \{k\varepsilon \mid k \in {}^*\mathbb{N} \cup 0, k \leq \varepsilon^{-1}\}$. Then st $\upharpoonright X$ is a surjection onto $[0,1]$. Let, for any internal subset A of X, $\nu(A) \equiv |A|/|X|$, with $|\ |$ denoting internal cardinality (so that $|X| = (\varepsilon)^{-1} + 1$). (X,A,ν) is a hyperfinite internal probability space (with A the algebra of internal subsets of X).

The corresponding Loeb space $(X,\sigma(A), L(\nu))$ is nothing but a hyperfinite model of the Lebesgue measure space associated with Lebesgue measure on $[0,1]$, in the sense that, for any measurable $A \subset [0,1]$, $\mathrm{st}^{-1}(A) \equiv \{x \in X \mid \mathrm{st}\, x \in A\}$ is Loeb measurable, and viceversa, and $\int_A dx = L(\nu)\,(\mathrm{st}^{-1}(A))$. This gives substance to the intuition that Lebesgue measure is uniform measure (in fact it is "near" the counting measure ν on X).

Similarly for any Lebesgue integrable f_0 there is a lifting F s.t. $f \equiv \mathrm{st}\, F = f_0 \circ \mathrm{st}$ is $L(\nu)$ integrable, and F is S-ν-integrable and viceversa, and we have

$$\int F d\nu = \sum_{x \in X} |X|^{-1} F(x) \approx \int_X f(x) dL(\nu)(x) = \int_0^1 f_0(y) dy \; .$$

We can thus approximate Lebesgue's integral by hyperfinite sums.

<u>Remark</u>: A similar construction holds for general Radon spaces, cf. [5]. We shall use below Loeb measures in the discussion of hyperfinite Dirichlet forms as well as in the discussion of stochastic processes.

IV.3 Return to Schrödinger operators. Dirichlet forms.

We shall now take up again the theme of IV.1, looking at the study of singular differential operators, of the form $-\Delta + \lambda \delta(x)$ e.g. In this study it is often useful to look at the basic mappings to be constructed as quadratic forms rather than operators (e.g., as operator in $L^2(\mathbb{R}, dx) \lambda \delta(x)$, $x \in \mathbb{R}$ does not make sense, it is however a small form perturbation of $-\Delta$, and hence $-\Delta + \lambda \delta(x)$ is well-defined as lower bounded quadratic form in this 1-dimensional case). Thus it is quite natural to try to develop a theory of hyperfinite quadratic forms. The fact that even in 3 dimensions $-\Delta + \lambda \delta(x)$, realized as H^α, (cf. Sect. IV.1), is, modulo a finite, α-dependent constant, unitary equivalent with a Dirichlet form (in the sense of [100]-[104],[109], see also e.g. [5],[111], and the approach to elliptic operators via Dirichlet forms is, at least in the symmetric case, the most powerful one, motivates the idea of trying to construct a theory of hyperfinite Dirichlet forms. This has been done [5], [99] and we shall here give some sketch of this approach.

Let $S = \{s_1,...,s_N\}$ be an internal subset of ${}^*\mathbb{R}^d$, $N \in {}^*\mathbb{N} - \mathbb{N}$ (think of S as a hyperfinite lattice "realization" of $\mathbb{R}^d$). Let $m: S \to {}^*\mathbb{R}_+$ be an internal function, which we shall think of as a measure on S. We then write for any internal $A \subset S$:

$$m(A) = \sum_{s_i \in A} m(s_i).$$

Let $L^2(S,m)$ be the internal space of all internal functions $f: S \to {}^*\mathbb{R}$ with the inner product

$$<f,g> \equiv \sum_{i=1}^{N} f(i)g(i)m_i.$$

Let Q be an $N \times N$ stochastic matrix ("transition matrix") s.t. $Q = ((q_{ij})), i,j, = 1,...N$, $q_{ij} \geq 0$, $\sum_{j=1}^{N} q_{ij} = 1$ and such that Q is symmetric with respect to m in the sense that $m_i q_{ij} = m_j q_{ji}$ for all i,j.

Let Δt be a positive infinitesimal. $Q^{\Delta t}$ is, by transfer, a well-defined operator on $L^2(S,m)$ given by $(Q^{\Delta t}f)(i) = \sum_{j=1}^{N} q_{ij} m_j f(j)$.

Let $A \equiv [I-Q^{\Delta t}]/\Delta t$. Then A is a symmetric positive operator which generates the "hyperfinite Dirichlet form"

$$E(f,g) = <A^{1/2}f, A^{1/2}g>$$

(this is in analogy with the standard theory of (discrete) Dirichlet forms developed originally by Beurling and Deny. $Q^{k\Delta t}$, $k \in \mathbb{N}$ is a hyperfinite version of the semigroup generated by A, with time replaced by the discretized version $k\Delta t$, $k \in \mathbb{N}$). The domain $D(E)$ of E is defined as the set of all $f \in L^2(S,m)$ s.t. $E_1(f,f) \equiv E(f,f) + <f,f>$ is finite and minimal among all $f' \approx f$. It turns out that E with domain $D(E)$ is always closed (in the norm given by $\|f\|_1 \equiv E_1(f,f)^{1/2}$). (This is based on the fact that for all $f \in D(E)$, $t = k\Delta t \approx 0$, $k \in \mathbb{N}$, $<([I-Q^t]/t)f,f> \approx E(f,f)$. In fact, if $\{f_n, n \in \mathbb{N}\}$ is an internal extension of an E_1-Cauchy sequence of elements f_n of

$D(E)$, let $N \in {}^*\mathbb{N}-\mathbb{N}$ so that $\mathrm{st}|f_n-f_N| \to 0$ as $n \to \infty$ in $\mathbb{N}$. Let $f = f_N$, we show $f \in D(E)$. In fact, if f were at absurdum not in $D(E)$, then by the above there would exist an infinitesimal t and an $\varepsilon > 0$, $\varepsilon \in \mathbb{R}$ s.t. $\|f-Q^t f\|_1 < \varepsilon/3$. Choose n so large that $\|f-f_n\|_1 < \varepsilon/3$, then, as easily shown, $\|Q^t f - Q^t f_n\|_1 < \varepsilon/3$. An $\varepsilon/3$-argument yields then, using $\|f_n - Q^t f_n\| \approx 0$, a contradiction).
We shall now see how one can associate to hyperfinite Dirichlet forms standard Dirichlet forms. Let $F(L^2(s,m))$ be the set of all elements in $L^2(S,m)$ with finite norm.

The set of equivalence classes modulo infinitesimals with representatives in $F(L^2(S,m))$ is a Hilbert space with scalar product given by the standard part of the one in $L^2(S,m)$. We define the standard form E_{st} induced by E to be the symmetric bilinear form on $F(L^2(S,m))/\approx$ given by $E_{st}(v,v) = \inf_{f \in v} \{\mathrm{st}\, E(f,f)\}$, with domain $D(E_{st}) = \{v \in F(L^2(S,m))/\approx \mid v \cap D(E) \neq \emptyset\}$. The argument giving the closedness of E also yields that E_{st} is closed. If $m\{s \in S \mid \|s\| < r\}$ is finite for all finite r, we can define a Radon measure $\tilde{m}$ on all measurable sets of $\mathbb{R}^d$ by $\tilde{m}(B) = L(m)(\mathrm{st}^{-1}(B))$, with $L(m)$ the Loeb measure to m.

The standard part of E is the Dirichlet form on $L^2(\mathbb{R}^d, \tilde{m})$ given by $E(f,f) = E_{st}(\tilde{f},\tilde{f})$, with $\tilde{f} = {}^*f \upharpoonright S$, for any bounded continuous real-valued f on $\mathbb{R}^d$. E is a standard Dirichlet form (in the sense of [100], i.e. E is a closed densely defined positive bilinear form with the contraction property $E(f^\#,f^\#) \leq E(f,f)$, $f^\# = (f \vee 0 \wedge 1)$. Equivalently the associated semigroup P_t in $L^2(\mathbb{R}^d,\tilde{m})$ is sub-Markov in the sense $0 \leq f \leq 1 \Rightarrow 0 \leq P_t f \leq 1$).

In this way from hyperfinite Dirichlet forms one produces standard Dirichlet forms, and viceversa one can associate to every standard Dirichlet form a hyperfinite Dirichlet form. One can then exploit the hyperfinite Dirichlet forms (which behave "by transfer" very similarly to finite Dirichlet forms) to give a new (and in a sense "more intuitive") construction of a strong Markov, Hunt process associated with the standard Dirichlet form (for the standard construction see [100]) and the development of a stochastic calculus for hyperfinite Dirichlet forms. For details on this see [5], [99]. A particular case is $S = \{z \in {}^*\!\mathbb{Z}^d, |z_i| < \delta^{-2}, i = 1,\ldots,d\}$, δ a positive infinitesimal. Let U be the set of unit vectors in ${}^*\!\mathbb{R}^d$ of the form $(0,\ldots,\pm1,0\ldots0)$ and define, for $s \in S, e \in U$ and any internal function $f: S \to {}^*\!\mathbb{R}$

$$D_e f(s) = \frac{f(s+\delta e)-f(s)}{\text{sign}(e)\delta},$$

with sign e the sign of the non-zero component of e.

Hyperfinite forms of the type

$$F(f,g) = \frac{1}{4} \sum_{s \in S} \sum_{e \in U} D_e g(s) D_e f(s) \nu(s),$$

with ν an internal measure on S are called <u>hyperfinite energy forms</u>. It is not difficult to show that, for some constant $\eta \in {}^*\!\mathbb{R}$, with $\Delta t = \eta \delta^2/d$ $F(f,g) = \eta^{-1} E(f,g)$, with E the hyperfinite Dirichlet form defined as above, with $m(s) = \nu(s) + (4d)^{-1} \sum_{e \in U} \nu(s+\delta e)$ (so that $L(m) \circ \text{st}^{-1} = L(\nu) \circ \text{st}^{-1}$!), $Q = ((q_{s,s'}))$, $s,s' \in S$, $q_{s,s'} = 0$ if $|s-s'| \neq \delta$ or $m(s) = 0$, $q_{s,s'} = [\nu(s) + \nu(s')]/[4dm(s)]$ if $|s-s'| = \delta$ and $m(s) \neq 0$.

The process associated with E has standard part which is a Hunt process with stationary measure $L(\nu) \circ \text{st}^{-1}$. In the case where $\eta = 1$ with a suitable choice of ν one gets the diffusion process associated with a standard energy form

$$\frac{1}{2} \int_{\mathbb{R}^d} <df,dg> \varphi^2 dx,$$

$\varphi > 0$ a.e., $\varphi \in L^2_{loc}(\mathbb{R}^d)$, in $L^2(\varphi^2 dx)$.

<u>Remark</u>: The same construction can be applied to the case where $\mathbb{R}^d$ is replaced by a differentiable manifold, or in a completely different direction, M is a (suitable) fractal set, see [99]. This is a study which deserves to be pushed much further, also in view of the recently discovered uses of diffusions on fractal sets.

Let us now come to the description of how the above hyperfinite approach copes with the problem of giving a meaning to "operators" of the form $-\Delta + \lambda\delta(x)$ in $L^2(\mathbb{R}^d, dx)$. More generally, we try to find operators associated with formal quadratic forms

$$E(f,g) \equiv E_0(f,g) + \int_C \lambda f g d\rho ,$$

with E_0 a known quadratic form (e.g. the one associated with $H_0 = -\Delta$), C a set of $\mathbb{R}^d$ of Lebesgue measure zero and ρ a measure concentrated on C. The point is to find hyperfinite quadratic forms $\tilde{E}, \tilde{E}_0$ on $L^2(S,m)$, S hyperfinite, m internal positive function, s.t. $\tilde{E}(\tilde{f},\tilde{g}) = \tilde{E}_0(\tilde{f},\tilde{g}) + \int_B \tilde{\lambda} \tilde{f} \tilde{g} \, d\tilde{\rho}$, with $B \subset S$ internal s.t. st $B = C$, $\tilde{\lambda}$ internal function resp. measure on B s.t. $\rho = L(\tilde{\rho}) \circ \mathrm{st}^{-1}$, f,g "versions" of $\tilde{f},\tilde{g}$ in $L^2(S,m)$. The generator $\tilde{L}$ of $\tilde{E}$ is given by $(\tilde{L}\tilde{f})(i) = \tilde{L}_0\tilde{f}(i) + \tilde{\lambda}(i)\tilde{f}(i)\frac{\tilde{\rho}(i)}{m(i)}$, with $\tilde{L}_0$ the generator associated with E_0.

With our case $C = \{0\}$, λ infinitesimal, in mind (corresponding to the interaction $\lambda\delta(x)$) we allow λ to be non-standard.

Here is a sketch of how to handle such singular hyperfinite perturbation problems. A computation using transfer applied to Neumann series yields for $\mathrm{Im}\, z \neq 0$

$$(\tilde{L} - z)^{-1} = G_z + \hat{G}_z^* (\tilde{\lambda}^{-1} - G_z')^{-1} \hat{G}_z ,$$

with $G_z \equiv (\tilde{L}_0 - z)^{-1}$, $\hat{G}_z, G_z', \hat{G}_z^*$ denoting G_z as operator $L^2(S) \to L^2(B)$, resp. $L^2(B) \to L^2(B)$ resp. $L^2(B) \to L^2(S)$.

The whole trick is now to find $\tilde{\lambda}$ s.t. the standard part of the operator $(\tilde{\lambda}^{-1} - G_z')^{-1}$ exists (in the sense that the kernel has a standard part on finite elements outside the diagonal) and the standard part $(H-z)^{-1}$ of $(\tilde{L}-z)^{-1}$ exists and is different from the one of $(\tilde{L}_0-z)^{-1}$ (which is of course $(H_0-z)^{-1}$). A sufficient condition for this is the existence of a $z_0 \in {}^*\mathbb{R}$ such that

$$\text{st } [\sum_i G_{z_0} G_z(\cdot - i) \, \tilde{\rho}(i)] \in L^2(\mathbb{R}^d, \rho) \, .$$

In such cases one gets a self-adjoint family of operators H in $L^2(\mathbb{R}^d, m)$ parametrized by internal functions $\tilde{\lambda}$ s.t.

$\tilde{\lambda}(x)^{-1} - \sum_i G_{z_0}(x-i)\tilde{\rho}(i)$ has a standard part. In the case $d = 3$ $C = \{0\}$, ρ = Lebesgue measure, in which case H is formally $-\Delta + \lambda\delta(x)$, we get that $\tilde{\lambda}$ is infinitesimal negative, in fact of the form $\tilde{\lambda}^{-1} = -G_0(0) + \tilde{\alpha}$, with $\tilde{\alpha} \in \mathbb{R}$ ($\tilde{\alpha}$ can be chosen constant).

We thus recover by the hyperfinite approach the result we obtained above by smooth non-standard perturbations concerning the definition of "$-\Delta + \lambda\delta(x)$".

Remark: The "hyperfinite" as well as the "smooth" non-standard approaches to $-\Delta + \lambda\delta(x)$ can be extended to the study of operators of the form $-\Delta + \sum_{y \in Y} \lambda_y \delta(x-y)$, with Y some finite subset of $\mathbb{R}^d$, see [122],[5]. The case of some infinite Y can also be handled, below we shall treat one example where Y is the path of a Brownian motion in $\mathbb{R}^d$. It would be very worthwile to have a systematic treatment of the case Y infinite, this has not yet been done (see however [5]).

Remark: The treatment of above singular Schrödinger operators $-\Delta + \sum_{y \in Y} \lambda_y \delta(\cdot - y)$ is just an example of the kind of problems which can, and partly have been, handled by non-standard analysis. In fact, many other problems can be treated, we have already mentioned detailed studies of dynamical systems ([90] and references therein), let us in addition mention the treatment of singular Sturm-Liouville problems with "measure coefficients" given in [60],[74],[109],[122], which has also been stimulus to new developments on the standard side, see [105],[131] and references therein (and [93]). Moreover, we would like to mention that also in the case of non-linear partial differential equations some of the methods of non-standard analysis have been useful, e.g. in classical boundary layers problems [1], in the asymptotic study of differential equations [70],[11],[88] and more recently in a breakthrough for the study of Boltzmann's equation of kinetic gas theory, through very important work by Arkeryd, see e.g. [97] and references therein, as well as [95], [5]. Here the decisive element is the construction of a suitable "Loeb" solution (a non-standard tool based on Loeb measures, playing a role somewhat similar to "generalized solutions" in the linear case, this joined with as much as possible control on standard parts to extract from such a solution a classical solution). It seems to us that more of these tools should prove to be useful in handling problems of non-linear functional analysis (turbulence problem in hydrodynamics, e.g.).

We shall now leave the domain of differential equations to take up the discussion of some examples of applications in stochastic analysis and quantum field theory.

IV.4 Some applications in probability theory and quantum field theory

Here is Anderson's hyperfinite model for Brownian motion (see [107], [5]). Let T be the above hyperfinite model of $[0,1]$, i.e. $T = \{k \Delta t \mid k \in {}^*\mathbb{N}, \ k \leq (\Delta t)^{-1}\}$ for a fixed $(\Delta t)^{-1} \in {}^*\mathbb{N} - \mathbb{N}$.

We look at T as "time set". Let $\Omega = \{-1,+1\}^T$. Let A be the internal subsets of the internal set Ω, let P be the product $\prod_{t \in T} P_t$ of symmetric Bernoulli measures (P_t the Bernoulli distribution of the component $\omega(t)$ of $\omega \in \Omega$).

Let $B(\omega,t) \equiv \sum_{s=0}^{t-\Delta t} \omega(s) \sqrt{\Delta t}$, $\omega \in \Omega$.

This is a hyperfinite random walk, moving by $\pm\sqrt{\Delta t}$ right or left at each infinitesimal time interval Δt. B is an <u>internal stochastic process</u>. It is a theorem then that st $B(\omega,t)$ exists ($L(P)$-a.s.) and calling it $b(\omega, \text{st } t)$ we have that b is a process indexed by $[0,1]$, with values in $\mathbb{R}$ (and underlying probability space $(\Omega, \sigma(A), L(P))$. b is a realization of Brownian motion on $\mathbb{R}$ and $(\Omega, \sigma(A), L(P))$ is a realization of Wiener measure.

The proof is by showing that the independent increments property of B
gives by transfer the same property for b (using the definition of
L(P)) and proving that b(u) - b(v), u ≥ v has mean zero and variance
u-v, by realizing that with u = st s, v = st s, α ∈ ℝ:

$$\int_\Omega e^{i[b(u)-b(v)]\alpha} \, dL(P) = st \int_\Omega e^{i[B(s)-B(t)]\alpha} \, dP$$

and computing the internal integral using the definition of B and P.
Of course a similar result holds for Brownian motion on $\mathbb{R}^d$. Moreover,
there is no difficulty extending to the infinite dimensional case, cf.
[5]. The above hyperfinite realization of Brownian motion (and other diffusion processes) has found many applications to stochastic analysis (see
e.g. [5],[17]-[19],[39]-[41],[46]-[49],[52],[57],[61],[72],[85],[99],[108],
[115],[120],[121],[123],[125]).

As particularly striking recent applications of non-standard methods in
stochastic analysis, let us mention Cutland's work on stochastic control, Hoover's, Keisler's, Lindstrøm's and Perkins' work on stochastic
differential equations, and Lindstrøm's diffusions on fractals, Perkins'
proof of Levy's formula for local times in terms of excursions, Stoll's
characterization of (higher dimensional time) Lêvy Brownian motion.

In the domain of random fields, important results have been obtained,
particularly by Kessler (on the problem of the global Markov property
of random fields and on a non-standard characterization of certain
generalized random fields).

In [76],[108],[120],[123] and more extensively in [5] we discussed an application of non-standard analysis to the study of certain stochastic Schrödinger operators which arise in the study of polymers. The problem is to give
a meaning to the operator $-\Delta + \lambda \int_0^t \delta(x-b)(s))ds$, with b Brownian
motion in $\mathbb{R}^d$, in $L^2(\mathbb{R}^d,dx)$ (formally this is of the type discussed
in IV.3, with Y a path of Brownian motion). By using the hyperfinite
model for b and the above hyperfinite Dirichlet form construction of
"$-\Delta + \lambda\delta(x)$", a meaning to the above operator can be found, for $d \leq 5$,
λ infinitesimal negative for d = 4,5.

In [5] we have given a hyperfinite version of Symanzik's polymer representation for quantum fields φ, with classical interaction depending only on φ^2 (like the $\lambda\varphi^4_d$ model). Using our result on above "polymer operator" for $d \leq 5$, we have given a (partial) (non-trivial) construction of the φ^4_d and $(\varphi^2_1\varphi^2_2)_d$ models, $d \leq 5$, with λ infinitesimal negative.

Hyperfinite lattice quantum field models are discussed in [5], with applications to exponential interactions and gauge fields. For a non-standard approach to perturbation theory in quantum electrodynamics see [43], [44].[7] We like to close by mentioning Stoll's work on hyperfinite random walks and polymer measures (obtaining in particular a new construction, with better control on approximations, of Symanzik's - Varadhan's result on Edwards' 2-dimensional polymer measure) ([85]).

We hope at this point the reader has at least gotten some impression of how rich the field of applications of non-standard analysis actually is and might wish to look further into some of the numerous more specific references we have given. If moreover she/he would start wondering about the possible use of non-standard tools in his own favourite subject, then our main goal with these lectures would have been achieved.

Acknowledgements

It is a special pleasure to thank Ludwig Streit for kindly inviting me to the ZiF-Project No. 2 and to write up these lectures. I also thank heartily Jens-Erik Fenstad, Raphael Høegh-Krohn and Tom Lindstrøm for the joy of collaboration over many years on our book on non-standard analysis, on which these lectures are based. I am grateful to Professors Keith Stroyan and Josi Bayod for sending me parts of their book [18], before publication. Skilfull typing by Mrs. Jegerlehner, Mischke and Richter is also gratefully acknowledged.

Footnotes

1) And also e.g. [1]-[3], [5], [8], [20], [23], [26], [33], [35], [42], [51], [53], [64], [70], [77], [78], [126].

2) Usually they are called simply "standard". To avoid the ensuing, at the beginning somewhat confusing, situation of having to call e.g. the set $^*\mathbb{R}$ of all hyperreal "standard" we replace here "standard" by "*-standard", reserving the name standard for the numbers r in $^*\mathbb{R}$ of the form $^*r$ (i.e. the real numbers as embedded in $^*\mathbb{R}$).

3) See e.g. [18], [5], [12], [14] for a precise, formalized statement.

4) We have based our formulation on a model, constructed using an ultrafilter U extending Cof ($\mathbb{N}$) ("ultrapower construction"). Alternative formulations of non-standard analysis exist. In particular, an axiomatic approach is possible and has some appealing features, in particular it has also been used in [17], [24] for "nonstandard" calculus. See also e.g. [1], [7], [25], [26], [29], [30], [31], [35], [59], [64]-[68], [69], [86], [106], [114], [126]. In particular, Edward Nelson has given a formulation of nonstandard analysis as "internal set theory", within a modified Zermelo-Fraenkel framework. This formulation has had great influence and has been taken as a basic "système de pensée" for the subsequent work, particularly on applications, see e.g. [31]-[35],[48], [54], [56], [64]-[67], [87], [90], [106], [112], [113]. Unfortunately, time and space do not permit us to enter a discussion of the relations between Nelson's approach and the one presented here, but going back and forth between Nelson's language and the one presented here causes little trouble once some working knowledge of one of the approaches has been acquired. Let us also mention that E. Nelson has subsequently also given fascinating and more radical approaches to portions of mathematics, somewhat related to nonstandard models, see [27], [28].

5) A question which often puzzles the beginners is the one of "uniqueness of the model" (dependence on the ultrafilter, e.g.). There are isomorphism theorems, see e.g. [16] and references therein, however

the main point is that for applications the flexibility of the construction of the nonstandard world is rather an advantage than a disadvantage, see [5] and references therein, and also [59], [69], [75], [81], [115].

6) More precisely, *-standard non-real.

7) For previous non-standard analysis work in quantum field theory see [38], [47], [50], [82].

References

[1] A. Robinson, Non-Standard Analysis, North-Holland, Amsterdam (1966)

[2] D. Laugwitz, Infinitesimalkalkül, Bibliographisches Institut, Mannheim (1978)

[3] D. Laugwitz, The Theory of Infinitesimals, Accademia Nazionale Lincei Roma (1980)

[4] Also e.g. a) Letters by A. Guggenheimer, D.D. Spalt, I. Graham-Guiness, and references therein in "The Mathematical Intelligencer" b) D. Laugwitz, Grundbegriffe der Infinitesimalmathematik bei Leonhard Euler, pp. 459-483 in Folkerts und Lindgren, Edts. Mathemata, Franz Steiner Verlag, Stuttgart (1985); [62],[63],[83],[118].

[5] S. Albeverio, J.E. Fenstad, R. Høegh-Krohn, T. Lindstrøm, Non standard methods in stochastic analysis and mathematical physics, Acad. Press, New York (1986)

[6] A. Robinson, Selected Papers, North-Holland, Amsterdam (1979)

[7] M. Machover, J. Hirschfeld, Lectures on Non-Standard Analysis, Lect. Notes in Maths. $\underline{94}$, Springer, Berlin (1969)

[8] W.A.J. Luxemburg, Non-standard Analysis, Lectures on A. Robinson's Theory of Infinitesimals and Infinitely Large Numbers, Caltech Bookstore, Pasadena, rev. 1964

[9] A. Hurd, P. Loeb, Victoria Symposium on Nonstandard Analysis, Lect. Notes in Maths. $\underline{3}$, 109, Springer (1974)

[10] W.A.J. Luxemburg, Ed., Applications of Model Theory to Algebra, Analysis and Probability, Holt, New York (1969)

[11] A.H. Lightstone, A. Robinson, Non-Archimedean Fields and Asymptotic Expansions, North-Holland (1975)

[12] M. Davis, Applied Nonstandard Analysis, J. Wiley, New York (1977)

[13] D.R. Johnson, Bibliography of Nonstandard Analysis, June 1975, Lect. Notes in Maths. $\underline{3}$, Dept. Maths., Univ. of Pittsburgh, Pittsburgh, PA

[14] A. Hurd, P.A. Loeb, An introduction to nonstandard real analysis, Academic Press, New York (1985)

[15] J.M. Henle, E.M. Kleinberg, Infinitesimal Calculus, MIT-Press, Cambridge (1979)

[16] H.J. Keisler, Foundations of infinitesimal calculus, Prindle, Weber and Schmidt, Boston (1976)

[17] H.J. Keisler, An infinitesimal approach to stochastic analysis, Mem. Am. Math. Soc. $\underline{297}$ (1984)

[18] K.D. Stroyan, J.M. Bayod, Foundations of infinitesimal stochastic analysis, North-Holland, Amsterdam (1986)

[19] A.E. Hurd, Ed., Nonstandard Analysis - Recent Developments, Lect. Notes in Maths. $\underline{983}$, Springer, Berlin (1983)

[20] K.D. Stroyan, W.A.J. Luxemburg, Introduction to the theory of infinitesimals, Academic Press (1976)

[21] G.H. Moore, Zermelos's axiom of choice, Springer, New York (1983)

[22] E. Zakon, Remarks on the nonstandard real axis, pp. 195-227 in Ref. [10]

[23] W.A.J. Luxemburg, What is nonstandard analysis? Amer. Math. Monthly 80, 38-67 (1973)

[24] J. Keisler, Elementary Calculus, Prindle, Weber & Schmidt, Boston (1976)

[25] Shih-Chao Lin, A proof theoretic approach to non-standard analysis with emphasis on distinguishing between constructive and non-constructive results I, pp. 391-414, in "The Kleene Symposium", 1980, North-Holland, Amsterdam (1980); II, pp. 281-296 in "Models and Sets", Eds. C.H. Müller, M.M. Richter, Lect. Notes in Maths. 1103, Springer, Berlin (1983)

[26] E. Nelson, Internal set theory, Bull. Am. Math. Soc. 83, 1165-1198 (1977)

[27] E. Nelson, Radically elementary probability theory, Princeton Preprint (1984)

[28] E. Nelson, Predicative Arithmetic, Princeton Preprint (1985)

[29] K. Hrbacek, Axiomatic foundations for non-standard analysis, Fund. Math. XCVIII, 1-19 (1978)

[30] P. Vopenka, Mathematics in the alternative set theory, Teubner, Leipzig (1979)

[31] M. Richter, Ideale Punkte, Monaden und Nichtstandard-Methoden, Vieweg (1982)

[32] B. Benninghofen, Superinfinitesimals and the calculus of the generalized Riemann integral, pp. 9-52 in Ref. [25]

[33] F. Diener, Cours d'Analyse Non-standard, Université d'Oran, Office Publ. Univ. (1983)

[34] T. Sari, Moyennisation dans les systèmes différentiels à solutions rapidement oscillantes, Thèse Mulhouse (1983)

[35] R. Lutz, M. Goze, Non standard analysis, Lect. Notes Math. 881, Springer (1981)

[36] A. Ostebee, P. Gambardella, M. Dresden, A "nonstandard approach to the thermodynamic limit II. Weakly tempered potentials and neutral Coulomb systems, J. Math. Phys. 17 1570-1578 (1976)

[37] Cz. Wôzniak, Non-standard analysis and its applications to mechanics, pp. 322-341 in "Trends in Appl. Pure Math. and Mech. II, Ed. H. Zorski, Pitman, London (1977)

[38] S. Nagamachi, T. Mishimura, Linear canonical transformations on Fermion Fock space with indefinite metric, Tokushima-Osaka Preprint (1984)

[39] N.J. Cutland, Simplified existence for solutions to stochastic differential equations, Stochastics 14 (1985)

[40] N.J. Cutland, Nonstandard measure theory and its applications, Bull. London Math. Soc. 15, 529-589 (1983)

[41] T.L. Lindstrøm, The structure of hyperfinite stochastic integrals, Z. Wahrsch. verw. Geb.

[42] K. Stroyan, Infinitesimal analysis of curves and surfaces, pp. 197-232 in Ed. J. Barwise, Handbook of Mathematical Logic, North-Holland, Amsterdam (1977)

[43] R. Fittler, Some nonstandard quantum electrodynamics, Helv. Phys. Acta 57. 579-609 (1984)

[44] R. Fittler, More nonstandard quantum electrodynamics, FU Berlin Preprint (1985)

[45] Li Bang-He, Nonstandard analysis and multiplication of distributions, Scientia Sinica 21, 561-   (1978)

[46] P.A. Loeb, Measure spaces in nonstandard models underlying standard stochastic processes, Proc. Int. Congress Math. (1983)

[47] Ph. Blanchard, J. Tarski, Renormalizable interactions in two-dimensions and sharp-time fields, Acta Phys. Austr. 49, 129-152 (1978)

[48] P. Cartier, Perturbations singulières des êquations différentielles ordinaires et analyse non-standard, Sêm. Bourbaki 81-82, Astêrisque 92-93 (1982)

[49] T.L. Lindstrøm, A Loeb-measure approach to theorems by Prohorov, Sazonov, and Gross, Trans. Am. Math. Soc. 269 (1982) 521-534

[50] P.J. Kelemen, A. Robinson, The nonstandard $:\varphi_2^4(x):$ modeld, I , J. Math. Phys. 13. 1870-1874 (1972); II, J. Math. Phys. 13, 1875-1978 (1972)

[51] A. Voros, Introduction to nonstandard analysis, J. Math. Phys. 14, 292-296 (1973)

[52] A. Stoll, A nonstandard construction of Lêvy Brownian motion, ZiF-Preprint 1984, to appear in Z. Wahrsch. verw. Geb.

[53] M.O. Farrukh, Applications of nonstandard analysis to quantum mechanics, J. Math. Phys. 16, 177-   (1975)

[54] M.A. Chaouch, Analyse non standard et théorêmes de stabilitê, Thèse 3e cycle, Strasbourg (1984)

[55] N.J. Cutland, Infinitesimal methods in control theory, deterministic and stochastic, Hull Preprint (1985), Acta Appl. Math.

[56] R. Bebbouchi, J.L. Callot, F. Diener, M. Diener, Phênomênes micro- et macroscopiques, ONRS, Mathêm., Oran (1983)

[57] J. Oikkonen, Harmonic analysis and nonstandard Brownian motion in the plane, Math. Scand.

[58] A. Sloan, The strong convergence of Schrödinger propagators, Trans. Am. Math. Soc. 264, 557-570 (1981)

[59] J.E. Fenstad, Is nonstandard analysis relevant for the philosophy of mathematics?, Fondamenti della Matematica: Attualitâ del Problema, Synthese 62 (1984)

[60] A.L. MacDonald, Sturm-Liouville theory via nonstandard analysis, Ind. Un. Math. J. 25 (1976)

[61] A. Stoll, A nonstandard construction of Lêvy Brownian motion with applications to invariance principles, Diplomarbeit, Freiburg (1982)

[62] G. Fisher, Cauchy's variables and the orders of the infinitely small, pp. 261-277, Brit. J. Phil. Soc. 30 (1979)

[63] J.P. Cleave, The concept of "variables" in nineteenth century analysis, pp. 266-278 in Ref. [62]

[64] J. Harthong, L'analyse non-standard, La Recherche 148, 1194-1201 (1983)

[65] G. Reeb, L'analyse non-standard, vieille de soixante ans?, Publ. IRMA, Strasbourg

[66] J. Harthong, G. Reeb, Intuitionisme 84, Preprint, Strasbourg (1983)

[67] J. Harthong, L'analyse non-standard, une rêvolution scientifique, Gazette des Math. 17, 25-54 (1981)

[68] L.A. Steen, New models of the real-number line, Scient. Amer., Aug. 71

[69] J.E. Fenstad, On the metaphysics of the real line, Oslo preprint

[70] V. Komkov, C. Waid, Asymptotic behavior of non-linear inhomogeneous equations via non-standard analysis, Ann. Pol. Math. 28, 67-87 (1973)

[71] M. Wolff, Spectral theory of group representation and their nonstandard hull, Trans. Am. Math. Soc.

[72] S.A. Kosciuk, Non-standard stochastic methods in diffusion theory, Ph.D. Thesis, Madison (1982)

[73] A. Alonso Coria, Shrinking potentials in the Schrödinger equation, Ph.D. Thesis, Princeton Univ. (1978)

[74] B. Birkeland, A singular Sturm-Liouville problem treated by non-standard analysis, Math. Scand. 47 (1980)

[75] J.E. Fenstad, Non-standard methods in stochastic analysis and mathematical physics, Jber. d. Dt. Math. Verein 82, 167,180 (1980)

[76] S. Albeverio, Non-standard analysis: Polymer models, quantum fields, Acta Phys. Austr. Supp. 26, 233-254 (1984)

[77] W.S. Hatcher, Calculus is algebra, Am. Math. Monthly 89, 362-370 (1982)

[78] D.H. Van Osdol, Truth with respecto to an ultrafilter or how to make intuition rigorous, Am. Math. Monthly 79, 355-363 (1972)

[79] S. Ward Henson, L.C. Moore Jr., Nonstandard analysis and the theory of Banach spaces, Preprint

[80] R.F. Hoskins, Standard and Nonstandard Mathematical Analysis, Ellis Horwood, Chichester (1986)

[81] J.E. Fenstad, The discrete and the continuum in the mathematics and natural sciences, Oslo Preprint, Nov. 1985

[82] S. Moore, Non-standard applications of non-standard analysis, Bogotá Preprint

[83] D. Laugwitz, Cauchy and infinitesimals, Darmstadt Preprint (1985)
[84] L.M. Pecora, A nonstandard infinite dimensional vector space approach to Gaussian functional measures, J. Math. Phys. $\underline{23}$, 969-982 (1982)
[85] A. Stoll, Self-repellent random walks and polymer measures in two dimensions, Diss., Bochum (1985); and to appear in Proc. BiBoS Symp. II, Lect. Notes Maths., Springer (Eds. S. Albeverio, Ph. Blanchard, L. Streit)
[86] W. Schnitzspan, Konstruktive Nonstandard-Analysis, Diss., Darmstadt (1976)
[87] A.K. Zvonkin, M.A. Shubin, Nonstandard analysis and singular perturbations of ordinary differential equations, Russ. Math. Surv. $\underline{39}$, 69-131 (1984)
[88] T. Kamae, A simple proof of the ergodic theorem using nonstandard analysis, Israel J. Math. $\underline{42}$, 284-290 (1982)
[89] P.A. Loeb, A functional approach to nonstandard measure theory, Contemp. Maths. $\underline{26}$, 251-261 (1984)
[90] III Rencontre de géometrie du Schnepfenried, Vol. 2, Astérisque 109-110 (1983)
[91] S. Albeverio, R. Høegh-Krohn, Schrödinger operator with point interactions and short range expansions, Physica $\underline{124A}$, 11-28 (1984)
[92] S. Albeverio, F. Gesztesy, R. Høegh-Krohn, H. Holden, Some exactly solvable models in quantum mechanics and the low energy expansions, pp. 12-28 in "Proceedings of the Second International Conference on Operator Algebras, Ideals and Their Applications in Theoretical Physics," Ed. H. Baumgärtel et al., Teubner, Leipzig (1984)
[93] S. Albeverio, F. Gesztesy, R. Høegh-Krohn, H. Holden, Solvable models in quantum mechanics, book in preparation
[94] L. Arkeryd, Loeb solutions of the Boltzmann equation, Arch. Rat. Mech. Anal. $\underline{86}$, 85-97 (1984)
[95] L. Arkeryd, J. Bergh, Some properties of Loeb-Sobolev spaces, Göteborg Preprint (1985)
[96] L.L. Hurd, P.A. Loeb, Bounds on the oscillation of spin systems, J. Math. Anal. Appl. $\underline{86}$, (1982)
[97] C. Keßler, Nonstandard methods in random fields, Diss., Bochum (1984)
[98] C.W. Henson, Unbounded Loeb measures, Proc. Am. Math. Soc. $\underline{74}$ (1979)
[99] T. Lindstrøm, Non-standard energy forms and diffusions on manifolds and fractals, to appear in Proc. Ascona Conf. "Stochastic Processes in Classical and Quantum Systems", Eds. S. Albeverio, G. Casati, D. Merlini, Lect. Notes Phys. Springer (1986)
[100] M. Fukushima, Dirichlet forms and Markov processes, North-Holland, Amsterdam (1980)
[101] M.L. Silverstein, Symmetric Markov Processes, Springer, Berlin (1974)

[102] S. Albeverio, R. Høegh-Krohn, Diffusion fields, quantum fields and fields with values in Lie groups, pp. 1-98 in M. Pinsky (Ed.) Stochastic Analysis and Applications, M. Dekker, New York (1985)

[103] M. Röckner, N. Wielens, Dirichlet forms - closability and change of speed measure. In "Infinite dimensional analysis and stochastic processes", Ed. S. Albeverio', Pitman, London (1985)

[104] M. Fukushima, Energy forms and diffusion processes, pp. 65-97 in "Mathematics + Physics, Lectures on Recent Results", Vol. I, Ed. L. Streit, World Scientific Publ. (1985)

[105] J. Persson, Fundamental theorems for linear measure differential equations, Lund Preprint (1985) (to appear in Math. Scand.)

[106] T. Kawai, Non-standard analysis by axiomatic method, pp. 55-76 in Logic Proc. Singapore 81, North-Holland, Amsterdam (1983)

[107] R.M. Anderson, A non-standard representation for Brownian motion and Ito integration, Isr. J. Math. $\underline{25}$, 15-46 (1976)

[108] S. Albeverio, J.E. Fenstad, R. Høegh-Krohn, W. Karwowski, T. Lindstrøm, Perturbations of the Laplacian supported by null sets, with applications to polymer measures and quantum fields, Phys. Letts. $\underline{104}$ (1984)

[109] S. Albeverio, R. Høegh-Krohn, L. Streit, Energy forms, Hamiltonians and distorted Brownian paths, J. Math. Phys. $\underline{18}$, 907-917 (1977)

[110] C. Keßler, Example of extremal lattice fields without the global Markov property

[111] S. Albeverio, Some points of interaction between stochastic analysis and quantum theory, BiBoS-Preprint, to appear in Proc. Conf. "Stochastic Systems and Applications", Ed. K. Helmes (1986)

[112] M. Diener, C. Lobry, Eds. Analyse non standard et représentation du réel, Actes de l'Ecole d'Eté, OPU (Alger - CNRS (Paris) (1984)

[113] M. Diener, G. Reeb, Champs polynomiaux: nouvelles trajectoires remarquables, Preprint (1985)

[114] H.J. Skala, Non-archimedean utility theory, D. Reidel (1975)

[115] J.E. Fenstad, Lectures on stochastic analysis with applications to mathematical physics, Proc. Simposio Chileno Log. Mat., Santiago (1986)

[116] C. Keßler, The global Markov property for lattice spin systems in the case of uniqueness (in preparation)

[117] C. Keßler, The global Markov property of a convex combination of GMP-states (in preparation)

[118] D. Laugwitz, Cauchy and infinitesimals, Darmstadt Preprint (1985)

[119] R. Vesly, An intuitionistic infinitesimal calculus, pp. in "Constructive Mathematics", Lect. Notes Maths. $\underline{873}$, Springer, Berlin (1981)

[120] T. Lindstrøm, Nonstandard analysis and perturbation of the Laplacian along Brownian paths, pp. 180-200 in "Stochastic Processes - Mathematics and Physics", Proc. BiBoS I, Eds. S. Albeverio, Ph. Blanchard, L. Streit, Lect. Notes Maths. 1158, Springer (1985)

[121] E. Perkins, A global intrinsic characterization of Brownian local time, Ann. of Prob. 9, 800-817 (1981)

[122] S. Albeverio, J.E. Fenstad, R. Høegh-Krohn, Singular perturbations and non-standard analysis, Trans. Am. Math. Soc. 252, 275-295 (1979)

[123] S. Albeverio, Ph. Blanchard, R. Høegh-Krohn, Newtonian diffusions and planets, with a remark on non-standard Dirichlet forms and polymers, pp. 1-24 in "Stochastic Analysis and Applications," Eds. A. Truman, D. Williams, Lect. Notes Maths. 1095, Springer (1984)

[124] C. Keßler, Hyperfinite representation of generalized random fields, Bochum Preprint (1984)

[125] G.F. Lawler, A self-avoiding random walk, Duke Math. J. 47, 655-692 (1980)

[126] A. Robert, Analyse non standard, Presses Polyt. Romandes XVII, Lausanne (1985)

[127] V.E. Lyantse, Can non standard analysis be ignored? (Jordan form of an operator in an infinite-dimensional space) (Russ.), General theory of boundary value problems, pp. 108-112, Naukova D., Kiew, 1983

[128] B. Birkeland, D. Normann, A non-standard treatment of the equation $y' = f(y,t)$, Mat. Sem. Oslo (1980)

[129] P.A. Loeb, Conversion from nonstandard to standard measure spaces and applications in probability theory, Trans. Am. Math. Soc. 211, 113-122 (1945)

[130] E. Perkins, Stochastic processes and nonstandard analysis, in Ref. [19]

[131] A.N. Kochubei, Elliptic operators with boundary conditions on a subset of measure zero, Funct. Anal. Appl. 16, 137-139 (1982)

[132] W.A.J. Luxemburg, A. Robinson, Contributions to Non-Standard Analysis, North-Holland, Amsterdam (1972)

[133] J. Hartong, Etudes sur la méchanique quantique, Astériques 111 (1984)

TOPOLOGY AND ANOMALIES

Luis ALVAREZ-GAUMÉ[*]
Lyman Laboratory of Physics
Harvard University
Cambridge, Massachusetts 02138
USA

I. Introduction

These lectures are intended to give an overview on some of the recent applications of algebraic and differential topology in the elucidation of the general structure of anomalies [1], especially in the case of the non-Abelian anomalies [2]. Since this subject is now evolving quite vigorously, it is likely that by the time these lectures appear, there will have been new developments not covered here. Our aim however is to provide the reader with a reasonably pedagogical exposition of the basic tools of the trade without necessarily entering into the detailed technical discussions. The main mathematical results that we will use are those given by the Atiyah-Singer index theorem for elliptic operators [3]. Section II is dedicated to a review of some basic formulae in differential geometry, and to a presentation of the Atiyah-Singer index theorem as applied to the Dirac operator on arbitrary manifolds and with arbitrary gauge fields. In Section III we will discuss the topological underpinnings of the anomaly (Abelian and non-Abelian). We will make particular emphasis on the Hamiltonian formulation of the anomaly, because it makes the physics quite clear. It is not possible to define the Fock vacuum for a fermionic theory in the presence of some arbitrary gauge field, satisfying Gauss' law if there is an anomaly [4].

---

[*] Research supported in part by the National Science Foundation under Grant No. PHY82-15249.

The analysis of this problem is greatly simplified using the Atiyah-Singer index theorem for families of Dirac operators. After we present the basic ideas in the Hamiltonian framework, we will quickly overview the changes required to obtain the Euclidean formulation of the origin of anomalies and how it is related to exactly the same topological obstructions [2].

The reference list is far from complete. It is intended to provide guidance for further reading and not an exhaustive listing of all the work that has been done on the subject in the last few years. I would thus like to apologize to those authors whose work is not mentioned in the reference list.

II. Preliminaries

We begin this section with an overview of gauge theories in the language of differential forms [5].

Let M be a d-dimensional manifold with coordinates x^μ; and let G be the gauge group. Geometrically the gauge field $A_\mu = A_\mu^i \lambda^i$ specifies a connection on a principal bundle P with base M and fiber G. We write the connection 1-form as:

$$A = A_\mu \, dx^\mu . \qquad (1)$$

A is a matrix-valued field taking values in the Lie algebra of G. We take the adjoint representation of this Lie algebra to be spanned by anti-hermitian matrices $(\lambda^i)^a_{\ b}$. The gauge field strength becomes

$$F = dA + A^2 = \tfrac{1}{2}(\partial_\mu A_\nu - \partial_\nu A_\mu + [A_\mu, A_\nu])dx^\mu \, dx^\nu = \tfrac{1}{2} F_{\mu\nu} \, dx^\mu \, dx^\nu . \qquad (2)$$

F is also antihermitian, and $d = dx^\mu \, \partial/\partial x^\mu$ is the standard exterior derivative. A gauge transformation is a map from M into G which assigns an element of G to every point of M. The change of A and F under a gauge transformation is

$$A^g \equiv g^{-1}(A+d)g$$
$$F^g = g^{-1}Fg.$$
(3)

If Φ is any other matrix valued field taking values in the adjoint representation of the Lie algebra of G, the covariant derivative of Φ with respect to A is

$$D\Phi \equiv d\Phi + [A,\Phi]$$
(4)
$$[A,\Phi] \equiv A\Phi - (-1)^P \Phi A.$$

P is the degree of Φ as a form. A consequence of (2) and (4) is the Bianchi identity for gauge fields:

$$dF = d(dA + A^2) = dAA - AdA = FA - AF = -[A,F]$$
(5)
$$DF = 0.$$

Similarly, if we want to include the possible effect of gravitational background fields in M, it is useful to recall briefly the standard formula of Riemannian geometry. Let $g_{\mu\nu}$ be the metric on M:

$$ds^2 = g_{\mu\nu}(x)dx^\mu \otimes dx^\nu$$
(6)

in terms of d-beins: $g_{\mu\nu} = \eta_{ab} e^a_\mu e^b_\nu$ where η_{ab} could be either the Minkowski or Euclidean metrics depending on the case considered.

$$ds^2 = \eta_{ab} e^a \otimes e^b$$
$$e^a \equiv e^a_\mu dx^\mu.$$
(7)

We can also introduce the inverse d-bein $E \equiv e^{-1}$; $E^\mu_a e^b_\nu = \delta^b_a$. From (7) it is clear that the orthonormal tangent frames specified by e^a_μ

are defined only up to local frame rotations, i.e. if e^a, $a = 1,d$, form an orthonormal frame basis, then so does $L^a_b(x)e^b$ where $\eta_{ab} L^a_c L^b_d = \eta_{cd}$. Thus L will be a space-time dependent group element of $SO(1,d-1)$ ($SO(d)$) for Minkowskian (Euclidean) signature.

Once the group of local frame redefinitions has been introduced (and this is necessary if one wants to have spinors, because $GL(d,\mathbb{R})$ does not admit any finite dimensional spinor representation); the formulae of Riemannian geometry can be written in a form entirely analogous to the same formulae in gauge theories. In order to define parallel transport, we have to introduce a connection. In terms of frames, the collection of all orthonormal frames over M forms the canonical frame bundle $F(M)$; $F(M)$ is a principal bundle whose structure group is $SO(1,d-1)$ or $SO(d)$. Thus a connection is a 1-form with values in the structure group of $F(M)$. Let this connection (the spin connection) be represented by $\omega^a_b = \omega^a_{\mu b} dx^\mu$, satisfying $\omega_{ab} \equiv \eta_{ac} \omega^c_b = -\omega_{ba}$. In analogy with the gauge theory case we can introduce a covariant differential acting on forms by $D\Phi = d\Phi + [\omega, \Phi]$. In terms of indices, let $\Phi^{a_1 \ldots a_m}_{b_1 \ldots b_n}$ be a tensor-valued form. Then

$$(D\Phi)^{a_1 \ldots a_m}_{b_1 \ldots b_n} = d\Phi^{a_1 \ldots a_m}_{b_1 \ldots b_n} + [\omega, \Phi]^{a_1 \ldots a_m}_{b_1 \ldots b_n}$$

$$[\omega, \Phi]^{a_1 \ldots a_m}_{b_1 \ldots b_n} \equiv \sum_{j=1}^{m} \omega^{a_j}_{c_j} \Phi^{a_1 \ldots c_j \ldots a_m}_{b_1 \ldots b_n} - (-1)^p \sum_{j=1}^{n} \Phi^{a_1 \ldots a_m}_{b_1 \ldots c_j \ldots b_n} \omega^{c_j}_{b_j}.$$

(8)

The torsion and curvature associated to the connection ω are defined by

$$T^a = de^a + \omega^a_b e^b, \quad \text{or} \quad T = de + \omega e$$

$$\Omega^a_b = d\omega^a_b + \omega^a_c \omega^c_b, \quad \text{or} \quad \Omega = d\omega + \omega^2.$$

(9)

The Bianchi identities now follow by acting on both equations in (9) with d and using the definition of T:

$$D\Omega = d\Omega + [\omega,\Omega] = 0$$

$$dT = R e - \omega T \ . \tag{10}$$

A well-known result of classical Riemannian geometry is that there is a unique metric connection which is torsion-free: if we define $de^a = \frac{1}{2}\xi^a_{bc} e^b e^c$, $\omega^a_b = \omega^a_{b,c} e^c$, the torsion-free condition ($T = 0$) implies

$$\xi_{a,bc} = \omega_{ab,c} - \omega_{ac,b} \ . \tag{11}$$

Adding and subtracting appropriate cyclic permutations of (11) one easily arrives at:

$$\omega_{ab,c} = \frac{1}{2}(\xi_{a,bc} + \xi_{b,ca} - \xi_{c,ab}) \ . \tag{12}$$

Had we used an ordinary coordinate basis dx^μ rather than an orthonormal basis, we would have obtained the standard Christoffel connection $\Gamma^\alpha_\beta = \Gamma^\alpha_{\gamma\beta} dx^\gamma$, $\Omega^\alpha_\beta = d\Gamma^\gamma_\beta + \Gamma^\alpha_\lambda \Gamma^\lambda_\beta$, $\Omega^\alpha_\beta = R^\alpha_{\beta\lambda\mu} dx^\lambda dx^\mu/2$, and $R^\alpha_{\beta\lambda\mu}$ is the Riemannian tensor. We can use Ω to stand for the curvature tensor referred to either orthonormal or coordinate frames, because $\Omega^a_b = e^a_\alpha \Omega^\alpha_\beta E^\beta_b$, i.e. $d\Gamma + \Gamma^2 = e^{-1}(d\omega + \omega^2)e$. In terms of $\Gamma^\alpha_{\gamma\beta}$, the torsion-free condition implies $\Gamma^\alpha_{\gamma\beta} = \Gamma^\alpha_{\beta\gamma}$, and the fact that $\omega_{ab} = -\omega_{ba}$ is equivalent to the fact that $g_{\mu\nu}$ is covariantly constant with respect to $\Gamma^\alpha_{\gamma\beta}$. These features are enough to prove that

$$\Gamma^\alpha_{\gamma\beta} = \frac{1}{2}g^{\alpha\rho}(-\partial_\rho g_{\gamma\beta} + \partial_\gamma g_{\rho\beta} + \partial_\beta g_{\rho\gamma}) \ . \tag{13}$$

It is also easy to show that $\omega^a_{b,\mu} = e^a_\nu \nabla_\mu E^\nu_b = e^a_\nu (\partial_\mu E^\nu_b + \Gamma^\nu_{\nu\rho} E^\rho_b)$. Finally, under local frame rotations, e, ω and Ω transform as follows:

$$e \rightarrow (L^{-1}(x))e$$
$$\omega \rightarrow L^{-1}(\omega + d)L$$
$$\Omega \rightarrow L^{-1}\Omega L \ . \tag{14}$$

As pointed out in the introduction, the basic tool which has provided the new and interesting links between topology and field theory is the Atiyah-Singer index theorem [3]. The rest of this section is dedicated to a brief review of this important piece of modern mathematics.

So far, we have been dealing exclusively with the local formulae of Riemannian geometry and gauge theories. Since we will eventually be interested in global topological properties of M_n and of the gauge fields defined on M_n, let us briefly review how to describe the global properties of M_n. Let M_n be the manifold we are interested in, and let B be a principal bundle with base space M_n and the group G as fibre. A curved M_n in general cannot be specified by a single coordinate patch, and requires instead a covering by several coordinate patches U_i, each diffeomorphic to $\mathbb{R}^n$, with well-defined transition rules on their overlap. The gauge field (a connection on the principal bundle B) is defined separately on each patch with the requirement that gauge fields $A_{U_i} = g_{ij}^{-1}(A_{U_j} + d)g_{ij}$ on any non-empty overlap $U_i \cap U_j$. The topological information is thereby encoded in the transition functions g_{ij}. For example, let $M_n = S_n$, the n-dimensional sphere, and consider a gauge group G. The simplest covering consists of two coordinate patches, the upper and lower hemispheres D_+ and D_-, intersecting on an equatorial S^{n-1}. Letting $A_\pm$ denote the gauge field on $D_\pm$, we see that the bundle is defined by a single transition function g_{+-}, which in turn is a map from S^{n-1} into G. The bundle may thus be non-trivial if and only if $\pi_{n-1}G \neq 0$, i.e. if the (n-1)'st homotopy group of G is non-trivial. A familiar example of this construction is the instanton bundle on a compactified space-time manifold S^4. In this case the topological non-triviality is encoded by $\pi_3 G$ which is $\mathbb{Z}$ for all compact simple Lie groups.

From an elementary point of view, characteristic classes are local forms on M_n, constructed from the curvature two-forms, whose integrals over the manifold are sensitive to the existence of non-trivial topology, i.e., to the presence of homotopically non-trivial transition functions. The characteristic classes are closed forms (a p-form

Ω_p is closed if $d\Omega_p = 0$), but not globally exact. By Poincaré's lemma, their closure property means that they can be written locally as exact forms $\alpha = d\phi_i$ on each patch U_i, so that Stokes' theorem $\int_B d = \int_{\partial B} \phi$ ensures that their integrals depend only on the transition functions. If the forms α_i piece together to form a global form, then Stokes' theorem ensures that their integrals over a manifold M_n without boundary vanish.

We now outline a general procedure which produces characteristic classes. (More details, and references to the original literature can be found in Ref. [5].) Let Ω be a matrix valued two-form taking values in $\hat{G}$, and let $P(\Omega)$ be a polynomial in Ω satisfying the invariance property

$$P(g^{-1}\Omega G) = P(\Omega) \tag{15}$$

for any $g \in G$. Ω will be taken to be the gauge field strength F or the curvature Ω so that G is either the gauge group or the frame rotation group. One can now show that any polynomial satisfying (15) is (i) closed: $dP = 0$, and (ii) has integrals which are topologically invariant under deformations of the gauge field or connection and depend only on the transition functions. To make the discussion simpler, we note that it suffices to consider polynomials of the form $P_m = \text{Tr}\Omega^m$, since all other invariant polynomials can be constructed as sums of products of the P_m's. The closure of P_m follows from the Bianchi identity:

$$dP_m = d\,\text{Tr}\Omega^m = m\,\text{Tr}\,d\Omega\,\Omega^{m-1} = m\,\text{Tr}(d\Omega + \omega\Omega - \Omega\omega)\Omega^{m-1} = 0 \tag{16}$$

where we have used the cyclicity of the trace. This is legitimate because Ω is a two-form and thus behaves like an even element of a Grassmann algebra. To show that the integral of P_m over M_n is independent of the connection, consider two connections ω_0, ω_1 with the same transition functions, and define the interpolation

$$\omega_t = \omega_o + t(\omega_1 - \omega_o)$$

$$\Omega_t = d\omega_t + \omega_t^2, \quad 0 \le t \le 1. \tag{17}$$

Using the relation $\partial\Omega_t/\partial t = d(\omega_1-\omega_o) + \omega_t(\omega_1-\omega_o) + (\omega_1-\omega_o)\omega_t \equiv D_t(\omega_1-\omega_o)$ gives

$$\frac{\partial}{\partial t} P_m = \frac{\partial}{\partial t} \operatorname{Tr}\Omega_t^m = m \operatorname{Tr} \frac{\partial\Omega_t}{\partial t} \Omega_t^{m-1}$$

$$= m \operatorname{Tr} D_t(\omega_1-\omega_o)\Omega_t^{m-1}$$

$$= m\, d\, \operatorname{Tr}(\omega_1-\omega_o)\Omega_t^{m-1}, \tag{18}$$

where in the last step we have used the chain rule and the Bianchi identity $D_t\Omega_t = d\Omega_t + \omega_t\Omega_t - \Omega_t\omega_t = 0$. Integrating in t from 0 to 1 gives

$$P_m(\Omega_1) - P_m(\Omega_o) = m\, d \int_0^1 dt\, \operatorname{Tr}(\omega_1-\omega_o)\Omega_t^{m-1}, \tag{19}$$

which by Stokes' theorem implies $\int_M P_m(\Omega_1) = \int_M P_m(\Omega_o)$ when M has no boundary. We will in particular consider bundles with $G = U(n)$ or $SO(n)$. Let's therefore analyze some of the most useful characteristic classes for these cases. When $G = U(n)$, the two-form Ω is an antihermitian matrix, and it can formally be diagonalized, with eigenvalues ix_j, $j = 1,n$. Then any polynomial (15) can be written in terms of symmetric functions of the x's. The symmetric and homogeneous function of the x's are known as the Chern classes, and they are generated by

$$c(\Omega) = \det(1 + \frac{1}{2\pi}\Omega) = 1 + c_1(\Omega) + c_2(\Omega) + \ldots . \tag{20}$$

Thus

$$(2\pi)c_1(\Omega) = \sum_k ix_k$$

$$(2\pi)^2 c_2(\Omega) = \sum_{i<j} (ix_i)(ix_j)$$

$$(2\pi)^j c_j(\Omega) = \sum_{i_1<\ldots<i_j} (ix_{i_1})(ix_{i_2})\ldots(ix_{i_j}), \text{ etc.} \tag{21}$$

the sum (20); $c(\Omega)$ is known as the total Chern class and it is multiplicative with respect to direct sums of bundles $c(E \oplus F) = c(E)c(F)$ where E and F are two U(n) vector bundles over some manifold M. Another useful set of characteristic classes are the Chern characters, which are generated by

$$ch(\Omega) = \sum_i e^{ix_i/2\pi} . \qquad (22)$$

(The Chern character is additive with respect to the direct sum of bundles: $ch(E \oplus F) = ch(E) + ch(F)$, and plays a central role in the index theorem.)

When the bundle group is SO(n), Ω is an antisymmetric matrix, and can be skew diagonalized:

$$\Omega = \begin{pmatrix} 0 & x_1 & & & & & \\ -x_1 & 0 & & & & & \\ & & 0 & x_2 & & & \\ & & -x_2 & 0 & & & \\ & & & & \ddots & & \\ & & & & & x_n & \\ & & & & & -x_n & 0 \end{pmatrix} . \qquad (23)$$

As before, the characteristic polynomials can be written in terms of the "eigenvalues" x_i. The Pontrjagin classes are defined by expanding $p(\Omega) = \det(1 + \Omega/2\pi)$. Since Ω is antisymmetric, only terms containing an even number of Ω's do not vanish:

$$p(\Omega) = \det(1 + \frac{\Omega}{2\pi}) = 1 + p_1 + p_2 + \ldots$$

$$(2\pi)^2 p_1 = \sum_i (x_i)^2$$

$$(2\pi)^4 p_2 = \sum_{i<j} x_i^2 x_j^2$$

$$(2\pi)^6 p_3 = \sum_{i<j<k} x_i^2 x_j^2 x_k^2 . \qquad (24)$$

This concludes our short review of characteristic polynomials.

After all these preliminaries, we can formulate the index theorem. In a generic index problem, we have some elliptic operator D from some bundle E to some bundle F. In order to make things more concrete, we can think of D as the Dirac operator defined on M_{2n} in the presence of some gauge field A with group G. Then E is the space of positive chirality spinors (note that we are now taking M to be even dimensional so that chirality is defined, or, to put it more precisely, there are two independent spinor representations of the frame group SO(2n)), and F is the space of negative chirality spinors.

$$\not{D} = E_a^\mu \gamma^a (\partial_\mu + A_\mu + \frac{1}{2} \omega_{cd,\mu} \sigma^{cd}) ,$$

$$\{\gamma^a, \gamma^b\} = 2\delta^{ab} , \quad \sigma^{ab} = \frac{1}{4} [\gamma^a, \gamma^b] ,$$

$$\hat{\Gamma} = i^n \prod_{a=1}^{2n} \gamma^a , \quad \hat{\Gamma}^2 = 1 . \tag{25}$$

E and F in this case are characterized by having $\hat{\Gamma} = +1$ or $\hat{\Gamma} = -1$; and since we are also including a gauge field A, E and F carry some representation of the gauge group G. In this setting, the Weyl operator is $D_+ \equiv \not{D}(1 + \hat{\Gamma}/2) : E \to F$, and its adjoint $\not{D}_- : F \to E$. The collection $(\not{D}_+, \not{D}_-, E, F)$ forms an elliptic complex, and the index of the complex is defined by

$$\text{Ind } \not{D} = \dim \ker \not{D}_+ - \dim \ker \not{D}_- . \tag{26}$$

In general, if we have an elliptic operator $\hat{O}$ acting on some bundle E (spinor fields, vector fields, tensor fields, etc.) $\hat{O} : E \to F$, and $\hat{O}^\dagger : F \to E$, then the index of $\hat{O}$ is defined as in (26) $\text{Ind } \hat{O} = \dim \ker \hat{O} - \dim \ker \hat{O}^\dagger$, where $\ker \hat{O}$ indicates the space of zero modes of the operator $\hat{O}$.

What the index theorem states is that $\text{Ind } \hat{O}$ is a topological invariant which depends on E, F and M_n, and which moreover can be written as the integral of a characteristic class which measures the topological twisting of E, F and M. One of the aims of the index theorem is to provide explicitly the characteristic class which determines the index of a given elliptic complex. There are several approaches to computing the index of an operator in the mathematical literature. The one which is the closest to methods used in physics is the heat kernel expansion [6]. This procedure is as follows.

Given the operators $\hat{O}$, $\hat{O}^\dagger$, we can construct two "Laplacians" $\Box_+ = \hat{O}^\dagger \hat{O}$, $\Box_- = \hat{O}\hat{O}^\dagger$. By construction, $\Box_\pm$ are elliptic, hermitian and self-adjoint, and since M_n is compact, the spectrum of $\Box_\pm$ is discrete, and the degeneracy of each eigenvalue is finite. Also $\Box_+ : E \to E$, $\Box_- : F \to F$ so that $\Box_\pm$ have well-defined eigenvalue problems. Furthermore, let λ be an eigenvalue of $\Box_+$ with eigenfunction $\phi_\lambda : \Box_+ \phi_\lambda = \lambda \phi_\lambda$, then $\hat{O}\phi_\lambda \equiv \psi_\lambda$ is an eigenfunction of $\Box_-$ with the same eigenvalue $\Box_- \psi_\lambda = \hat{O}\hat{O}^\dagger \hat{O} \psi_\lambda = \lambda \hat{O} \psi_\lambda$, and vice-versa. We thus learn that the spectrum of non-zero eigenvalues of $\Box_+$ and $\Box_-$ are identical. This is not true however of the zero eigenvalues. The zeros of $\Box_+$ are the same as the zeros of $\hat{O}$, i.e., $\ker \Box_+ = \ker \hat{O}$ and $\ker \Box_- = \ker \hat{O}^\dagger$. These properties imply that

$$\text{Ind } \hat{O} = \text{Tr } e^{-t\Box_+} - \text{Tr } e^{-t\Box_-}. \tag{27}$$

The first trace is taken over E, and the second over F. In the standard heat kernel expansion, one computes the asymptotic expansion in $1/t$ for $t \to 0^+$ for both terms on the right-hand side of (27), and then picks out the piece which is independent of t.

Rather than elaborating on the proof of the index theorem in the form that will be used, we will simply state the result and calculate a few interesting examples. For an elliptic operator $\hat{O}$ interpolating between the bundles E and F, the general formula of Atiyah and Singer [3] gives:

$$\text{Ind } \hat{O} = \int_{M_{2n}} \left[\left(\text{ch}(E) - \text{ch}(F) \right) \frac{\text{td}(TM \otimes C)}{e(TM)} \right]_{\text{vol}} \tag{28}$$

ch(E), ch(F) represent the Chern characters of the bundles E and F. td(TM ⊗ C) is a characteristic class known as the Todd genus for the complexified tangent bundle TM ⊗ C' of the manifold M. In terms of eigenvalues of $\Omega/2\pi$, x_i, it is represented by

$$\mathrm{td}(TM \otimes C) = \prod_i \left(\frac{x_i}{1-e^{-x_i}}\right)\left(\frac{x_i}{1-e^{x_i}}\right). \quad (29)$$

e(TM) is the Euler class, i.e., it is the invariant polynomial represented by $\prod_i x_i$. The division by e(TM) in (28) simply means that one has to expand the characteristic polynomial appearing in (28) and consider only the term proportional to the volume form of M. As a first example of (28), let us consider the Dirac operator without gauge fields on M. From a physical point of view, this means considering only the interaction of fermions with some external gravitational background. From (28) we notice that the bundles E, F are the positive and negative chirality spinor bundles Δ_+, Δ_- are just those induced by the standard spin connection. In other words, ch(E) is obtained by computing $\mathrm{Tr}\, \Omega_{ab} \sigma_+^{ab}/4\pi$ where σ_+^{ab} are the generators of SO(2n) in the spinor representation of positive chirality. Similarly, ch(F) is obtained by computing $\mathrm{Tr}\, \exp 1/4\pi\Omega_{ab}\, \sigma_-^{ab}$, σ_-^{ab} being the generators in the spinor representation of negative chirality. If for a moment we forget that $\Omega_{ab}/2\pi$ is a 2-form, and consider it just as an infinitesimal element of SO(2n), the traces to be computed are nothing but the characters of a generic element of SO(2n) in the two spinor representations. This can be done as follows. SO(2n) is a group of rank n, which means that the space of weights has n-generators e_i, $i = 1,n$. Then the weights of the spinor representations (i.e. the eigenvalues of the elements in the Cartan subalgebra of SO(2n) are given by $\frac{1}{2}\sum_{a=1}^{n}\varepsilon_a e_a$, $\varepsilon_a = \pm 1$, and the two spinor representations are distinguished by the conditions $\prod_{a=1}^{n}\varepsilon_a = +1$ ($\hat{\Gamma} = +1$) or $\prod_{a=1}^{n}\varepsilon_a = -1$ ($\hat{\Gamma} = -1$). In general, the weights Λ of a given representation of a group of rank r, G_r, are r-component vectors, and the characters of an arbitrary group element g is

$$\text{Tr}_\Lambda g = \Sigma \exp(i \sum_{i=1}^{r} \Lambda_i \phi_i).$$

The angles ϕ_i are determined by noticing that by a similarity transformation in G_r we can write $g = S(\exp i\phi_i H_i)S^{-1}$ where the H_i's are the generators of the Cartan subalgebra of G_r: the maximal set of commuting operators. Since the character involves the trace of the matrix representation of a group element, it is invariant under similarity transformations, and therefore only depends on the angles ϕ_i and on the weights of the representation (the eigenvalues Λ_i of the H_i's, $H_i|\Lambda\rangle = \Lambda_i|\Lambda\rangle$). After this small digression in group theory, we can now compute the Chern characters of the spinor bundles. If x_α, $\alpha = 1, n$, are the skew eigenvalues of $\Omega_{ab}/2\pi$, one can easily show that

$$\text{Tr}_{\Delta_+}(\exp \frac{1}{4\pi}\Omega_{ab}\sigma^{ab}) + \text{Tr}_{\Delta_-}(\exp \frac{1}{4\pi}\Omega_{ab}\sigma^{ab}) = \prod_\alpha 2\cosh(\frac{x_\alpha}{2}) \quad (30)$$

$$\text{Tr}_{\Delta_+}(\exp \frac{1}{4\pi}\Omega_{ab}\sigma^{ab}) - \text{Tr}_{\Delta_-}(\exp \frac{1}{4\pi}\Omega_{ab}\sigma^{ab}) = \prod_\alpha 2\sinh(\frac{x_\alpha}{2}). \quad (31)$$

(31) is the desired expression for $\text{ch}(\Delta_+) - \text{ch}(\Delta_-)$. Thus the characteristic polynomial which determines the index of the Dirac operator is:

$$\hat{A}(M) = \prod_\alpha 2\sinh(\frac{x_\alpha}{2}) \frac{1}{(\prod_\alpha x_\alpha)} \cdot \prod_\alpha \left(\frac{x_\alpha}{1-e^{-x_\alpha}}\right)\left(\frac{-x_\alpha}{1-e^{x_\alpha}}\right) = \prod_\alpha \frac{x_\alpha/2}{\sinh x_\alpha/2}. \quad (32)$$

The characteristic polynomial ($\hat{A}$-genus, or Dirac genus) determines the index of $\slashed{D}$ when there are no gauge fields. If we also want to consider fermions coupled to some gauge field with gauge group G, let V be the vector space carrying the representation of the fermions with respect to G. Then $E = \Delta_+ \otimes V$, $F = \Delta_- \otimes V$. Using the properties of Chern characters alluded to before: $\text{ch}(\Delta_\pm \otimes V) = \text{ch}\Delta_\pm \text{ch}V$, and the characteristic polynomial which determines the index in this case is simply

$$\text{ch } V \hat{A}(M) = (\text{Tr } e^{F/2\pi i}) \hat{A}(M). \tag{33}$$

F is the gauge field strength, and the trace is taken over the representation of G carried by the fermions.

This concludes our review of some classical results in index theory. As we willl see later on, (33) is all one needs to understand anomalies in global currents. If we are concerned also with anomalies in gauge currents, then (33) is not enough, and one needs an extension of the index theorem also derived by Atiyah and Singer [3], which corresponds to the study of families of operators. It is this more general form of the index theorem for families of elliptic operators that allows a geometrical and topological interpretation of many qualitative features of the fermion-gauge field dynamics. Though the rigorous derivation of the general form of this theorem requires rather sophisticated mathematical techniques, one can understand some of the ideas in simple examples. We know that the index of an operator is an invariant under continuous deformations. Thus given some elliptic complex with operator $A = E \to F$, E,F bundles over M, we know how to compute ind $A = \dim \ker A - \dim \ker A^+$. Let us for the time being imagine that $\ker A^+ = 0$, and $\ker A = n$ (like in the case of a Dirac operator in the presence of an instanton with Pontrjagin number equal to n). Rather than considering an operator A or M, we could consider a family of continuous deformations of A; A_p where p takes values of some parameter space P itself a compact manifold. Assuming that $\ker A_p^+ = 0$, we can construct a smooth vector bundle over P where the fiber at each point $p \in P$ is $\ker A_p$. Since the index is a topological invariant, we know that $\text{Ind } A_p = \dim \ker A_p = \text{constant}$. The fact that $\dim \ker A_p = n$, means that there is a lot more topological information in this problem than the standard index theorem. The standard index theorem is the simplest topological invariant of the bundle over P defined by $\ker A_p$. In order to obtain the maximal amount of topological information, we have to compute the Chern classes of this "index" bundle over P. These Chern classes are computed by the index theorem for families of operators [3]. The index for families of operators can be defined even in those situations where there are jumps in the dimension of ker A and ker A^+, and this is indeed the case appearing in the treatment of the non-Abelian anomaly.

III. Gauss' Law and the Topological Origin of Anomalies

The connection between topology and field theory in the presence of gauge fields is basically due to the necessity of imposing Gauss' law as a constraint in the space of physical states. There are two ways of analyzing this problem, one is Hamiltonian and the other is Euclidean. Though both methods are equivalent, each has its own merits and drawbacks. In this section we present the basic Hamiltonian analysis of the problem. Although the computation of the anomaly will be carried out to full completion in the next section using the Euclidean treatment, it is helpful to understand how to detect gauge anomalies in a purely Hamiltonian framework [4].

In a pure gauge theory, the easiest way to obtain the Hamiltonian formulation is to work in the $A_0 = 0$ gauge. In this form the theory is manifestly Hamiltonian with

$$H = \frac{1}{2} \int d^3x \, \text{Tr}(\vec{E}^2 + \vec{B}^2). \tag{34}$$

$\vec{E}$ and $\vec{B}$ are matrix fields valued in the Lie algebra of G, and they represent the non-Abelian analogues of the usual electric and magnetic fields. H is still invariant under arbitrary dependent gauge transformations? If we compactify space to a three sphere S^3, the space of gauge transformations is equivalent to the spaces of maps from S^3 into G defined so that the north pole (the point at infinity in the uncompcatified space) is always mapped into the unit element of G. Let $\mathfrak{g}^{(3)}$ stand for the three-dimensional gauge group; and let $A^{(3)}$ be the space of all possible three-dimensional gauge field configurations on S^3. The naive space of configurations associated to the dynamics generated by H in (34) is $A^{(3)}$. However, because of gauge invariance, the true configurations $A^{(3)}/\mathfrak{g}^{(3)} \equiv M^{(3)}$. Geometrically $A^{(3)}$ is the total bundle with base space $M^{(3)}$ and fiber $g^{(3)}$. $A^{(3)}$ however is topologically trivial, because given any two configurations $A_1, A_2 \in A^{(3)}$ there is always a line joining them, namely $A(s) = A_1 + s(A_2-A_1)$: $A(0) = A_1, A(1) = A_2$. Hence all the topology of $M^{(3)}$ is induced by the topology of $g^{(3)}$ using the exact homotopy sequence:

$$\pi_n(A^{(3)}) \to \pi_n(A^{(3)}/g^{(3)}) \to \pi_{n-1}(g^{(3)}) \to \pi_{n-1}(A^{(3)})$$

and the fact that $\pi_n(A^{(3)}) = 0$ for all n, we get $\pi_n(M^{(3)}) \approx \pi_{n-1}(g^{(3)})$. In particular, for $n=1$, we find that $M^{(3)}$ is not simply connected if G is a non-Abelian group. For any such group, $\pi_3(G) = \mathbb{Z}$ (take G to be a simple group); then $\pi_0 g^{(3)} \approx \pi_3 G = \mathbb{Z}$. It is also easy to explicitly construct the non-trivial loops in $M^{(3)}$. They are given by one-parameter families of gauge fields interpolating between any A and A^g where g is a map of S^3 into G with winding number $\nu \in \mathbb{Z}$. The existence of these non-trivial gauge transformations is the origin of the θ-parameter in gauge theories [8]. The quantization of the theory defined by (34) is accomplished by imposing Gauss' law on the space of wave functionals. This means that the operator $\vec{D} \cdot \vec{E}$ annihilates any physical state:

$$\vec{D} \cdot \vec{E} \, \psi[A] = 0 \qquad (35)$$

where $\vec{D}$ is the three-dimensional gauge covariant derivative. Recall that the generator of the infinitesimal gauge transformations is

$$Q_\theta = \frac{1}{g} \int d^3\vec{x} \, \text{Tr}(\theta \cdot \vec{D} \cdot \vec{E}) . \qquad (36)$$

θ is an arbitrary space-dependent element of the Lie algebra of G. Equation (35) is the quantum version of Gauss' law. By making infinitesimal transformations, we can construct a finite gauge transformation $g(\vec{x})$. However these types of gauge transformations have zero winding because there is a continuous deformation $g(\vec{x},t)$ such that $g(\vec{x},0) = g(\vec{x})$ and $g(\vec{x},1) = 1$. These gauge transformations are trivial, and these are the ones appearing in Gauss' law. In finite form we can write (35) as:

$$\psi[A^g] = \psi[A] \tag{37}$$

and as a consequence of the gauge invariance of H, $[Q_\theta, H] = 0$. Since $\pi_0 g^{(3)} = \pi_3 G = \mathbb{Z}$, there are still gauge transformations which cannot be deformed continuously to the identity, those with non-zero winding number

$$\nu(g) = \frac{1}{24\pi^2} \int_{S^3} \mathrm{Tr}(g^{-1} \, dg)^3 \, . \tag{38}$$

Let U be the operator which implements a gauge transformation with unit winding number: $U\psi[A] = \psi[A^U]$, $\nu(U) = 1$. The Hamiltonian still commutes with U, $[H,U] = 0$. However Gauss' law cannot be used to require that physical states be eigenstates of U with eigenvalue 1. In general we have to expect that

$$U\psi[A] = e^{-i\theta}\psi[A] \qquad 0 \leq \theta \leq 2\pi \, . \tag{39}$$

The new parameter θ is the celebrated vacuum angle in gauge theories. If $\theta \neq 0$ one can show that the theory violates CP symmetry. This implies in particular that QCD with a θ-parameter violates CP (strong CP-violation). Once the existence of the θ-parameter has been established there remains the non-trivial dynamical question of computing the dependence of the vacuum energy on θ. A partial answer to this question can be given by computing the instanton contribution to the vacuum energy. One finds $E_0(\theta) \sim e^{-8\pi^2/g^2} \cos\theta$ (see [8] for details and references).

Next we want to analyze what happens when we include fermions. There are several possibilities depending on the type of representation of G carried by the fermions. There are three types or representations for an arbitrary group: real, pseudoreal and complex. If T^a is the generator of G in some representation, one says that the representation is real (pseudoreal), if $T^a = R \, {}^tT^a \, R^{-1}$ and R is symmetric (resp. antisymmetric). In the first case, it is possible to find an explicit parametrization where the T^a's are purely real and antisymmetric. This however is not possible in the second case. A representation T^a is complex whenever its complex conjugate representation ${}^tT^a$ (recall that $T^{a+} = -T^a$, $T^{a+} = {}^tT^{a*}$) is not equivalent to T^a. For each of these three cases one can use the index theorem to gain information on the behavior of fermions in the presence of gauge fields.

Let us first consider a purely vector-like theory like QCD. We consider a theory with a Dirac fermion in the fundamental representation of G which for definiteness we will take to be $SU(N)$. A Dirac fermion is equivalent to two Weyl fermions $\lambda_L^i, \lambda_R^i, i = 1,N$ of opposite chirality in the same representation of G, or by using charge conjugation in four dimensions (or more generally in any 4k-dimensions) we can represent a Dirac fermion in terms or two left-handed fermions ($\gamma_5 = +1$) in complex conjugate representations of the gauge group λ_L^i, λ_{L_j} (the upper index label of the N of $SU(N)$, whilst the lower j label the $\bar{N}$). In terms of creation and annihilation operators we can say that λ_L^i is the field which annihilates left-handed quarks, and creates right-handed antiquarks. On the other hand, λ_{L_j} annihilates left-handed antiquarks and creates right-handed quarks. All these representations clearly correspond to the same physics. The Hamiltonian in the presence of left-handed fermions takes the form:

$$H = \frac{1}{2} \int d^3x \, \text{Tr}(\vec{E}^2 + \vec{B}^2) + \int d^3x \, \lambda^+{}_i(i\sigma^a D_a)\lambda^i + \int d^3x \eta^{+i}(i\sigma^a D_a)\eta_i \quad (40)$$

$\lambda^i \equiv \lambda_L^i, \eta_i \equiv \lambda_{R_i}$, and D_a is the gauge covariant derivative in the appropriate representation. The analysis of fermions in the presence of gauge fields is carried out usually in two steps. We first quantize the fermions in the presence of gauge fields, and then quantize the gauge fields. The advantage of this procedure is that the issues concerning the quantum conservation of global or gauge symmetries are most transparent in this way. In the rest of this section, we will follow the philosophy of the first entry of Ref. [4]. From this point of view we have an infinite family of fermionic quantum theories parametrized by the elements of $A^{(3)}$. For each point in $A^{(3)}$, we can construct the full fermionic quantum theory in terms of the appropriate Fock vacuum and creation and annihilation operators. This way we erect a Hilbert space on top of each point of $A^{(3)}$, or rather, a Hilbert bundle over $A^{(3)}$. If we want to eventually quantize the gauge field, we know that the true configuration space is $M^{(3)} = A^{(3)}/g^{(3)}$. Thus to ask whether the Hilbert bundle over $A^{(3)}$ naturally extends down to $M^{(3)}$, or whether there are any obstructions to this procedure. If we consider two gauge fields which only differ by a trivial gauge transformation A, A^g, we have in principle two different Hilbert spaces H_A, H_A^g for the fermionic theories on top of A and A^g. Gauge invariance of the complete theory is now expressed by saying that there is a smooth way of defining a unitary transformation U_g between H_A and H_A^g, in such a way that

$$\psi[\vec{A}^g] = U_g \, \psi[\vec{A}] \, . \tag{41}$$

To see that this is equivalent to Gauss' law for the full theory, simply take g to be infinitesimal $g \simeq 1 + \theta$, then (41) becomes:

$$(\vec{D} \cdot \vec{E}^a + \lambda^\dagger T^a \lambda)\psi[\vec{A}] = 0 \tag{42}$$

where $\lambda^\dagger T^a \lambda$ is the true component of the fermionic gauge current. It can be shown that for real representations it is always possible to quantize the theory in such a way that (41) is satisfied. (We will present the argument in the next section.)

After this long preamble, we can present the first application of the index theorem to the anomaly in global axial currents. We start by assuming for the time being that the representations carried by the left-handed fermions are such that (41) is satisfied. A cursory look at (40) shows that the theory is naively invariant under a global axial rotation:

$$\lambda \to e^{-i\alpha} \lambda \, . \tag{43}$$

The generator of this symmetry is

$$j_5^\mu = \sum_r \lambda_r^\dagger \sigma^\mu \lambda_r \qquad (44)$$

where r is a given representation of G, $\sigma^\mu = (1,-\vec{\sigma})$ and the σ^a's are the standard Pauli matrices. Let Q_5 be the associated charge $Q_5 = \int d^3x\, j_5^0$. We want to study whether Q_5 is truly conserved at the quantum level. This turns out not to be the case, and one can use the index theorem to understand quite easily why not. The argument (essentially the same as in [2]) goes as follows: Let $|0\rangle_{\vec{A}}$ be the Fock vacuum for the fermionic Hilbert space at $\vec{A}$. This can be constructed by solving the three-dimensional Dirac equation $i\sigma^i D_i \psi_E = E\psi_E$ and separating the Hilbert space into a set of positive energy states H_A^+ and a set of negative energy states H_A^-. The Fock vacuum $|0\rangle_A$ is constructed by filling all the negative energy states. If Q_5 were a well-defined conserved quantity, we should be able to assign the same value of Q_5 for two Fock vacua related by a gauge transformation. Since we are insisting on quantizing the theory in a gauge-invariant manner, i.e., we are diagonalizing simultaneously the Hamiltonian and the generators of small and large gauge transformations, if we cannot assign consistently a Q_5 charge to gauge-equivalent Fock vacua, then it follows that Q_5 is not going to be conserved. The vacuum of the full theory will not maintain Q_5 as a generator of symmetry. We can analyze what happens as follows: Let U be the generator of a large gauge transformation U of unit winding number, and let A_t be a one-parameter configuration which interpolates adiabatically between $\vec{A}$ and $\vec{A}^U$. Since $\vec{\sigma}\cdot\vec{D}(A^U) = U^{-1}\vec{\sigma}\cdot\vec{D}(A)U$, the spectrum of $i\vec{\sigma}\cdot\vec{D}(A)$ and $i\vec{\sigma}\cdot\vec{D}(A)$ have the same eigenvalues, and if ψ_E is an eigenfunction of $i\vec{\sigma}\cdot\vec{D}(A)$ with eigenvalue E, then $U^{-1}\psi_E$ is the eigenfunction of $i\vec{\sigma}\cdot\vec{D}(A^U)$ with the same eigenvalue. We start at $t = -\infty$ with $|0\rangle_A$, which can be given zero charge with respect to Q_5, then we slowly interpolate between A and A^g. This way we can compare the Q_5 assignment between $|0\rangle_A$ and $|0\rangle_{A}U$. If we follow in detail the behavior of the spectrum of $i\vec{\sigma}\cdot\vec{D}(A(t))$, we have a set of eigenvalues $E_n(t)$. Since the spectrum of $i\vec{\sigma}\cdot\vec{D}(A)$ and $i\vec{\sigma}\cdot\vec{D}(A^U)$ are the same, there are several possibilities for the behavior of $E_n(t)$. We could have $E_b(-\infty) = E_b(\infty)$ in wich case the Fock vacuum $|0>_A$ would evolve into $|0>_{A^g}$ without further change. It could also be that there is a net spectral flow as we interpolate between A and A^G, i.e. $E_n(1) = E_{n+\nu}(0)$. (ν is some integer to be determined presently.) The difference between both situations can be posed as an index problem. We can solve the following equations

$$\left(\frac{\partial}{\partial t} - i\,\vec{\sigma}\cdot\vec{D}(A_t)\right)\psi_L = 0$$
$$\left(\frac{\partial}{\partial t} + i\,\vec{\sigma}\cdot\vec{D}(A_t)\right)\psi_R = 0 \tag{45}$$

adiabatically. (45) can be written more concisely as the equation for zero modes of the four-dimensional Dirac operators $\displaystyle{\not}D\psi = 0$, where ψ_L, ψ_R stand for $\gamma_5 = +1$ or -1 from the four-dimensional point of view. If the four-dimensional Dirac operator has a left-handed normalizable zero mode, it can be expressed in the adiabatic approximation as $\psi_L(t,\vec{x}) = f(t)\lambda_n^t(\vec{x})$, where $\lambda_n^t(\vec{x})$ is the eigenfunction of $i\,\vec{\sigma}\cdot\vec{D}(A_t)$ with eigenvalue $E_n(t)$. Then $f(t)$ satisfies the equation:

$$\frac{df(t)}{dt} - E_n(t)f(t) = 0$$
$$f(t) = f(0)\exp\int_0^t E_n(t')dt'. \tag{46}$$

Since $\lambda_n^t(x)$ is normalizable for all t, $\psi_L(t,\vec{x})$ will be normalizable if $f(t)$ is. This requires that as $t \to +\infty$, $E_n(t) \to -a$ ($a > 0$) and as $t \to -\infty$, $E_n(t) \to +b$ ($b > 0$), i.e. there is a spectral flow, $E_n(t)$ looks like an antikink. To determine the integer ν, we have to figure out how many four-dimensional zero modes there are for the Dirac operator. Note however that of the four-dimensional problem (45) we are interested in the net flow, i.e. $n_L - n_R$ zero modes; $n_L - n_R = \text{ind } i\,\displaystyle{\not}D$. Topologically, the four-dimensional configuration is equivalent to a gauge field configuration on a four sphere where in the upper hemisphere

D_+ we have a configuration A_+ (in our case A_t) and in the lower hemisphere D_-, A_- (in our case $A_- = 0$) so that the transition function in the overlap $D_+ \cap D_- = S^3$ (the sphere at infinity) $A_+ = U^{-1}(A_- + d)U$. Since in the four sphere the $\hat{A}(S^4) = 1$ (this can easily be proven by considering the standard maximally symmetric metric on S^4 and the topological invariance of $\hat{A}(M)$), we get

$$\text{ind } i \not{D} = -\frac{1}{(2\pi)^2 2!} \int_{S^4} \text{Tr } F^2$$

$$= -\frac{1}{(2\pi)^2 2!} \left[\int_{D_+} \text{Tr } F_+^2 + \int_{D_-} \text{Tr } F_-^2 \right]. \tag{47}$$

Since $\text{Tr } F_\pm^2 = d \, \text{Tr}(A_\pm dA_\pm + \frac{2}{3} A_\pm^3)$, a short computation leads to

$$\text{ind } i \not{D} = -\frac{1}{24\pi^2} \int_{S^3} \text{Tr}(U^{-1} dU)^3 = -1. \tag{48}$$

Hence the net spectral flow $\nu = -1$, and we have $E_n(t = +\infty) = E_{n-1}(t = -\infty)$, and we find that a positive energy left-handed state $t = -\infty$ becomes a negative energy state at $t = +\infty$, thus the difference in the Q_5 charges is:

$$Q_5(t = -\infty) - Q_5(t = -\infty) = -1 = (\text{ind } i \not{D}) \tag{49}$$

and we learn that the axial charge is not conserved if we want to define the theory in a manifestly gauge invariant way. This result can be re-expressed in terms of currents by:

$$\partial_\mu j_5^\mu(x) = \frac{1}{32\pi^2} \text{Tr } F_{\mu\nu} {}^*F^{\mu\nu}$$

$${}^*F^{\mu\nu} = \varepsilon^{\mu\nu\alpha\beta} F_{\alpha\beta}. \tag{50}$$

This is the celebrated axial U(1) anomaly [1] which led to an understanding of π_o decay and to the resolution of the U(1) problem [9]. If we had more left-handed fermion species, then the right-hand side of (49) would have to inlcude the contribution of each fermion. Thus we can write in full generality:

$$\Delta Q_5 = \text{ind}(i\not{D}) \qquad (51)$$

where the index has to be taken for the reducible rerpesentation of G generated by the direct sum of the representations of all the fermions involved.

Next, let us analyze the more difficult problem of understanding whether Gauss' law in the form given by Equation (41) holds in the presence of complex representations of fermions. From the previous paragraphs we learn that the key issue in constructing a manifestly gauge covariant Hilbert bundle over $A^{(3)}$ lies in the construction of the Fock vacuum. This can be seen more explicitly as follows: We can write down the expansion of the field $\lambda(\vec{x})$ in terms of creation and annihilation operators. Let $\psi_E(\vec{x})$ be the eigenstates of $i\vec{\sigma}\cdot\vec{D}(A)$ for some three dimensional configuration A. For simplicity we can assume that $i\vec{\sigma}\cdot\vec{D}(A)$ has no zero eigenvalues:

$$i\vec{\sigma}\cdot\vec{D}(A)\psi_E = E\psi_E . \qquad (52)$$

Thus we can split $H_A = H_A^+ \oplus H_A^-$ the space of positive and negative energy states. As usual in second quantization, we can expand $\lambda(\vec{x})$ in terms of ψ_E by

$$\lambda(\vec{x}) = \sum_{E>0} a_A(E)\psi_E(\vec{x}) + \sum_{E<0} b_A^\dagger(E)\psi_E(\vec{x}). \qquad (53)$$

a_A, $b_A^\dagger$ are the corresponding creation and annihilation operators: and the Fock vacuum $|0\rangle_A$ is that state annihilated by all the annihilation operators. Any state in the Fock space at $\vec{A}$ can be construced now in terms of creation operators acting on $|0\rangle_A$. Given the transformation properties of $\psi_E(x)$ under gauge transformation, if $U(g)$ is the operator implementing gauge transformation g, it follows that $U a_A U^\dagger = a_A^g$, $U b_A U^\dagger = b_A^g$. Thus the covariance properties of any state (41) can be translated into the statement of whether the Fock vacuum can be smoothly defined to satisfy (41). This simplifies things considerably. In the first quantized picture the Fock vacuum is the state with all the negative energy states filled. This gives a single state $|0>_A$ for every A. In other words, the Fock vacuum defines a complex line bundle over $A^{(3)}$. Gauss' law forces this Fock vacuum to be equivariant under gauge transformations (41) so that we can reduce this Fock line bundle from $A^{(3)}$ to $M^{(3)}$. Failure to satisfy (41) implies that there is no consistent way of quantizing the theory maintaining gauge invariance. Thus, rather than working with $A^{(3)}$, we can simply analyze the behavior of $|0>_A$ on $\mathfrak{g}^{(3)}$ for given $\vec{A}$. The end point of this analysis is therefore that the smooth implementation of Gauss' law depends on whether the line bundle over $\mathfrak{g}^{(3)}$ defined by the Fock vacuum satisfies (41). However, the triviality of a line bundle can be determined by restricting it to the non-contractible two spheres on $\mathfrak{g}^{(3)}$ (see [7] for more details). We have to compute $\pi_2 \mathfrak{g}^{(3)}$, or the homotopy classes of maps from $S^2 \times S^3$ into $\mathfrak{g}$. If we take $\mathfrak{g}$ to be $SU(N)$, $N > 2$, and consider only those gauge transformations which have zero winding number on S^3 (the only ones that Gauss' law can deal with), the maps from $S^2 \times S^3 \to G$ are classified by $\pi_5(SU(N)) = \mathbb{Z}$, $N > 2$. Thus, let $g(\theta,\phi,\vec{x})$ be the two-parameter family of gauge transformations $(0 \le \theta \le \pi, 0 \le \phi \le 2\pi)$ generating the non-trivial two-sphere in $\mathfrak{g}^{(3)}$. The final step is to compute the first Chern class of the Fock line bundle over the S^2 constructed. Intuitively this means that we consider the Fock vacuum at some fiducial point (say the north pole on S^2, $\theta = 0$) and transport it to any other point (θ,ϕ) on S^2. This can be represented as a parallel transport of the Fock vacuum from the north pole to (θ,ϕ) in terms of a $U(1)$ gauge field on S^2, $\omega = \omega_\theta d\theta + \omega_\phi d\phi$, if the magnetic field is associated to ω, $B = d\omega$ has a non-trivial flux over S^2, $\int_{S^2} d\omega = 2\pi n$, $n \ne 0$, then the Fock vacuum behaves like the wave function of a charged particle moving on S^2 with a monopole in the "center" of the sphere. Thus there will be string singularities, and the Fock vacuum will twist in a rather non-trivial way over S^2. This twisting in turn provides an obstruction to the reduction of the Fock bundle from $A^{(3)}$ to $M^{(3)}$ satisfying (41).

To see more explicitly what happens, let $\vec{A}$ be the three-dimensional gauge field at $\theta = 0$. Since we are assuming for simplicity that $\vec{A}$ has no zero modes, we can consistently define H_A^+, H_A^- and the Fock vacuum at A. Next we consider an interpolation between A and $A^{(\theta,\phi)}$ where $A^{(\theta,\phi)} = g^{-1}(\theta,\phi)(A+d)g(\theta,\phi)$ $(d = dx^i \partial/\partial x^i)A_t$, $A_{t=0} = A$, $A_{t=1} = A^{(\theta,\phi)}$. Hence (t,θ,ϕ) becomes geometrically a three ball whose boundary is the non-contractible two sphere in $g^{(3)}$. Since we want to compare the Fock vacuum at A and $A^{(\theta,\phi)}$, we consider the flow of eigenvalues between A and $A^{(\theta,\phi)}$. Since $i\vec{\sigma}\cdot D(A)$ has no zero modes, it is clear that for points (θ,ϕ) close to the north pole there will be no eigenvalue flow. In terms of the four-dimensional Dirac operator, this means that none of the two equations

$$\left[\frac{\partial}{\partial t} \pm i\vec{\sigma}\cdot\vec{D}(A_t^{(\theta,\phi)})\right]\psi_\pm = 0 \tag{54}$$

has a solution, or equivalently that ind $i\rlap{/}D(A_t^{(\theta,\phi)}) = 0$. Hence the interpolating field $A_t^{(\theta,\phi)}$ defines for each point (θ,ϕ) on S^2 a four-dimensional gauge configuration, i.e. an element of $A^{(4)}$. As we change (θ,ϕ) we are simply making a continuous deformation of this four-dimensional configuration, and consequently ind $i\rlap{/}D(A_t^{(\theta,\phi)})$ is still equal to zero. It may however be possible that for some value of (θ_0,ϕ_0) there is a solution to both Equations (54). There has to be a solution to both equations because the ordinary index of $i\rlap{/}D(A_t^{(\theta,\phi)})$ vanishes. If this were the case, then at the end of the interpolation we would end up not with the Fock vacuum but with a particle-antiparticle pair, and there would be no smooth way of defining the vacuum state consistent with Gauss' law. There is a net flow of gauge charge in the interpolation from $\theta = 0$ to (θ_0,ϕ_0). To check that this is indeed the case, we have to compute the index of the two-parameter family of four-dimensional Dirac operators $i\rlap{/}D(A_t^{(\theta,\phi)})$. This can be done as a simple variation on the Euclidean approach taken in [2]. In this problem we have six variables $t,\theta,\phi,\vec{x}$, and we want to consider for a moment the six-dimensional operator on $S^3 \times S^3$ in the presence of the following gauge configuration: we split the first S^3 into upper and lower hemispheres $D_+ \times S^3$, $D_- \times S^3$. On $D_\pm \times S^3$ we define the following six-dimensional gauge fields.

$$A_+ = A + t(g^{-1}Ag + g^{-1}dg - A)$$

$$A_- = A + sg\delta g^{-1} \tag{55}$$

$$d = dx^i \frac{\partial}{\partial x_i}, \quad \delta = d\theta \frac{\partial}{\partial \theta} + d\phi \frac{\partial}{\partial \phi} \tag{56a}$$

$$d\delta + \delta d = d^2 = \delta^2 = 0. \tag{56b}$$

Thus in the overlap region $(D_+ \times S^3) \cap (D_- \times S^3) = S^2 \times S^3$ we have $t = 1$, $s = 1$, and

$$A_+(t=1) = g^{-1}(A_-(s=1) + d + \delta)g \tag{57}$$

i.e. A_+, A_- define a six-dimensional bundle over $S^3 \times S^3$ with transition function $g(\theta,\phi,\vec{x})$. Using the methods discussed in the previous section, we can compute the index theorem for the six-dimensional Dirac operator:

$$\text{ind } \not{D}^{(6)} = \frac{1}{(2\pi)^3 3!} \left(\int_{D_+ \times S^3} \text{Tr } F_+^3 + \int_{D_- \times S^3} \text{Tr } F_-^3 \right)$$

$$F_+ = (d+\delta)A_+ + A_+^2$$

$$F_- = (d+\delta)A_- + A_-^2. \tag{58}$$

Let $d + \delta = \hat{d}$, then using

$$\text{Tr } F^3 = 3\hat{d} \int_0^1 du \text{ Tr } A(u\hat{d}A + u^2 A^2)^2$$

$$= \hat{d} \text{ Tr } (A\hat{d}A\hat{d}A + \frac{3}{2}\hat{d}A A^3 + \frac{3}{5} A^5) \tag{59}$$

and the explicit form of $A_\pm$, we get:

$$\text{ind } \not{D}^{(6)} = \frac{1}{480\pi^3} \int_{S^2 \times S^3} \text{Tr}[g^{-1}(d+\delta)g]^5 \tag{60}$$

the five-dimensional analog of (48). The right-hand side of Eq. (60) is the five-dimensional winding number of $g(\theta,\phi,\vec{x})$, choosing for simplicity $f(\theta,\phi,\vec{x})$ to have unit winding number (the generator of $\pi_5 SU(N)$, we conclude that ind $\not{D}^{(6)} = 1$, i.e. there is a zero mode of $\not{D}^{(6)}$ with positive six-dimensional chirality. Let this mode be $\eta(t,\theta,\phi,\vec{x})$. Since the index is invariant under continuous deformations let us consider the six-dimensional operator $\not{D}_\varepsilon^{(6)}$:

$$i \not{D}_\varepsilon^{(6)} \equiv i \Gamma^a D_a + \frac{1}{\varepsilon} i \Gamma^i D_i . \tag{61}$$

The a-indices, a = 1,2,3, refer to the auxiliary variables (t,θ,ϕ) or (s,θ,ϕ), and i = 1,2,3 to ordinary space. By topological invariance, ind $i\not{D}_\varepsilon^{(6)} = $ ind $\not{D}^{(6)}$. If we make ε very small, we can compute $\eta(t,\theta,\phi,\vec{x})$ and the spectrum of $\not{D}_\varepsilon^{(6)}$ in the Born-Oppenheimer approximation. Let $H_\varepsilon = (\not{D}_\varepsilon^{(6)})$

$$H_\varepsilon = -D_a D^a + \frac{1}{\varepsilon^2}(i \Gamma^i D_i)^2 - \frac{1}{\varepsilon} \Gamma^a \Gamma^i \partial_a A . \tag{62}$$

The first two terms on the right hand side of (62) are manifestly positive definite. If we consider H_ε in the lower hemisphere, $i \Gamma^i D_i = i \Gamma^i (\partial_i + A_i)$. Since $\sigma^i D_i(A)$ has no zero modes, by making ε small enough, the positive term $(i \Gamma^i D_i)^2/\varepsilon^2$ can be chosen to overcome the third term. Thus, in the adiabatic approximation we can write

$$\eta(t,\theta,\phi) = f^{(1)}(t,\theta,\phi)\psi_{E_1}^{(t,\theta,\phi)}(\vec{x}) + f^{(2)}(t,\theta,\phi)\psi_{E_2}^{(t,\theta,\psi)}(\vec{x}). \quad (63)$$

To make the argument more transparent, we can take t to run from $-\infty$ to $+\infty$. Substituting (63) in (62) for ε small, we obtain an operator acting only on $f^{(1)}, f^{(2)}$. It is then easy to convince oneself [2] that the only way one can find a normalizable solution of $D_{\varepsilon,\eta}^{(6)}$ is if for some (t_o,θ_o,ϕ_o) $E_1(t_o,\theta_o,\phi_o) = E_2(t_o,\theta_o,\phi_o) = 0$. Put differently, there is a solution to (54) for both $\psi_+\psi_-$ at some (θ_o,ϕ_o), which from previous arguments implies that $E_1(t_o,\theta_o,\phi_o)$ must behave like a kink and $E_2(t_o,\theta_o,\phi_o)$ as an antikink as we interpolate from the north pole of S^2 to (θ_o,ϕ_o).

In order to connect our approach with the standard formulation of non-Abelian anomalies, let us translate our argument into the Euclidean functional integral approach. From this point of view one starts with the fermionic effective action for fermions in the presence of some gauge field $A \in A^{(4)}$:

$$e^{-\Gamma_{eff}^{(R)}[A]} = \int d\bar{\psi}\, d\psi \,\exp - \int d^4x \, x\bar{\psi}\, i\rlap{/}D_+\psi$$

$$i\rlap{/}D_+ = i\rlap{/}D(1+\Gamma)/2 \quad (64)$$

with some suitable regulator. From the Hamiltonian point of view (64) is computing the vacuum expectation value of the time evolution operator $U(\infty,-\infty,A)$ in the presence of some time-dependent gauge field configuration. The superscript (R) in (64) stands for the representation of G carried by the fermions. The gauge currents become $\bar{\psi}\gamma^\mu \frac{\lambda^a}{2}(1+\Gamma)\psi$, and the λ^a's are the anti-Hermitean generators of G in a complex representation R. If we make an infinitesimal gauge transformation $A_\mu \to A_\mu - D_\mu v$, $D_\mu v = \partial_\mu v + [A_\mu,v]$, the effective action changes by

$$\Gamma^{(R)}[A] \to \Gamma^{(R)}[A] + \int dx\, v^a(x)\, D_\mu \frac{\delta \Gamma^{(R)}[A]}{\delta A_\mu^a}. \quad (65)$$

Hence gauge invariance would require $D_\mu \delta\Gamma^{(R)}/\delta A_\mu^a = 0$ which is equivalent to the current conservation condition $D_\mu \langle j^{\mu a}\rangle = 0$, since $\delta\Gamma^{(R)}/\delta A_\mu^a = \langle j^{\mu a}\rangle$. General arguments [10] allow us to determine when $\Gamma^{(R)}[A]$ is potentially gauge non-invariant. The real part of $\Gamma^{(R)}[A]$ is always gauge invariant: $2\,\text{Re}\,\Gamma^{(R)} = \Gamma^{(R)} + \Gamma^{(\bar{R})} = \Gamma^{(R+\bar{R})}$. ($\bar{R}$ is the representation complex conjugate to R.) Since a Weyl fermion in the representation $R+\bar{R}$ is equivalent to a Dirac fermion in the representation R, we can always define $2\,\text{Re}\,\Gamma^{(R)}$ in a gauge invariant way via, for example, a Pauli-Villars regulator. Thus we learn that only complex representations can give a gauge non-invariant $\Gamma^{(R)}[A]$, and only the imaginary part of $\Gamma^{(R)}[A]$ may pick up an anomalous variation under gauge transformations. Naively, one would like to regard (64) as $\det(i\!\!\not{\!D}_+)$. This is problematic because $i\!\!\not{\!D}_+$ maps spinors of positive chirality into spinors of negative chirality, and consequently does not have a well-defined eigenvalue problem. This can be fixed [2] by defining the following elliptic operator:

$$i\hat{D} \equiv i(\not{\partial} + \not{A}\,\frac{1+\Gamma}{2}) = i(\not{\partial}_- + \not{D}_+). \qquad (66)$$

From a perturbative and non-perturbative point of view we can define $e^{-\Gamma^{(R)}[A]}$ as $\det i\hat{D}(A)$; $i\hat{D}(A)$ is elliptic and admits a well-defined eigenvalue problem. Even though $\det i\hat{D}$ is not in general gauge invariant, it follows from rather simple manipulations that $|\det \hat{D}|$ is gauge invariant. In a basis where Γ is diagonal

$$i\hat{D} = \begin{bmatrix} 0 & iD_+ \\ i\partial_- & 0 \end{bmatrix}$$

and

$$\det|iD|^2 = \det i\hat{D}^\dagger \det i\hat{D} = \det(i\!\!\not{\!D}_- i\!\!\not{\!D}_+)\det(i\!\!\not{\!\partial}_+ i\!\!\not{\!\partial}_-) \qquad (67)$$

which is equal to the ordinary $\det i\!\!\not{\!D} = \det(i\!\!\not{\!D}_- i\!\!\not{\!D}_+)$ up to a gauge field independent normalization factor. In order to compute the anomalous variation of $\text{Im}\,\Gamma^R[A]$, we consider now a one-parameter family of four-dimensional gauge transformations $g(\theta,x)$, $g(0,x) = g(2\pi,x) = 1$, $g(\theta,x)$ is a map from $S^1 \times S^4 \to G$. (We compactify space-time to S^4.) Once again these maps are classified topologically by $\pi_5 G = \mathbb{Z}$ (for $G = SU(N)$, $N > 2$). Now we define a circle in $A^{(4)}$ by:

$$A^\theta = g^{-1}(A+d)g. \qquad (68)$$

Assuming that A defines a toplogically trivial bundle over S^4, we can thicken up this circle to become a disc in the form $A^{(t,\theta)} = tA^\theta$. If A were topologically non-trivial, we could use a different definition of the disc: $A^{(t,\theta)} = A + t(A^\theta - A)$. The fermion determinant $\det(i\hat{D}^{t,\theta})$ can now be considered as a complex function on this disc. Notice that at $t=1$ $A^{(t,\theta)} = A^\theta$, and thus this disc in $A^{(4)}$ becomes a two-sphere in $M^{(4)} = A^{(4)}/\mathfrak{g}^{(4)}$. The analogy with the Hamiltonian analysis is clear, the line bundle which will measure the breakdown of gauge invariance is provided by the determinant $\det(i\hat{D}^{(t,\theta)})$, and the two-parameter family of gauge transformations has become a one-parameter family for the Euclidean, four-dimensional point of view. Along the boundary of the disc

$$\det i\hat{D}(A^\theta) = [\det i\rlap{/}{D}(A)]^{1/2} e^{iw(A,\theta)}. \qquad (69)$$

Thus the phase of the determinant restricted to the boundary of the disc defines a mapping $e^{iw(\theta,A)}: S^1 \to S^1$. Such mappings are characterized by our integer winding numbers

$$n = \frac{1}{2\pi} \int_0^{2\pi} d\theta \; \frac{\partial w}{\partial \theta}(\theta,A). \qquad (70)$$

Let us now take the family $g(\theta,x)$ to have unit winding number from the five-dimensional point of view. The integer n in (70) measures the integrated change of $\operatorname{Im} \Gamma^{(R)}(A,\theta)$:

$$\int_0^{2\pi} d\theta \; \frac{\partial}{\partial \theta} \operatorname{Im} \Gamma^{(R)}(A,\theta) = 2\pi n. \qquad (71)$$

Since n is a topological invariant, we can deform the θ-circle inwards into the disc. Even though we start with a configuration A such that $i\slashed{D}(A)$ has no zero modes, it may happen that at some points (t_o, θ_o), the fermion determinant may vanish. Then the winding number (70) can be written as the sum of the winding numbers of Im $\Gamma^{(R)}$ for arbitrarily small circles around those points (t_o, θ_o) where $i\slashed{D}$ has zero modes. These zeroes generically occur at isolated points since an arbitrary perturbation of a Dirac operator with zero index typically mixes any positive and negative chirality zero modes to produce only non-zero modes. Regarding the determinant as a regularized product of complex eigenvalues, its winding number around any contour is equal to the sum of the winding numbers of the individual eigenvalues. But by continuity, it is clear that the winding number vanishes for any eigenvalue which does not vanish within the contour. The index at the zeroes of det $i\hat{D}^{(t,\theta)}$ is thus determined entirely by the winding number of the smallest eigenvalues in the neighborhood of the zeroes and the contribution to the total winding number of each one of these zeroes will tpyically be ±1. Thus we have reduced the problem of calculating the winding number of det $\hat{D}(A^\theta)$ to characterizing the behavior of the smallest eigenvalues of $i\hat{D}^{(t,\theta)}$ locally about their zeroes. This is done by considering instead a six-dimensional operator defined on $S^2 \times S^4$ given by the following gauge configurations:

$$D_+ \times S^4 \qquad A_+ = t\, g^{-1}(A + d_x)g$$
$$D_- \times S^4 \qquad A_- = A - s\, d_\theta\, g g^{-1} \qquad (72)$$

on the overlap $(D_+ \times S^4) \cap (D_- \times S^4) = S^1 \times S^4$, $A_+ = g^{-1}(A_- + d + d_g)g$. Using now the index theorem and the adiabatic approximation in a similar manner to the Hamiltonian case, it can be shown after a not too lengthy argument that (see entry 2 in [2]):

$$i \int d\theta\, \frac{\partial}{\partial \theta} \text{Im } \Gamma^{(R)}[A,\theta] = 2\pi\, \frac{1}{(2\pi)^3 3!} \int_{S^2 \times S^4} \text{Tr } F^3 \qquad (73)$$

where F is the six-dimensional gauge field strength associated with (72). An explicit computation shows that

$$\int_0^{2\pi} d\theta \frac{\partial}{\partial \theta} \text{Im } \Gamma[A,\theta] = \frac{1}{24\pi^2} \int_0^{2\pi} d\theta \text{ Tr}(g^{-1} \frac{\partial}{\partial \theta} g) d(A^\theta dA^\theta + \frac{1}{2} A^{\theta 3}). \quad (74)$$

From (74) we can identify the anomaly to be:

$$D_\mu <j^{\mu a}> = \frac{i}{24\pi^2} \epsilon^{\mu\nu\alpha\beta} \text{ Tr } \lambda^a \partial_\mu (A_\nu \partial_\alpha A_\beta + \frac{1}{2} A_\nu A_\alpha A_\beta) \quad (75)$$

in agreement with standard perturbative computations [1].

This concludes our overview of the use of topological methods to understand the structure of non-Abelian anomalies. It should be clear from the arguments presented in this section, that all the non-trivial twisting of either the Fock vacuum or the fermionic determinant were given by the topological non-triviality of $A^{(3)}/g^{(3)}$ or $A^{(4)}/g^{(4)}$. The results described in these lectures only explore the lowest non-trivial topological invariants of these spaces; namely the θ-parameter can be understood from the fact that $\pi_1(A^{(3)}/g^{(3)}) = \mathbb{Z}$, and the non-Abelian anomaly essentially follows from $\pi_3(A^{(3)}/g^{(3)})$. However, $M^{(3)}$ or $M^{(4)}$ contains a much richer geometry and topology. Which role (if any) these more complicated structures will play in the gauge field-fermion dynamics, remains an open question.

Acknowledgement

I would like to thank all the members of ZiF and especially Professor L. Streit for their very kind hospitality, and for giving me the opportunity to present this material in such a stimulating environment.

References

[1] S.L. Adler, Phys. Rev. 177 (1969) 2426;

J.S. Bell and R. Jackiw, Nuovo Cim. 60A (1969) 47;

S.L. Adler and W.A. Bardeen, Phys. Rev. 182 (1969) 1517;

W.A. Bardeen, Phys. Rev. 184 (1969) 1848.

See also: S.L. Adler in Lectures on Elementary Particles and Quantum Field Theory, S. Deser et al. eds. (MIT Press, 1971), and R, Jackiw in Lectures in Current Algebra (Princeton Univ. Press, 1972).

[2] M.F. Atiyah and I.M. Singer, Proc. Nat. Acad. Sci. 81 (1984) 2597;

L. Alvarez-Gaumé and P. Ginsparg, Nucl. Phys. B243 (1984) 449;

C. Gomez, Salamanca Preprint (1983);

J.O. Alvarez, I.M. Singer, and B. Zumino, Comm. Math. Phys. 96 1984) 409.

[3] M.F. Atiyah and I.M. Singer, Ann. of Math. 87 (1968) 485, 546; 93 (1971), 1, 119, 139;

M.F. Atiyah and G.B. Segal, Ann. of Math. 87 (1968) 531.

[4] L. Alvarez-Gaumé and P. Nelson, Comm. Math. Phys. 99 (1985) 103;

L.D. Faddeev, Phys. Lett. 145B (1984) 81;

L.D. Faddeev and S. Shatashvili, Theor. Math. Phys. 60 (1984) 206;

B. Zumino, Nucl. Phys. B253 (1985) 477;

R, Jackiw, MIT preprint CTP 1298.

[5] In this section we are mainly following L. Alvarez-Gaumé and P. Ginsparg, Ann. of Phys. 161 (1985) 423.

See also B. Zumino in Relativity, Groups and Topology II, B. DeWitt and R. Stora eds. (North Holland, 1984);

T. Eguchi, A.H. Hanson, and P.B. Gilkey, Phys. Rep. 66 (1980) 243.

[6] For a review and references see P.B. Gilkey, The Index Theorem and the Heat Equation (Publish or Perish, Boston, 1974).

[7] See, for instance, N. Steenrod, The Topology of Fiber Bundles (Princeton University Press, 1951).

[8] R. Jackiw and C. Rebbi, Phys. Rev. Lett. 37 (1976) 172;

C. Callan, R. Dashen, and D. Gross, Phys. Lett. 63B (1976) 334.

For review see, for instance, R. Jackiw in Relativity Groups and Topology II, B. DeWitt and R. Stora eds. (North Holland, 1984); S. Coleman, The Uses of Instantons in The Whys of Subnuclear Physics, A. Zichichi ed. (Plenum Press, N.Y., (1979) 805.

[9] G. 't Hooft, Phys. Rev. Lett. $\underline{37}$ (1976) 8 and Phys. Rev. $\underline{D14}$ (1976) 3432.

[10] L. Alvarez-Gaumé and E. Witten, Nucl. Phys. $\underline{B234}$ (1983) 269.

SUM RULES IN SCATTERING THEORY AND APPLICATIONS TO STATISTICAL MECHANICS

D. Bollé[*]

Instituut voor Theoretische Fysica
Universiteit Leuven
B-3030 Leuven, Belgium

CONTENTS

I. INTRODUCTION

II. GENERALIZED LEVINSON'S THEOREM
    A. Some Results on Two-Body Scattering Theory
    B. Jauch's Proof of Levinson's Theorem
    C. Levinson's Theorem "à la carte"

III. HIGHER-ORDER LEVINSON'S THEOREM
    A. Introductory Remarks
    B. Derivation
    C. Discussion

        1. Rules in dimension 1 and 2; relation with the Korteweg-de-Vries invariants
        2. Local sum rules; sum rule dynamics

IV. LOW-ENERGY SUM RULES
    A. Sum Rules for three dimensions
    B. Scattering on the Line
    C. Two-Dimensional Results

V. RESULTS FOR OTHER SCATTERING SITUATIONS
    A. Non-Local Interactions

---

[*]Onderzoeksleider N.F.W.O., Belgium

B. Coulomb Plus Short-Range Interactions
　　C. Impurity Scattering, Friedel's Sume Rule
　　D. Classical Scattering
　　E. Remarks on Relativistic and Many-Particle Scattering

VI. APPLICATIONS
　　A. Statistical Mechanics
　　B. Other Fields
　　　　1. Nuclear physics
　　　　2. Field theory

ACKNOWLEDGMENTS

APPENDIX

I. <u>INTRODUCTION</u>

　　The study of sum rules in scattering theory has a long history. It started in fact in 1949 when Levinson [1] studied the uniqueness of the potential in the Schrödinger equation for a given asymtotic phase. Thereby he first proved rigorously that for scattering in a given partial wave, the value of the phase shift at the threshold energy zero is proportional to the number of distinct bound states in this partial wave. (This fact was apparently known to other people.) The first formulation of this Levinson theorem for general scattering systems was made by Jauch [2]. For later studies for non-local, singular ... interactions we refer to the literature cited in Newton [3], Beregi et al. [4] and Buslaev [5]. Especially non-local interactions received some attention around the question what exactly the role is played by the so-called positive energy eigenvalues. This problem was completely settled by Dreyfus [6]. Around this time different approaches were worked out to prove the theorem under very general conditions. These approaches will be discussed in some detail here. Simultaneously, it was recognized that this theorem is the first one of a whole set of sum rules or trace relations, as realized already earlier in the Russian literature (see ref. [5]).

Besides these recent generalizations of this theorem as an expression of spectral stability, new applications to different areas of physics have been given. It is exactly the purpose of these lectures to elaborate on these two main topics.

In the first part, we present the proof of Jauch and discuss the different forms under which Levinson's theorem appears in the literature, depending on the context it is used in.

In a second part, we discuss the higher-order rules in $d \leq 3$ dimensions in arbitrary regions of configuration space. In particular, we study the relation with the invariants of the Korteweg-de Vries equation.

In a third part, we treat the sum rules appropriate for low energies in $d \leq 3$ dimensions, taking into account explicitly the possibility of zero-energy resonances and/or zero-energy bound states for the underlying Hamiltonian. This requires a detailed study of the low-energy properties of this Hamiltonian.

The extension of these rules to other scattering situations like classical scattering, impurity scattering etc., is the subject of a fourth part. Finally, the fifth part reviews some applications of these sum rules in statistical mechanics, in particular in connection with the theory of virial coefficients, and, very shortly, in nuclear physics and field theory.

Quite a lot of material on this subject has been published more recently, ranging from abstract mathematical to pure phenomenological work. We will present here a selection out of this material to give an overview of the ideas and methods used in this field. (The material treated in Sections II to IV and partly in Section V is completely rigorous.) No technical details will be presented but as we go along we will try to build up an extensive reference list of the more recent works.

II. GENERALIZED LEVINSON THEOREM

At first sight, a lot of different forms of Levinson's theorem exist in the literature. In this part, we study the precise relationship between them. We start by shortly reviewing some results on two-body scattering we need in the following.

A. Some Results on Two-Body Scattering Theory

Let V be a real measurable function and assume $V \in R$, where R denotes the Rollnik class, i.e.

$$\|V\|_R^2 = \int d^3x \int d^3y \; \frac{|V(\underline{x})||V(\underline{y})|}{|\underline{x}-\underline{y}|^2} < \infty . \tag{2.1}$$

Let $H_0 = -\Delta$ denote the usual self-adjoint realization of the kinetic energy operator. (We use natural units $\hbar = 2m = 1$.) Then, the Schrödinger Hamiltonian can be defined on $H = L^2(\mathbb{R}^3)$ through the method of forms [7], viz.

$$H = H_0 \dotplus V . \tag{2.2}$$

Then it is well-known that the wave operators

$$\Omega^{\pm} = \underset{t \to \mp\infty}{\text{s-lim}} \; e^{itH} \, U(t), \quad U(t) = e^{-iH_0 t} \tag{2.3}$$

satisfy the following properties

$$\Omega^{\pm *} \Omega^{\pm} = \mathbb{1} , \tag{2.4}$$

$$\Omega^{\pm} \Omega^{\pm *} = P_{in} = P_{out} , \tag{2.5}$$

$$H\Omega^{\pm} = \Omega^{\pm} H_0 , \tag{2.6}$$

where $P_{in,out}$ are the projections onto ran $\Omega^{\pm}$. The first equation (2.4) constitutes conservation of probability, the second (2.5) completeness of scattering states, the third (2.6) expresses conservation of energy. Introducing

$$v(\underline{x}) = |V(\underline{x})|^{1/2}, \quad u(\underline{x}) = v(\underline{x}) \, \text{sign} \, V(\underline{x}) ,\quad (2.7)$$

the symmetrized free resolvent operator $uR_0(k)v$ defined as the norm limit

$$uR_0(\pm k)v = \underset{\varepsilon \to 0^+}{\text{n-lim}} \; u(H_0 - k^2 \mp i\varepsilon)^{-1}v, \quad k \geq 0 \quad (2.8)$$

in Hilbert-Schmidt and satisfies

$$u R_0(-k)v = (v R_0(k)u)^* . \quad (2.9)$$

The scattering wave functions $\Phi^\pm$ obey the following inhomogeneous Lippmann-Schwinger equations

$$\Phi^\pm(k\underline{\omega},\underline{x}) = \Phi_0^\pm(k\underline{\omega},\underline{x}) - (u R_0(\mp k)v\Phi^\pm)(k\underline{\omega},\underline{x}), \quad k^2 \notin E , \quad (2.10)$$

which can also be written in the form

$$\Phi^\pm(k\underline{\omega},\underline{x}) = ((1 + u R_0(\mp k)v)^{-1}\Phi_0^\pm)(k\underline{\omega},\underline{x}), \quad k^2 \notin E , \quad (2.11)$$

where the $\Phi_0^\pm$ are defined by

$$\Phi_0^-(k\underline{\omega},\underline{x}) = u(\underline{x}) e^{ik\underline{\omega}\cdot\underline{x}}, \quad \Phi_0^+(k\underline{\omega},\underline{x}) = v(\underline{x}) e^{ik\underline{\omega}\cdot\underline{x}}, \quad (2.12)$$

and where $\underline{\omega} \in S^2$, the unit sphere in $\mathbb{R}^3$. The set E is defined as

$$E = \{k^2 \geq 0 \mid u R_0(k)v\psi = -\psi \text{ for some } \psi \in L^2(\mathbb{R}^3), k \geq 0\}. \quad (2.13)$$

E is a closed subset of $[0,\infty)$ with Lebesgue measure zero containing the singular continuous spectrum and positive eigenvalue of H [7].

Then one can prove the following results:
Let $V \in L^1(\mathbb{R}^3) \cap R$ then the scattering operator

$$S = \Omega^{-*} \Omega^{+} \qquad (2.14)$$

associated with the pair (H, H_0) is unitary, commutes with H_0 and in the spectral representation of H_0, the corresponding on-shell operator $S(k)$ reads

$$(S(k)\phi)(\underline{\omega}) = \phi(\underline{\omega}) - \frac{k}{2\pi i} \int_{S^2} d\underline{\omega}' \, f(k,\underline{\omega},\underline{\omega}')\phi(\underline{\omega}'), \quad k^2 \notin E, \, \phi \in L^2(S^2), \qquad (2.15)$$

where $f(k,\underline{\omega},\underline{\omega}')$ represents the on-shell scattering amplitude. An explicit characterization of $f(k,\underline{\omega},\underline{\omega}')$ is obtained from

$$f(k,\underline{\omega},\underline{\omega}') = -(4\pi)^{-1} \, (\Phi_0^+(k\underline{\omega}), \Phi^-(k\underline{\omega}')) \qquad (2.16)$$

$$= -(4\pi)^{-1} \, (\Phi^+(k\underline{\omega}), \Phi_0^-(k\underline{\omega}'))$$

$$= -(4\pi)^{-1} \, (\Phi_0^+(k\underline{\omega}), T(k)\Phi_0^-(k\underline{\omega}')) \, . \qquad (2.17)$$

Here $T(k)$ is the T-matrix

$$T(k) = (u \, R_0(k) v + 1)^{-1} \, . \qquad (2.18)$$

It satisfies the relation

$$T(k) = 1 - u \, R_0(k) v \, T(k). \qquad (2.19)$$

For $k^2 \notin E$, $(S(k)-1) \in B_1(L^2(S^2))$ and continuous in trace norm with respect to k. For proofs we refer to [7], [8]. Finally, the averaged total cross-section $\bar{\sigma}(k)$ is defined by [9]

$$\bar{\sigma}(k) = (4\pi)^{-1} \int_{S^2 \times S^2} d\underline{\omega} \, d\underline{\omega}' \, |f(k,\underline{\omega},\underline{\omega}')|^2$$

$$= \frac{\pi}{k^2} \|S(k) - 1\|_2^2 = -\frac{2\pi}{k} \, \text{Re}(\text{Tr}(S(k)-1)). \qquad (2.20)$$

If V is spherically symmetric then H and S commute with the generator of the rotation group, i.e. the angular momentum operators. One can then define all scattering observables onto the subspace of angular momentum, e.g. the scattering amplitude is represented by

$$f(k,\underline{\omega},\underline{\omega}') = 4\pi \sum_{\ell=0}^{\infty} \sum_{m=-\ell}^{\ell} \frac{e^{2i\delta_\ell(k)} - 1}{2ik} Y^*_{\ell m}(\underline{\omega}) Y_{\ell m}(\underline{\omega}'), \quad k > 0, \quad (2.21)$$

$$e^{2i\delta_\ell(k)} = S_\ell(k), \qquad (2.22)$$

where δ_ℓ denotes the partial wave phase shift.

B. Jauch's Proof of Levinson's Theorem

For s-wave scattering, Levinson rigorously proved [1], while working on the inverse problem, the following theorem using the theory of ordinary differential equations. If

$$\int_0^\infty dr\, r\, |V(r)| + \int^\infty dr\, r^2\, |V(r)| < \infty \qquad (2.23)$$

then $\delta(k)$ is continuous in k, $\delta(\infty) = \lim_{k \to \infty} \delta(k)$ exists and

$$\delta(0) - \delta(\infty) = \pi(n + q\frac{1}{2}), \qquad (2.24)$$

where n is the number of s-wave bound states, $q = 1$ if there is a zero-energy resonance, $q = 0$ otherwise. This result turned out not only to provide some deep theoretical insight into the scattering process but also to be very useful in applications (see Sections V and VI). Therefore, it is undoubtedly interesting to generalize this theorem to very general scattering systems. In the rest of this Section, we will assume no positive or zero-energy bound states and no zero-energy resonances. They can, however, be included (see Section IV).

The first in the literature to realize this was Jauch [2]. He gave a formal proof based on the orthogonality and completeness relation for the wave operators. He considered systems where V is not necessarily central and where the spectrum of H differs from that of H_0 by a finite number of bound states below the continuum. The condition on the potential is replaced by a condition for the behavior of the scattering state wave function at high-energy. We want to give an outline of Jauch's proof because it implicitly gives the crucial indication where the partial wave and non-central Levinson theorem exactly differ. One had to work about five years when Buslaev [10] proved the theorem for non-central potential in the Schwarz class of rapidly decreasing C^∞-functions. Then, in the seventies modern abstract scattering theory was used to extend these results to more general classes of potentials (see below).

The method of Jauch is simple in principle. From Eq. (2.5) we immediately obtain for the system under consideration

$$\mathrm{Tr}[\Omega^{\pm *}, \Omega^\pm] = N \qquad (2.25)$$

and, restricted to a fixed partial wave subspace,

$$\mathrm{Tr}_\ell [\Omega^{\pm *}_\ell, \Omega^\pm_\ell] = n_\ell \quad .$$

Here N is the total number of bound states, n_ℓ is the number of distinct bound states for the partial wave ℓ (each bound state has a $(2\ell+1)$ degeneracy). In the following, we forget up to a certain point in the derivation the partial wave index ℓ since the method works for both the restricted and full problem. The relations (2.25) and (2.26) are in fact the basic forms of Levinson's theorem. The only thing we have to do is to find a suitable relation between the wave operators and scattering observables like the S-matrix, phase shift, etc.

To realize this, we recall the known representation of the wave operators in terms of the T-matrix [2], [11], viz.

$$\Omega^\pm = \mathbb{1} - K^\pm , \qquad (2.27)$$

where the kernel of K is given by the following generalized function (denote $vT(k)u \equiv t$)

$$K^{\pm}(\underline{p},\underline{p}') = \frac{t(\underline{p},\underline{p}';p'^2 \pm i0)}{p^2 - p'^2 \mp i0} . \qquad (2.28)$$

Then Eq. (2.25) becomes

$$\text{Tr } [K^{\pm *}, K^{\pm}] = N . \qquad (2.29)$$

Since the kernel of K is a generalized function, the trace needs to be computed through a limiting process. It is convenient to introduce the operator $K^{\pm}(k)$ corresponding to $K^{\pm}$ in the spectral representation of H_0. We then get after some manipulations (see [2], [12]) in terms of the $S(k)$ operator

$$N = \frac{1}{4\pi} \int_0^\infty dk^2 \left\{ \text{Tr}[i \, S^*(k) \, \frac{d}{dk^2} \, S(k) - i \, \frac{d \, S^*(k)}{dk^2} \, S(k)] \right. \qquad (2.30)$$

$$\left. + \text{Tr}[i \, \frac{d}{dk^2} \, (S^*(k) - S(k))] \right\}$$

where Tr is taken on the appropriate space $L^2(S^2)$. A similar equation holds on the subspace of angular momentum ℓ. The condition that is needed to make these detailed arguments work boils down to the integrability of the r.h.s. of (2.30) at ∞. For partial wave scattering we immediately get, using Eq. (2.22),

$$n_\ell = -\frac{1}{4\pi} \int_0^\infty dk^2 \left\{ 4 \, \frac{d}{dk^2} \, \delta_\ell(k) - 8\pi \, \frac{d}{dk^2} \, \sin(2\delta_\ell(k)) \right\} , \qquad (2.31)$$

or

$$\pi n_\ell = \delta_\ell(0) - \delta_\ell(\infty) - 2\pi[\sin(2\delta_\ell(0)) - \sin(2\delta_\ell(\infty))] . \qquad (2.32)$$

This has the wanted solution

$$\pi n_\ell = \delta_\ell(0) - \delta_\ell(\infty) . \qquad (2.33)$$

For non-spherically symmetric scattering, a heuristic study of the behavior of the integrand of (2.30) at ∞ leads to the following [12].

Using Eqs. (2.15) - (2.19), we can write

$$i \frac{d}{dk^2} \text{Tr} [S^*(k) - S(k)] = -\frac{1}{2\pi k} \int d^3x\, V(\underline{x}) - 4\pi \frac{d}{dk^2} \text{Re Tr}[kVR_0(k)t(k)]. \quad (2.34)$$

Then Eq. (2.30) reads

$$2\pi N = \int_0^\infty d\,k^2 \left\{ \text{Tr}[i\, S^*(k) \frac{d}{dk^2} S(k)] - \frac{1}{4\pi k} \int d^3x\, V(\underline{x}) \right\}$$

$$- 2\pi \text{ Re Tr } [kVR_0(k)t(k)] \Big|_{k^2 = 0}^{k^2 = \infty} . \quad (2.35)$$

The fact that the zero-energy on-shell t-matrix $t(k)$ is proportional to the scattering length (Section IV) means that the last term on the r.h.s. of (2.35) disappears in the $k^2 = 0$ limit. Furthermore, the symmetry property of the forward scattering amplitude under the transformation $k \to -k$, i.e. $f^*(k,\underline{\omega},\underline{\omega}) = f(-k,\underline{\omega},\underline{\omega})$ [3, sect. 10.3.3] implies that Re $f(k,\underline{\omega},\underline{\omega})$ is an even function of k. This, together with the fact that [13]

$$|(V R_0(k)\, t(k))(\underline{k},\underline{k})| \xrightarrow[k \to \infty]{} 0 \quad (2.36)$$

means that the last term in the r.h.s. of (2.35) also disappears for $k^2 = \infty$. So finally we arrive at

$$2\pi N = \int_0^\infty d\,k^2 \left\{ \text{Tr } [i\, S^*(k) \frac{d}{dk^2} S(k)] - \frac{1}{4\pi k} \int d^3x V(\underline{x}) \right\} . \quad (2.37)$$

The details of this derivation indicate why the extended Levinson theorem must have the potential term present. For sufficiently high energies we expect that the t-matrix will be dominated by the Born term $V(\underline{k},\underline{k})$. If we replace the t-matrix by this Born term in the expression (2.19) for $S(k)$, then the first order contribution is exactly this potential term, meaning that $\text{Tr}[i\, S^*(k) \frac{d}{dk^2} S(k)]$ is not integrable at

∞ with respect to k^2. By subtracting then this potential term we cancel this singular behavior.

Buslaev [10] was the first to give a rigorous proof of the extended Levinson theorem (2.37) for $V \in S(\mathbb{R}^3)$, the Schwarz class of rapidly decreasing C^∞-functions. Alternative proofs for more general classes of scattering systems will be discussed now.

C. Levinson's Theorem "à la carte"

Eqs. (2.37) and (2.33) give Levinson's theorem in terms of the S-matrix or phase shift. In the latter form it relates two observables in a scattering problem: the number of bound states and the phase shift. In more complicated scattering situations like e.g. non-spherically symmetric scattering, break-up scattering etc., a phase shift representation does no longer exist. An alternative in that case is to write the theorem in terms of time delay.

Let us introduce the notion of time delay heuristically in the simple case of partial wave scattering. Consider scattering of a spherical wave packet at time $t=0$. For $t<0$ we have an incoming packet of the form

$$\int_0^\infty dk\, f(k)\, e^{-i(kr + \frac{k^2}{2m}t)} \qquad (2.38)$$

where the function $f(k)$ is peaked around $k = k_0$. By the method of stationary phase we find the most important contribution to this integral at

$$\frac{d}{dk}\left(kr + \frac{k^2}{2m}t\right)\bigg|_{k=k_0} = 0 \Rightarrow r = -v_0 t,\ v_0 = \frac{k_0}{m}. \qquad (2.39)$$

After the scattering, for $t>0$, we have a scattered wave packet of the form

$$\int_0^\infty dk\, f(k)\, e^{ikr - \frac{ik^2}{2m}t + 2i\delta} \qquad (2.40)$$

such that the most important contribution is now obtained from

$$r = v_0 t - 2 \frac{d\delta}{dk}\bigg|_{k=k_0} . \qquad (2.41)$$

This means that the center of the outgoing wave packet is delayed due to the presence of the scatterer by

$$\tau = \Delta t = \frac{2}{v_0} \frac{d\delta}{dk}\bigg|_{k_0} = 2 \frac{d\delta}{dk^2}\bigg|_{k_0} = -i \, S^*(k) \frac{dS(k)}{dk^2}\bigg|_{k_0} . \qquad (2.42)$$

This notion of time delay and its connection with the S-matrix has been studied very extensively for general scattering situations (multichannel, absorptive, few-body, acoustical scattering ...) in the 70'ies and beginning of the 80'ies (see e.g. [14], [15] and the references cited therein).

For a general scattering system we proceed as follows. Let P be a projection on some subspace PH of states in H. In particular for two-body scattering, we take $P = P_\Sigma$ to be the projection on a bounded region $\Sigma \subset \mathbb{R}^3$ in configuration space. (Most of the time, we take this region to be a sphere.) The probability of finding the scattering state Φ_t in Σ at time t is $(\Phi_t, P_\Sigma \Phi_t) = \|P\Phi_t\|^2$. The total time spent by Φ_t in Σ, called the transit time T_Σ, is then

$$\begin{aligned} T_\Sigma(\phi) &= \int_{-\infty}^{+\infty} dt \, \|P_\Sigma \Phi_t\|^2 \\ &\equiv \int_{-\infty}^{+\infty} dt \, \|P_\Sigma \, e^{-itH} \, \Omega^- \phi\|^2 \qquad (2.43) \\ &= \int_{-\infty}^{+\infty} dt \, \|P_\Sigma \, \Omega^- \, e^{-itH_0} \phi\|^2 , \end{aligned}$$

where ϕ is the incoming state and the last equality is obtained by intertwining (see Eq. (2.6)). The difference between the transit time $T_\Sigma(\phi)$ of the scattered particle and the transit time $T_\Sigma^0(\phi)$ of the freely (asymptotically) moving particle defines the time delay for the region Σ and state ϕ, viz.

$$\tau_\Sigma(\phi) = T_\Sigma(\phi) - T_\Sigma^0(\phi) \ . \tag{2.44}$$

As Σ approaches the whole space $\mathbb{R}^3$ both T_Σ and T_Σ^0 diverge. However, one can show using different methods ([14] - [17] and references cited therein) that the limit $\Sigma \to \mathbb{R}^3$ of the difference (2.44) exists and that it is connected with the scattering matrix for potentials decreasing faster than $|x|^{-2-\varepsilon}$ at ∞.

One way to establish this connection with the S-matrix is to use trace methods. We briefly report on this here since the method is useful later on. It is based on the following theorem [17].

Let U_t be a unitary group with self-adjoint generator H_0 having absolutely continuous spectrum Λ, and Γ a trace class operator. Then

(i) $\displaystyle\int_{-\infty}^{+\infty} dt (f, U_t^* \Gamma U_t g) < \infty$ for f,g in a dense set D in H. (2.45)

(ii) There exists an essentially unique family of trace class operators Γ_λ acting on the components H_λ of the direct integral $H = \displaystyle\int^\oplus H_\lambda \, d\lambda$ which diagonalizes H_0, such that

$$\int_\Lambda d\lambda \, (\phi_\lambda, \Gamma_\lambda, \psi_\lambda) = \int_{-\infty}^{+\infty} dt \, (\phi, U_t^* \Gamma U_t \psi), \tag{2.46}$$

$$\frac{1}{2\pi} \int_\Lambda d\lambda \, \mathrm{Tr}_\lambda \, \Gamma_\lambda = \mathrm{Tr} \, \Gamma \ , \tag{2.47}$$

$$\frac{1}{2\pi} \int_\Lambda d\lambda \, \|\Gamma_\lambda\|_{\lambda,1} \leq \|\Gamma\|_1 \qquad \lambda = k^2 \ . \tag{2.48}$$

Here, Tr, Tr_λ, $\|\cdot\|_1$, $\|\cdot\|_{\lambda,1}$ are respectively the trace-norm in H and H_λ (we always omit the index λ in the Tr when there is no confusion possible).

We can use this theorem directly on $T_\Sigma(\phi)$ written down in Eq. (2.43) by choosing ϕ of compact support on Λ and taking $\Gamma = E_0(\Delta)\Omega^{-*} P_\Sigma \Omega^- E_0(\Delta)$ with $E_0(\Delta)$ being a spectral projection of H_0 on a bounded subset Δ of Λ. In this way, we obtain a transit time $T_\Sigma(k)$ and time delay $\tau_\Sigma(k)$ defined in the spectral representation of H_0, i.e. on the energy shell. The connection between $\tau(k) = \lim_{\Sigma \to \mathbb{R}^3} \tau_\Sigma(k)$ and the $S(k)$-matrix then reads

$$\tau(k) = -i\, S^*(k)\, \frac{dS(k)}{dk^2}, \qquad (2.49)$$

such that the extended Levinson theorem can also be written as

$$\int_0^\infty dk^2 \left\{ \text{Tr } \tau(k) + \frac{1}{4\pi k} \int d^3x\, V(\underline{x}) \right\} = -2\pi N. \qquad (2.50)$$

A direct proof of this form (2.50) for $V \in L^1 \cap L^2$ has been given in [12], [14] on the basis of analyticity properties for the resolvent difference $[R(k) - R_0(k)]$ in the complex k^2-plane (here $R(k) = (k^2 - H)^{-1}$) and a contour integration around the spectrum of H using the relation

$$\text{Im Tr } [R(k+i\delta) - R_0(k+i\delta)]$$

$$= \left\{ \frac{1}{2\pi} \int_0^\infty dk'^2 \text{ Tr } \tau(k') \frac{\delta}{(k'^2-k^2)+\delta^2} \right\} + \sum_j \frac{\delta}{(k'^2_j - k^2)+\delta^2}, \qquad (2.51)$$

where k'^2_j are the eigenvalues of H, repeated according to their multiplicities. We call this relation, which we will also use later on, the spectral property of time delay.

We remark that, besides its generality, the nice feature of (2.50) is that it is close to experiment. Indeed, a form of time delay can be measured in crystal blocking experiments [18] - [20]. In these experiments one shoots particles at a nucleus sitting on a lattice point of a crystal. A compound nucleus is formed. It moves through the crystal and decays after a certain time by emitting a particle. The emitted particle is forced to move in a certain symmetry direction of the crystal by the interactions with the atoms on the lattice points. Consequently, the angular distribution one measures outside the crystal shows a typical pattern around the symmetry axis. Since the place of emission gives the travelling distance of the compound nucleus, one can extract information about the lifetime of the compound nucleus. These measurements easily detect lifetimes of 10^{-17} sec. [18].

A third form of Levinson's theorem can be stated in terms of the density of states or the spectral displacement function of the system. This is interesting because this is the reason why this theorem is important in e.g. statistical mechanics applications (see Section VI). To arrive at such a form, we note that, as we explained before in Eq. (2.49), an application of the trace theorem (2.45) - (2.48) to Eq. (2.43) gives in particular

$$\begin{aligned}\frac{1}{2\pi} \int_\Delta dk^2 \; \text{Tr} \; T_\Sigma(k) &= \text{Tr} \; E_0(\Delta)\Omega^{-*} P_\Sigma \Omega^- E_0(\Delta) \\ &= \text{Tr}(P_\Sigma \Omega^- E_0(\Delta))^* (P_\Sigma \Omega^- E_0(\Delta)) \\ &= \text{Tr} \; P_\Sigma \Omega^- E_0(\Delta) \Omega^{-*} P_\Sigma \\ &= \text{Tr} \; P_\Sigma E(\Delta) P_\Sigma \; , \end{aligned} \qquad (2.52)$$

where we have used the cyclicity of the trace and the intertwining relation (2.6). Since $E(\Delta)$ is now a spectral projection of H, (2.52) shows that $\frac{1}{2}\text{Tr}_{k^2}T_\Sigma(k)$ is the density of states of H at the value k^2 of the total energy and lying in the subspace $P_\Sigma H$. Of course a similar conclusion is valid for $T_\Sigma^0(k)$ such that the trace of the time delay operator is the change in state density at the value k in the subspace $P_\Sigma H$ due to the interaction V, viz.

$$\frac{1}{2\pi} \text{Tr} \; \tau(k) = \frac{d}{dk^2} \xi(k) \; . \qquad (2.53)$$

In the theory of Krein and Birman [21]-[25], $\xi(k)$ is called the spectral displacement function for the pair (H,H_0). There the following has been shown. Let (H,H_0) be a pair of selfadjoint operators on H with $[R(k) - R_0(k)]$ trace-class, $\operatorname{Im} k \neq 0$. Then there exists a function $\xi(k)$ on R such that

$$\xi(k)(1+k^2)^{-1} \in L^1(-\infty, +\infty), \tag{2.54}$$

$$\operatorname{Tr}[R(k) - R_0(k)] = -\int_{-\infty}^{+\infty} dk'^2 \frac{\xi(k')}{(k^2-k'^2)^2}. \tag{2.55}$$

Furthermore, $\xi(k)$ is also connected with the phase of the S-operator through the relation obtained by combining (2.53) and (2.49). Finally,

$$\xi(k) = \lim_{\varepsilon \to 0} \frac{1}{\pi} \arg \det_2(1 + u R_0(k+i\varepsilon)v), \tag{2.56}$$

where $\det_2$ is the modified Fredholm determinant.

Formula (2.56) leads us to a last method for obtaining the extended Levinson theorem [26] (see also [3], [27]). Suppose $V \in L^1 \cap L^2$. Then the kernel $u R_0(k) v$ is Hilbert-Schmidt, and we can introduce

$$D(k) = \det_2(1 + uR_0(k)v). \tag{2.57}$$

By the analytic Fredholm theorem, $D(k)$ is analytic in the upper half-plane and approaches 1 as $|k| \to \infty$. We then write

$$D(-k) = \det_2(1 + uR_0(-k)v) \tag{2.58}$$

$$= \det_2\left\{(1+uR_0(k)v)[1 - (1+uR_0(k)v)^{-1}(uR_0(k)v - uR_0(-k)v)]\right\}.$$

Employing in (2.58)

$$\det_2(1-A)(1-B) = \det_2(1-A)\det_2(1-B)\exp(\operatorname{Tr}(1-A)B), \tag{2.59}$$

(the meaning of A and B being clear), valid if B is trace class, which is satisfied in our case, it is straightforward to show that (see Eqs. (2.10) - (2.19))

$$(uR_o(k)v - uR_o(-k)v)(\underline{x},\underline{y}) = -\frac{ik}{8\pi^2}\int_{S^2} d\omega \; \Phi_o^+(k\underline{\omega},\underline{x})\Phi_o^{-*}(k\underline{\omega},\underline{y}) \quad (2.60)$$

such that

$$Tr(uR_o(k)v - uR_o(-k)v) = Tr(1-A)B = -\frac{ik}{2\pi}\int d^3x \; V(\underline{x}) \; . \quad (2.61)$$

Furthermore,

$$B(\underline{x},\underline{y}) = \frac{ik}{8\pi^2}\int_{S^2} d\omega \; \Phi_o^+(k\underline{\omega},\underline{x})\Phi_o^-(k\underline{\omega},\underline{y}) \; . \quad (2.62)$$

This quantity has the same trace as the scattering operator ([26], [27]) such that (by using the Plemelj-Smithies formulas for the Fredholm determinant, see [3], [26], [27])

$$\det(1-B) = \det S \; . \quad (2.63)$$

Combining these results in (2.58), we finally get

$$D(-k) = D(k) \det S(k) \exp\left[-\frac{ik}{2\pi}\int d^3x \; V(\underline{x})\right] \; . \quad (2.64)$$

This is a generalization of the well-known partial wave result [3]

$$D_\ell(-k) = D_\ell(k) \det S_\ell(k) \; . \quad (2.65)$$

It again illustrates very nicely the structural change due to the non-spherical symmetry of the interaction.

Levinson's theorem is then obtained on the basis of the analyticity properties of $D(k)$ by integrating the total derivative $d(\log D(k))$ along a contour which extends along the real axis from $-R$ to $+R$, avoiding the origin by a small semicircle in the upper half-plane, and which closes in the upper half-plane with a large semicircle of radius R. The result is [26]

$$2\pi i N = \log \det S(0) - \lim_{k \to \infty} [\log \det S(k) - \frac{ki}{2\pi} \int d^3x \, V(\underline{x})]. \qquad (2.66)$$

In this form, the theorem can be generalized to very abstract scattering systems (H, H_0)

$$N = \frac{1}{2\pi i} \ln \det S(k) \Big|_{\partial \sigma (H_0)} \qquad (2.67)$$

where $\partial \sigma(H_0)$ denotes the boundary of the spectrum $\sigma(H_0) = [0, \infty)$. For a proof and detailed conditions on the interaction we refer to the literature [27] - [31].

This completes our discussion of the relations between the different forms of Levinson's theorem.

III. HIGHER-ORDER LEVINSON THEOREMS

A. Introductory Remarks

From Section II, it is clear that Levinson's theorem connects the bound state poles and the scattering phase shift, the point spectrum and the continuous spectrum of the Schrödinger Hamiltonian. One can say that this theorem expresses spectral stability. An interesting question that can be asked immediately then is if other relations of this type exists. If one thinks about a simple quantum mechanical system with a finite number N of independent states, of energy λ_j, $j = 1, \ldots, N$, one knows e.g. that

$$\text{Tr } H^n = \sum_{j=1}^{N} \lambda_j^n, \quad n = 1, 2, 3, \ldots \qquad (3.1)$$

These trace relations provide information about general properties of the spectrum without the need for evaluating the individual energy levels. They were actually already used in this sense in the study of magnetic properties of crystals in 1932 [32].

Recently, [33] there appeared some work on the trace relations connected with the θ-function of the Schrödinger Hamiltonian $H = -\frac{1}{2}\Delta + \frac{1}{2}(\underline{x}, A^2 \underline{x}) + V(\underline{x})$, A^2 being a symmetric strictly positive matrix in $\mathbb{R}^d$, $\underline{x} \in \mathbb{R}^d$ and $V(\underline{x})$ a bounded continuous potential on $\mathbb{R}^d$, viz.

$$\theta(t) = \text{Tr } e^{-itH} = \sum_{j=0}^{\infty} e^{-it\lambda_j}. \qquad (3.2)$$

In particular, the asymptotic behavior of $\theta(t)$ for small $\hbar$ is discussed in terms of classical periodic orbits and the relation with the Bohr-Sommerfeld quantization formula is given.

The type of relations we would like to discuss here are those involving higher-order energy moments of the phase shift or, more general, the trace of the resolvent difference $\text{Tr}[R(k) - R_0(k)]$ (see Eqs. (2.50) - (2.51)). These relations are in fact generalizations of (3.1) to systems where continuous spectra are allowed. The first derivation of some of these relations goes back to work by Buslaev, Dikii, Gelfand and Levitan (see references in [5]). Around the same time, there appeared some discussions on the partial-wave version of these relations [34] - [36]. Here we want to present a systematic discussion of the latest developments. As we will see in the following sections, these relations are not only important for a basic, mathematical understanding of the scattering system, they also lead to interesting physical applications.

B. Derivation

The method we use is based on a study of the analyticity properties of the resolvent difference, its high-energy behavior and contour integration in the complex energy (or momentum) plane around the spectrum of the Hamiltonian.

For potentials satisfying $V \in L^1(\mathbb{R}^3) \cap R$, we know that (see e.g. [7] - [9], [12])

$$\|u\, R_0(k)\|_2^2 = (8\pi \, \text{Im } k)^{-1} \|V\|_1, \quad \text{Im } k > 0, \qquad (3.3)$$

$$\| u \, R_0(k) v \|_2^2 \leq (4\pi)^{-2} \|V\|_R^2 \, , \quad \text{Im} \, k \geq 0 \, , \tag{3.4}$$

$$\| u \, R_0'(k) v \|_2^2 \leq (4\pi)^{-2} \|V\|_1^2 \, , \quad \text{Im} \, k \geq 0 \, , \tag{3.5}$$

where $R_0(k)$ is defined by Eq. (2.8), u and v by Eq. (2.7) and $R_0'(k)$ is the derivative of $R_0(k)$ with respect to k. Defining

$$\tilde{T}(k) = 1 - T(k) = u \, R_0(k) v \, (1 + u \, R_0(k) v)^{-1} \, , \tag{3.6}$$

where we have used Eq. (2.17), we have that

$$\tilde{T}(k) \in B_2(L^2(\mathbb{R}^3)), \quad \text{Im} \, k \geq 0, \quad k^2 \notin E \, , \tag{3.7}$$

$$[R(k) - R_0(k)] \in B_1(L^2(\mathbb{R}^3)), \quad \text{Im} \, k > 0, \quad k^2 \notin E, \tag{3.8}$$

with B_2 the Hilbert-Schmidt operators, B_1 the trace-class operators and E the exceptional set (see Eq. (2.13)). Under the conditions for V we assume in this section, E can be shown to be a compact subset of $\mathbb{R}$ of Lebesgue measure zero. Furthermore,

$$\text{Tr}\,[R(k) - R_0(k)] = -\text{Tr}[R_0(k) v \, u \, R_0(k)] + \text{Tr}[R_0(k) v \tilde{T}(k) u R_0(k)]. \tag{3.9}$$

The first term on the r.h.s. of (3.9) can be calculated in a straight-forward way. For the second term we use cyclic invariance properties of the trace and $R_0^2(k) = (2k)^{-1} R_0'(k)$. Then

$$\text{Tr}[R(k) - R_0(k)] = (8\pi i k)^{-1} \int d^3x \, V(x) + (2k)^{-1} \text{Tr}[\tilde{T}(k) u R_0'(k) v]. \tag{3.10}$$

By dominated convergence, $\text{Tr}[R(k) - R_0(k)]$ can be shown to be continuous with respect to k for $\text{Im} \, k \geq 0$, $k^2 \in E$, $k \neq 0$. If one further assumes that $e^{2a|\cdot|} V \in R$, $a > 0$, one can even show that $\text{Tr}[R(k) - R_0(k)]$ has a meromorphic continuation into $\text{Im} \, k > -a$.

On the basis of the formulas (3.3) - (3.10) one can also prove that for $|k| > |k_B|$ such that $\|u R_0(k_B)\|_2^2 < 1$ and $\|[u R_0(k_B)v]^2\|_2^2 < 1$

$$|Tr(VR_0(k))^n| \leq C < \infty, \quad |Re\, k| > |Re\, k_B|, \quad n \geq 2, \qquad (3.11)$$

$$\lim_{|Re\, k| \to \infty} Tr\, (V R_0(k))^n = 0, \quad n \geq 2, \qquad (3.12)$$

both uniform with respect to $Im\, k$. Consequently, for $|k| > |k_B|$, $Tr[R(k) - R_0(k)]$ allows a Born series expansion convergent in trace norm i.e.

$$Tr\,[R(k) - R_0(k)] = \sum_{n=1}^{\infty} (-1)^n Tr\,\{R_0(k)\,[VR_0(k)]^n\}. \qquad (3.13)$$

For more details see e.g. [7] - [9], [12], [14].

These results lead us to conclude that the function defined by

$$Q_N(k) = k^{2N} Tr\{R(k) - R_0(k) - \sum_{n=1}^{N+1} (-1)^n [R_0(k)(VR_0(k))^n]\}, \quad N = 0,1,2,\ldots \qquad (3.14)$$

is analytic in a region Π_δ of the complex energy plane, with Π_δ the set of points in the $z = k^2$-plane a distance δ or greater away from the positive real axis. (The symbol Π_0 will denote the cut z-plane obtained by letting $\delta \to 0$).

At this point, one could ask why we introduce the N+1 subtraction terms in (3.14). From Section II, we know that in the non-spherically symmetric scattering problem, a correction term appears already in Levinson's theorem itself (see Eq. (2.37) which corresponds to taking N = 0 in (3.14)) due to the (too) slow high-energy behavior of the resolvent difference (or time delay or S-matrix). For the same reason we anticipate that there will be N correction terms in the N-th order Levinson theorem.

A careful study of the high-energy behavior of the first N+1 Born terms in the expansion of the resolvent difference confirms this. To pursue that study, different methods can be followed (for more details see [37]-[39]). Let us consider potentials that are represented by Fourier transforms of bounded measures, and that have bounded partial derivatives up to order 2N, i.e.

$$V(\underline{x}) = \int_{\mathbb{R}^3} e^{i\,\underline{\alpha}\cdot\underline{x}}\, d\mu(\underline{\alpha}); \quad \int_{\mathbb{R}^3} |\alpha|^n \, d|\mu(\alpha)| < \infty, \quad n = 0,1,\ldots,2N. \tag{3.15}$$

Then, in the "heat-kernel ([39]) or time evolution kernel ([38])" method one first writes down the following formula

$$R(k) = \int_0^\infty d\beta\, e^{\beta k^2}\, e^{-\beta H}, \quad \text{Re } k^2 < -\|\mu\|, \tag{3.16}$$

$$(e^{-\beta H})(\underline{x},\underline{y}) = (e^{-\beta H_0})(\underline{x},\underline{y})\, F(\beta,\underline{x},\underline{y}). \tag{3.17}$$

One then shows that $F(\beta,\underline{x},\underline{y})$ has an asymptotic series representation in β up to order M that is uniform in $\underline{x}$ and $\underline{y}$. One finally reinserts this expansion into (3.16)-(3.17) and uses analytic continuation to obtain e.g. (for more details see [38])

$$\sum_{n=1}^{M+1} (-1)^n\, (R_0(k)[VR_0(k)]^n)(\underline{x},\underline{x})$$

$$= \frac{i}{8\pi} \sum_{j=1}^{N+1} \frac{(2j-2)!}{2^{2j-2}\, j!\,(j-1)!} \frac{P_j(\underline{x},\underline{x})}{k^{2j-1}} + R_{N+1}(\underline{x},k), \tag{3.18}$$

where the $P_j(\underline{x},\underline{x})$ are known functions ([37], [40], [41]), e.g.

$$P_1(\underline{x},\underline{x}) = V(\underline{x}),$$
$$P_2(\underline{x},\underline{x}) = V^2(\underline{x}) - \frac{1}{3}\Delta V(\underline{x}), \tag{3.19}$$
$$P_3(\underline{x},\underline{x}) = V^3(\underline{x}) - V(\underline{x})\Delta V(\underline{x}) - \frac{1}{2}[\underline{\nabla}\cdot V(\underline{x})]^2 + \frac{1}{10}\Delta^2 V(\underline{x})$$

and where

$$|R_{N+1}(\underline{x},k)| < \frac{C(\|\mu\|, N)}{|k^2|^{N+1+1/2}}. \tag{3.20}$$

The easy control of the rest-term is a particularly nice aspect of this method. A more detailed discussion of the further meaning of these corrections will be given at the end of this section. It is worthwhile to notice at this point that a recent general survey of the properties of $\exp(-tH), t > 0$ has been given in [42].

In order to derive the higher-order Levinson theorems we apply contour integration techniques and integrate $Q_N(k)$ along the following path in the complex energy plane: C_R around the positive real axis form $\Gamma-i\delta$ to $\Gamma+i\delta$, $\Gamma > 0$ avoiding the origin by a circle $C_\delta = \{\delta e^{i\theta} | \theta \in [\frac{3\pi}{2}, \frac{\pi}{2}]\}$, along the circle $C_\Gamma = \{\Gamma e^{i\theta} | \theta \in [\arcsin\frac{\delta}{R}, 2\pi - \arcsin\frac{\delta}{R}]\}$ and finally encircling all bound state energy positions λ_j clockwise $C_j = \{\epsilon_j e^{i\theta_j} | \theta_j \in [0, 2\pi], \epsilon_j$ sufficiently small $\}$, $1 \leq j \leq N_b$.

We then briefly indicate - leaving out the technical details - how the different contributions to this contour integral are obtained:

(i) Because the exact resolvent $R(k)$ has a first-order pole at the bound-state energy positions λ_j we get from C_j:

$$\sum_{j=1}^{N_b} \oint_{C_j} dk^2 \, Q_N(k) = 2\pi i \sum_{j=1}^{N_b} \lambda_j^N . \tag{3.21}$$

(ii) Assuming no zero-energy bound states and/or no zero-energy resonances, we know that $\mathrm{Tr}[R(k) - R_0(k)]$ behaves like k^{-1} around $k^2 = 0$ such that the C_δ-piece of the contour does not contribute in the limit $\delta \to 0$. For more details and a treatment of the zero-energy bound states and zero-energy resonances we refer to Section IV.

(iii) The function $Q_N(k)$ is constructed in such a way that the C_Γ-piece of the contour, $\Gamma \to \infty$ does not contribute.

(iv) The contribution from the two segments above and below the real axis can be written as

$$\lim_{\Gamma\to\infty}\lim_{\delta\to 0}\int_0^\Gamma dk^2\ 2i\ \text{Im}\ Q_N(k+i\delta)$$

$$=\int_0^\infty dk^2\ \Big\{2i\ \text{Im}\ \text{Tr}\ [R(k)-R_0(k)]$$ (3.22)

$$+\sum_{n=1}^{N+1}\frac{i}{4\pi}\frac{(2n-2)!}{2^{2n-2}n!(n-1)!}\frac{1}{k^{2n-1}}\int d^3x\ P_n(\underline{x},\underline{x})\Big\}.$$

Putting these results in Cauchy's theorem, viz.

$$\oint_C dk^2\ Q_N(k)=0,\quad C=C_\Gamma\cup C_R\cup C_\delta\cup\Big\{\bigcup_{j=1}^{N_b}C_j\Big\},$$ (3.23)

we obtain

$$\int_0^\infty dk^2\ k^{2N}\Big\{2\ \text{Im}\ \text{Tr}[R(k)-R_0(k)]+\sum_{n=1}^{N+1}\frac{(2n-2)!}{4\pi 2^{2n-2}n!(n-1)!}\frac{1}{k^{2n-1}}\int d^3x\ P_n(\underline{x},\underline{x})\Big\}$$

$$=2\pi\sum_{j=1}^{N_b}\lambda_j^N,\quad N=0,1,2,\ldots\ .$$ (3.24)

We remark that by using Eq. (2.51) in (iv), these rules can also be written in terms of time delay or the S-matrix. In their S-matrix form they were first discussed by Buslaev [10] for the restricted class of potentials possessing derivatives of any order and decreasing together with those derivatives faster than any power as $|\underline{x}|\to\infty$. For partial wave scattering ([34]–[36], [43], [44]) only N correction terms appear due to the better behavior of the S-matrix at high-energies. For potentials like the Yukawa potential, that have a singularity at the origin, it is straightforward to check that only a finite number ($N=2$ for Yukawa) of the rules (3.24) are valid, since the integrals over the P_n, $n>2$ do no longer exist.

C. Discussion

1. Rules in dimension 2 and 1, relation with the Korteweg-de Vries Invariants

The derivation presented here is not restricted to positive integer N and three dimensions. For negative N, one needs to know more details about the low-energy behavior of the resolvent difference. These details and the corresponding rules which have interesting physical applications will be presented in the next section. Also rules for complex N can be derived ([45] - [47]).

A similar treatment as above can be given for two-dimensional scattering and scattering on the line. In two dimensions, important e.g. for surface problems, the equivalent of Eqs. (3.18) - (3.20) describing the corrections can be written as

$$R(k,\underline{x},\underline{x}) - R_0(k,\underline{x},\underline{x}) = \sum_{n=1}^{N+1} \frac{1}{4\pi n} \frac{P_n(\underline{x},\underline{x})}{k^{2n}} + O\left(\frac{1}{k^{2N+4}}\right) , \qquad (3.25)$$

where the P_n are again given by (3.19) (where of course ∇ and Δ are now the two-dimensional gradient and Laplacian). The higher-order Levinson theorems then read [48]

$$\int_0^\infty dk^2 \, k^{2N} \, \mathrm{Im} \, \mathrm{Tr}[R(k) - R_0(k)] = -\pi \sum_{j=1}^{N_b} \lambda_j^N - \frac{1}{4(N+1)} \int d^2x \, P_{N+1}(\underline{x},\underline{x}),$$

$$N = 0,1,2,\ldots \quad . \qquad (3.26)$$

We immediately see that in contrast with $d = 3$, where a number of correction terms appeared in the integral (see (3.24)), we now have one "surface" correction. The $N = 0$ rule of (3.26) has also been derived by Cheney [27]. In higher dimensions, we expect to have a combination of both [47].

Concerning one dimension, we write down the theorems for half integer N, i.e. $N = j + \frac{1}{2}$

$$\frac{1}{2\pi i} \int_{-\infty}^{+\infty} dk \, k^{2j} \, \ln(1 - |r(k)|^2) = \frac{2}{2j+1} \sum_{n=1}^{N_b} (\lambda_n)^{j+\frac{1}{2}} - (-1)^j \left(\frac{1}{2i}\right)^{2j+1} \int_{-\infty}^{+\infty} dx \sigma_{2j+1}(x),$$

$$j = 0,1,2,\ldots \quad . \qquad (3.27)$$

Here $r(k)$ is the reflection coefficient and, similar to Eq. (3.19), the σ_{2j+1} are given by

$$\sigma_1(x) = V(x) ,$$
$$\sigma_3(x) = V^2(x) , \qquad (3.28)$$
$$\sigma_5(x) = 2V^3(x) + (\tfrac{d}{dx} V(x))^2 ,$$

(where we now have omitted total derivative terms). These rules appear in the work of Zakharov and Fadeev [49] proving that the one-dimensional Korteweg-de Vries (KdV) equation $u_t = 6uu_x - u_{xxx}$ is a completely integrable Hamiltonian system. They express the fact that the infinite set of integrals of motion for this equation, given by the second term on the r.h.s. of (3.27), can be expressed in terms of scattering data for a Schrödinger equation.

This leads us to a point which was only fully realized in the late 70'ies [37], [49], [50]. The correction terms in the higher-order Levinson theorems are nothing but the invariant densities of the one-dimensional KdV equation or a higher-dimensional generalization of it [51], [41]. This observation is based upon the deep relationship between the Schrödinger equation spectrum and the KdV equation, first discovered by Gardner, Greene, Kruskal and Miura [52]: Consider $V(x,t)$ to be the one-parameter family of potentials satisfying the KdV equation, then the Schrödinger equation spectrum for these potentials is t-independent. If the evolution of V in the Schrödinger equation is governed by any equation whatsoever which leaves the eigenvalues invariant in time, then that equation possesses the same conserved densities as the KdV equation. These invariants are unique [53]. In Ref. [52] it has also been shown how an explicit solution of the KdV equation can be obtained by using (inverse) scattering techniques for the Schrödinger equation. Nowadays, this has become a standard method for studying other non-linear evolution equations important e.g. in quantum field theory [54].

Furthermore, it is clear by looking at the Θ-function for the

Schrödinger equation (see Eq. (3.2)) that the problem of determining the asymptotic behavior of the Schrödinger spectrum is reduced to finding the asymptotic behavior of the solution of the heat equation satisfied by $\theta(t)$ for $t \to 0$. This has been used by Perelomov [51] (see also [50], [41] to find recursion relations for the generalized ($d>1$) KdV-type invariants. (We recall that the spectrum is t-independent.) In fact, when discussing the correction terms in the Levinson theorems in IIIB we have followed a closely related method to find the P_j's themselves (Eqs. (3.16) - (3.20)). The recursion relations they satisfy can be constructed by plugging the expansion in β of $F(\beta,\underline{x},\underline{y})$, occurring in Eq. (3.17), directly into the heat equation $((\partial/\partial\beta + H) \exp(-\beta H))(\underline{x},\underline{x}') = 0$. This same procedure has led to a renewed discussion of the off-diagonal thermal kernel for systems with magnetic field, in particular a new equation for its WKB approximation has been derived together with recursion relations for the coefficients in its Wigner-Kirkwood and high-temperature expansion (see e.g. Refs. [55], [56] and reference cited therein).

2. Local sum rules, sum rule dynamics

If we study the proof of the higher-order Levinson theorems in detail, we see that we are allowed to write down so-called "local sum rules", viz. (take e.g. $d=2$, $N=0$)

$$\int_0^\infty dk^2 \, \text{Im Tr } P_\Sigma[R(k)-R_0(k)]P_\Sigma = -\pi \sum_{j=1}^{N_b} \|P_\Sigma \psi_j\|^2 - \frac{1}{4} \int_\Sigma d^2x \, V(\underline{x}) \,, \quad (3.29)$$

where P_Σ is the projection operator onto a bounded region $\Sigma \subset \mathbb{R}^2$, as introduced before (see Section II, after Eq. (2.43)) and where ψ_j are the eigenfunctions of H, i.e. $H\psi_j = \lambda_j \psi_j$, $\psi_j \in L^2(\mathbb{R}^2)$, $j = 1,2,\ldots,N_b$. This rule has been derived very recently [57] under the very weak restriction on the potential $V \in L^{4/3}(\mathbb{R}^2)$. (As shown in this work, also the contribution from the singular continuous spectrum and the zero-energy bound states can be taken into account in a rather straightforward way.) As we will discuss in the next section, the contribution from possible zero-energy resonances does not seem to show up explicitly in these local rules, in contrast to the global results. This observation is made in dimension 3 but we expect this to be independent on the dimension. So one has to be extremely careful if one wants e.g. to take the limit $\Sigma \to \mathrm{I\!R}^2$ in Eq. (3.29) to recover Eq. (3.26) for $N = 0$.

The l.h.s. of Eq. (3.29) can be written in terms of (local) time delay for a finite space region [58]. For every space region Σ having finite Lebesgue measure and for all integers $N \geq 0$ these rules give then the precise inter-relationship between the N-th energy moment of time delay and the energy-weighted probability for finding the bound-state wave functions in the region Σ. The existence of this type of local sum rules was first discovered in classical scattering [59]

The region Σ can even be contracted to one space point. In that case the sum rules can be shown to take the form of one-dimensional power-moment problem and so provides an alternative description of dynamics [60].

IV LOW ENERGY SUM RULES

In this section, sum rules for negative energy moments of the phase shift or, more general, the resolvent difference are considered. Besides their fundamental importance, they can be fruitfully used to study e.g. low-energy nuclear scattering, the low-temperature behavior of the second virial coefficient etc. For more details about these applications, we refer to Section VI.

A systematic study of negative-moment sum rules requires the low-energy expansion of the trace of the resolvent difference because we now expect low-energy correction terms. It is clear that such an analysis depends very strongly on the zero-energy spectral properties of the underlying Schrödinger Hamiltonian H. In particular, the possibility of having zero-energy bound-states and/or zero-energy resonances will influence these results.

Some time ago, the first negative energy moment rules for three-dimensional scattering by spherically symmetric potentials satisfying $\int_0^\infty dr\, r|V(r)|e^{2ar} < \infty$, $a > 0$ have appeared [46]. The first moment rule for a rather restricted class of non-central potentials has been written down in Ref. [62]. These works explicitly assumed no zero-energy resonances and no zero-energy bound states. For a recent unified treatment of bound states and resonances of Hamiltonians defined as quadratic forms, we refer to [61]. (This work also contains an extensive list of references to related works.)

Quite recently there have appeared a number of new discussions of the spectral properties of H in hte low-energy limit in d-dimensions ([63] - [67] and references therein). These results strongly depend on the dimensionality. Using these results, a systematic treatment of low-energy sum rules for general potentials has been given in [68] for $d = 3$ and in [69] - [71] for $d = 1$. An account of such a study for $d = 2$ is in preparation [72].

A. Sum Rules for Three Dimensions

In order to obtain a low-energy expansion for the trace of the resolvent difference, we need such an expansion for the transition operator $T(k)$ (cf. Eqs. (3.10) and (3.6)) and thus for $(1 + uR_0(k)v)^{-1}$. It turns out that the explicit form of the latter expansion strongly depends on whether

$$u\, R_0(0)v\, \phi_j = -\phi_j \quad \text{for some} \quad \phi_j \in L^2(\mathbb{R}^3),\ j = 1,2,\ldots,N_0. \quad (4.1)$$

The functions

$$\psi_j(\underline{x}) = (R_0(0)v\, \phi_j)(\underline{x}), \quad j = 1,\ldots,N_0 \quad (4.2)$$

are called zero-energy resonance functions. They satisfy

$$H\psi_j = (-\Delta + V)\psi_j = 0 \quad \text{for} \quad V \in R,\ j = 1,\ldots,N_0 \quad (4.3)$$

in the sense of distributions. Since $uR_0(0)v \in B_2(L^2(\mathbb{R}^3))$ (Estimate (3.4)) N_0 is necessarily finite. In general, ψ_j need not be in $L^2(\mathbb{R}^3)$. We now distinguish the following cases (see also [26], [63], [65], [73]).

Case I: There exist no resonance functions ψ_j or, equivalently, -1 is not an eigenvalue of $uR_0(0)v$.

Case II: There exists only 1 resonance function ψ (i.e. -1 is a simple eigenvalue of $uR_0(0)v$) and $\psi \notin L^2(\mathbb{R}^3)$.

Case III: There exist $N_0 \geq 1$ resonance functions ψ_j, $j = 1,\ldots,N_0$, which are all in $L^2(\mathbb{R}^3)$.

Case IV: There exist $N_0 \geq 2$ resonance functions ψ_j, $j = 1,\ldots,N_0$, and at least one of them is not in $L^2(\mathbb{R}^3)$.

There is a simple criterion which helps to decide whether a resonance function ψ is in $L^2(\mathbb{R}^3)$ or not. If $V \in L^1(\mathbb{R}^3) \cap R$ and $|\underline{x}|V(\underline{x}) \in L^1(\mathbb{R}^3)$, then

$$\psi \in L^2(\mathbb{R}^3) \text{ is equivalent to } (v,\phi) = 0. \qquad (4.4)$$

One can check this by decomposing, following Newton [26],

$$\psi(\underline{x}) \equiv (4\pi)^{-1} \int d^3y \, |\underline{x}-\underline{y}|^{-1} v(\underline{y})\phi(\underline{y}) \qquad (4.5)$$
$$= \frac{1}{4\pi|\underline{x}|} \int d^3y \, v(\underline{y})\phi(\underline{y}) + \frac{1}{4\pi} \int d^3y \left(\frac{1}{|\underline{x}-\underline{y}|} - \frac{1}{|\underline{x}|} \right) v(\underline{y})\phi(\underline{y}).$$

This implies that in Case IV we can always choose a particular set of linear combinations of the resonance functions ψ_j such that only $(v,\phi_{j_0}) \neq 0$ and all other $(v,\phi_j) = 0$, $j = 1,\ldots N_0$, $j \neq j_0$. For spherically symmetric interactions, Case II corresponds to a s-wave resonance, Case III to a p- or higher-wave bound state.

We then have the following results. First, let $V \in L^1(\mathbb{R}^3) \cap R$ and $\varepsilon \in \mathbb{C}\setminus\{0\}$ small enough. Then $(uR_0(0)v + 1 + \varepsilon)^{-1}$ has a norm convergent Laurent expansion around $\varepsilon = 0$ [65]

$$(uR_o(0)v + 1+\varepsilon)^{-1} = \frac{P_o}{\varepsilon} + \sum_{m=0}^{\infty} (-\varepsilon)^m T_o^{m+1}, \qquad (4.6)$$

where P_o is the projector onto the N_o-dimensional eigenspace of $uR_o(0)v$ to the eigenvalue -1

$$P_o = 1 \quad \text{in Case I},$$

$$P_o = \sum_{j=1}^{N_o} \frac{(\tilde{\phi}_j, \cdot)\phi_j}{(\tilde{\phi}_j, \phi_j)} \quad \text{in Cases II - IV}, \qquad (4.7)$$

with ϕ_j the solutions of Eq. (4.1) and

$$\tilde{\phi}_j = (\text{sgn } V)\phi_j, \ (\tilde{\phi}_j, \phi_\ell) = 0, \ j \neq \ell, \ j = 1,,\ldots,N_o, \qquad (4.8)$$

and where T_o is a bounded operator denoting the reduced resolvent

$$T_o = \text{n-}\lim_{\varepsilon \to 0^+} (uR_o(0)v + 1 + \varepsilon)^{-1} Q_o, \ Q_o = 1 - P_o. \qquad (4.9)$$

Secondly, for V satisfying

$$e^{2a|\cdot|} V \in R, \text{ for some } a > 0 \qquad (4.10)$$

$uR_o(k)v$ is analytic with respect to k in the region $\text{Im } k > -a/2$, and

$$uR_o(k)v = \sum_{n=0}^{\infty} (ik)^n r_n \qquad (4.11)$$

where the $r_n \in B_2(L^2(\mathbb{R}^2))$. They have the following kernel representation

$$r_n(\underline{x},\underline{y}) = \frac{1}{4\pi n!} u(\underline{x}) |\underline{x}-\underline{y}|^{n-1} v(\underline{y}). \qquad (4.12)$$

The expansion (4.11) converges in B_2-norm [63].

This allows us to show that for potentials satisfying condition (4.10), the transition operator $T(k)$ has the following Laurent expansion in k around $k = 0$ in Cases I to IV [68] (see also [65])

$$T(k) = \sum_{n=-2}^{\infty} (ik)^n t_n , \qquad (4.13)$$

where

$$t_{-2} = t_{-1} = 0 \quad \text{in Case I} ,$$

$$t_{-2} = 0 \quad \text{in Case II.}$$

For the coefficients t_n we can derive recursion relations in all cases. They are based on a decoupling $T(k) = P_0 T(k) + Q_0 T(k)$, the expansion (4.11) and the Lippmann-Schwinger equation (2.18). For more details we refer to [68]. Using this $T(k)$ expansion, we finally arrive at the wanted expansion for the trace of the resolvent difference:

Assume condition (4.10). Then $\text{Tr}[R(k) - R_0(k)]$ has a Laurent expansion in k around $k = 0$ in all Cases I to IV, viz.

$$\text{Tr}[R(k) - R_0(k)] = \sum_{n=q-1}^{\infty} (ik)^n \Delta_n , \qquad (4.14)$$

where $q = 0$ in Case I, $q = -1$ in Case II and $q = -2$ in Cases III and IV. The coefficients Δ_n are real and given by

$$\Delta_n = \frac{\delta_{n,-1}}{8\pi} (v,u) - \text{Tr} \sum_{\ell=0}^{n-q+1} \sum_{j=0}^{n-\ell-q+1} \frac{j+1}{2} r_{j+1} t_{n-\ell-j+1} r_\ell . \quad (4.15)$$

The following remarks are in order when comparing with Ref. [68]. We have replaced the "old" $2^{-1}(-1)^n \Delta_{n+1}$ by Δ_n for convenience. Furthermore, the expression for Δ_n differs from the formal expression derived before because we have developed a new proof based upon the relation (3.10), which clearly allows us to write the coefficients Δ_n in (4.15) as a product of B_2-operators. It turns out that the first two coefficients are the same, the "old" Δ_1 requires a minor modification (see Appendix).

We now have all information to finally derive the sum rules. This can be done by contour integration, as done in Section III, but starting from the function

$$F_N(k) = k^{-2N} \left\{ \text{Tr}[R(k)-R_o(k)] - \frac{\delta_{N,0}}{8\pi i k} - \sum_{n=-2}^{2N-2} (ik)^n \Delta_n \right\}. \quad (4.16)$$

We recognize the second term on the r.h.s. as a necessary high-energy correction. We immediately write down the result (V satisfies (4.10))

$$\int_0^\infty dk^2 \, k^{-2N} \left\{ 2 \, \text{Im} \, \text{Tr}[R(k)-R_o(k)] + \frac{\delta_{N,0}}{4\pi k} (v,u) + 2 \sum_{n=0}^{N-1} (-1)^n k^{2n-1} \Delta_{2n-1} \right\}$$

$$= -2\pi \sum_{j=1}^{N_b} (\lambda_j)^{-N} + 2\pi(-1)^{N+1} \Delta_{2N-2}, \quad N = 0,1,\ldots \quad (4.17)$$

where N_b is as in Section III.

We are especially interested in the low-energy sum rules (4.17) for $N=0$ and $N=1$. For $N=0$ we obtain an extension of Levinson's theorem discussed in Sections II and III to allow for possible zero-energy resonances and/or zero-energy bound states

$$\int_0^\infty dk^2 \left\{ 2 \, \text{Im} \, \text{Tr}[R(k)-R_o(k)] + \frac{(v,u)}{4\pi k} \right\} = -2\pi [N_b + \Delta_{-2}], \quad (4.18)$$

where $\Delta_{-2}^{I} = 0$, $\Delta_{-2}^{II} = 1/2$, $\Delta_{-2}^{III} = N_o$, $\Delta_{-2}^{IV} = (N_o-1)+1/2$. This relation generalizes the well-known partial wave result (2.24) and completely agrees with [26].

For $N=1$ the sum rule reads

$$\int_0^\infty dk^2 \, k^{-2} \left\{ 2 \, \text{Im} \, \text{Tr}[R(k)-R_o(k)] + 2k^{-1}\Delta_{-1} \right\} = -2\pi \sum_{j=1}^{N_b} \lambda_j^{-1} + 2\pi \Delta_o. \quad (4.19)$$

The exact expressions for Δ_{-1} and Δ_o are given in the Appendix for all cases. To discuss the further meaning of Δ_{-1}, it is interesting to note that an appropriate generalization of the spherically symmetric low-energy parameters like scattering length a and effective range r (see Refs. [3] and [74] for a recent survey) to the non-spherically symmetric case ([65], [68], [75]), can be given directly in terms of the threshold behaviour of the scattering amplitude $f(k,\underline{\omega},\underline{\omega}')$ defined in Eq. (2.16) [75]. Some of the results of this work are

Cases I and III

$$a \equiv -\lim_{k \to 0+} (4\pi)^{-2} \int_{S^2 \times S^2} d\omega\, d\omega'\, f(k,\underline{\omega},\underline{\omega}')$$
$$= (4\pi)^{-1} (v, T_0 u) \quad . \tag{4.20}$$

Cases II and IV

$$r \equiv \lim_{k \to 0+} 2(4\pi)^{-2} \int_{S^2 \times S^2} d\omega\, d\omega'\, [f(k,\underline{\omega},\underline{\omega}') - ik^{-1}], \quad a^{-1} = 0$$
$$= 8\pi |(v,\phi)|^{-2} (\tilde{\phi}, r_2 \phi) \quad . \tag{4.21}$$

The relation with the Δ_{-1} given in the Appendix can be read off immediately in all cases. For a more detailed discussion of these low-energy parameters and their extensions to Coulomb scattering and d-dimensions, we refer to the literature ([74] - [78] and references therein). We already see that this rule (4.19) will be particularly interesting because it connects different observables in a general scattering situation, namely the bound state energies, the resolvent difference or time delay or S-matrix, and the (generalized) scattering length with the potential.

We end this discussion in $d = 3$ with three further remarks. First, the use of the strong condition (4.10) leads to strong results in the sense that the expansions we get for $T(k)$ and $\text{Tr}[R(k)-R_0(k)]$ are Taylor or Laurent expansions. If we only assume $V \in R$ and $(1+|\underline{x}|)^n V(\underline{x}) \in L^1(\mathbb{R}^3)$, $n > 4$, we then obtain asymptotic expansions, the order of which depends on the value of n ([63] - [68]). Consequently, we get a finite number of sum rules in that case. Secondly, the local version of sum rule (4.18) (see Section III,C) can also be derived. In that case the equivalent of Eq. (3.1o), which has to be used to obtain an expansion of the resolvent difference reads

$$\text{Tr } P_\Sigma [R(k) - R_0(k)] P_\Sigma = \frac{1}{8\pi i k} \int_{x \in \Sigma} d^3 x \; V(\underline{x}) + \text{Tr}[\tilde{T}(k) u R_0(k) P_\Sigma R_0(k) v]. \quad (4.22)$$

The appearance of the projection P_Σ in the second term on the r.h.s. prevents the collaps of $u R_0(k) R_0(k) v$ into $(2k)^{-1} u R_0'(k) v$. Consequently, zero-energy resonances, causing a k^{-1} behaviour of $\tilde{T}(k)$ no longer lead to a k^{-2} behaviour in the trace, whereas zero-energy bound states keep their k^{-2} behaviour. So $\Delta_{-2}(\Sigma)$ only contains the latter and we do not see explicitly the influence of the zero-energy resonances in the r.h.s. of the local version of (4.18) [57]. Thirdly, an analogous study to the one presented here can be made for non-local potentials $W = W_1 W_2 \in B_1(L^2(\mathbb{R}^3))$ with $W_j \in B_2(L^2(\mathbb{R}^3))$, $j = 1,2$. We refer to [6], [79], [80] and Section V for further details.

B. Scattering on the Line

Different aspects of the one-dimensional scattering problem on the line have received much attention in the past, especially in connection with inverse scattering techniques, which are used extensively in quantum mechanical problems and quantum field theory [54]. For an extensive list of references of earlier one-dimensional results, see [69], [70], [81].

Here we want to present an analysis similar to the one above (IV.A). We start by realizing that in one dimension the kernel of the free Green function

$$R_0(k,x,y) = \frac{i}{2k} e^{ik|x-y|} \quad (4.23)$$

has a first order pole as $k \to 0$. In order to isolate this pole we follow Ref. [82] and decompose

$$u R_0(k) v = \frac{i}{2k} (v, \cdot) u + M(k), \quad \text{Im } k \geq 0, \quad k \neq 0, \quad (4.24)$$

where $M(k) \in B_2(L^2\mathbb{R}))$ for all $\text{Im } k \geq 0$. If

$$\int_{\mathbb{R}} dx \, e^{a|x|} |V(x)| < \infty \quad \text{for some} \quad a > 0, \quad (4.25)$$

then $M(k)$ is analytic with respect to k in the region $\text{Im } k > -a/2$, and

$$M(k) = \sum_{n=0}^{\infty} (ik)^n M_n, \quad (4.26)$$

where the M_n are Hilbert-Schmidt operators with kernels

$$M_n(x,y) = -\frac{1}{2} u(x) \frac{|x-y|^{n+1}}{(n+1)!} v(y). \quad (4.27)$$

The expansion (4.26) converges in B_2-norm. Defining

$$P = (v,u)^{-1} (v,\cdot)u, \quad Q = 1-P, \quad (4.28)$$

the transition operator $T(k)$ becomes

$$T(k) = [1 + (i(v,u)/2k)P + M(k)]^{-1}, \quad \text{Im } k \geq 0, \, k \neq 0. \quad (4.29)$$

Here and throughout this analysis we assume $(v,u) \neq 0$. For a treatment when $(v,u) = 0$, we refer to [71]. To get an idea what happens when $k \to 0$, we take out the $(1+P)$-part in Eq. (4.29) and invert

$$T(k) = \left\{1 + \left[1 - \frac{i(v,u)P}{2k+i(v,u)}\right] M(k)\right\}^{-1} \left[1 - \frac{i(v,u)P}{2k+i(v,u)}\right] \quad (4.29)$$

$$= \left[1 + QM_0 + O(k)\right]^{-1} \left[Q + O(k)\right].$$

Furthermore, we use

$$(1 + QM_0)^{-1} = 1 - (1 + QM_0Q)^{-1} QM_0. \quad (4.30)$$

This makes it plausible that, in analogy with Eq. (4.1), one has to look at the equation

$$QM_0 Q \phi_0 = -\phi_0, \quad \phi_0 \in L^2(\mathbb{R}) \qquad (4.31)$$

to find out if there are any zero-energy resonances and/or zero-energy bound states. It turns out that one can show the following [70] (see also [73], [83], [84]).

Let V satisfy

$$\int_\mathbb{R} dx\, (1+|x|^2)|V(x)| < \infty, \quad \int_\mathbb{R} dx\, V(x) \neq 0. \qquad (4.32)$$

Assume that Eq. (4.31) has a solution and define the zero-energy resonance function ψ by

$$\psi_0(x) = -(v,u)^{-1}\, (v, M_0\phi_0) - 2^{-1} \int_\mathbb{R} dy\, |x-y|\, v(y)\, \phi_0(y). \qquad (4.33)$$

Then

(i) $\psi_0 \in L^\infty(\mathbb{R})$ and $H\psi = 0$ in the sense of distributions

(ii) $\psi_0 \notin L^2(\mathbb{R})$

(iii) $u(x)\psi_0(x) = -\phi_0(x)$ a.e.

(iv) $[\psi_0 + (v,u)^{-1}(v, M_0\phi_0) - 2^{-1}\, \text{sign}(\cdot)\, ((\cdot)v, \phi_0)] \in L^2(\mathbb{R})$

$\psi_0(\pm\infty) = -(v,u)^{-1}(v, M_0\phi_0) \pm 2^{-1}\, ((\cdot)v, \phi_0)$

(v) ψ_0 is unique and thus the non-zero eigenvalues of QM_0Q are simple.

This result leads to the following case distinction in one dimension:

Case I: -1 is not an eigenvalue of QM_0Q i.e. H has no zero-energy resonance

Case II: -1 is a simple eigenvalue of QM_0Q, $QM_0Q\phi_0 = -\phi_0$

for some $\phi_o \in L^2(\mathbb{R})$ i.e. H has a zero-energy resonance, and

a) $c_1 \equiv (v,u)^{-1}(v,M_o\phi_o) = 0$, $c_2 \equiv 2^{-1}((\cdot)v,\phi_o) \neq 0$

or

b) $c_1 \neq 0$, $c_2 = 0$ (4.34)

or

c) $c_1 \neq 0$, $c_2 \neq 0$.

In all the Cases II we have that $(v,\phi_o) = 0$. In these cases there exists precisely one resonance function $\psi_o \notin L^2(\mathbb{R})$, up to multiplicative constants, given by Eq. (4.33). There exist no zero-energy bound states or, equivalently, c_1 and c_2 do not vanish simultaneously (cf. (ii) and (iv)). It is straightforward to realize all these cases in the example of an asymmetric square well.

We can then follow the same procedure as above in three dimensions and write down low-energy (Laurent or Taylor) expansions for the transition operator $T(k)$ and consequently also for the scattering amplitude $f(k)$ and the scattering matrix $S(k)$. As is well-known (e.g. [84]-[86]), the latter quantities are two by two matrices. These low-energy expansions lead to some new results on the reflection and transmission coefficients when zero-energy resonances are present [69]-[70]. They also allow us to state the following: Let V satisfy (4.25). Then $\text{Tr}[R(k)-R_o(k)]$ has the following Laurent expansion in k around $k = 0$ in all Cases I, IIa)-c)

$$\text{Tr}[R(k) - R_o(k)] = \sum_{n=q-2}^{\infty} (ik)^n \Delta_n, \quad (4.35)$$

where $q = 0$ in Case I and $q = -1$ in Cases II, and where the coefficients Δ_n are given in terms of the M_n ((4.27)) and the expansion coefficients t_n of $T(k)$, for which recursion relations can be derived. This result is based upon the equivalent of (3.10), viz.

$$\text{Tr }[R(k) - R_0(k)] = -\text{Tr }[R_0(k) v\, T(k)\, u\, R_0(k)]$$

$$= -(2k)^{-1} \text{Tr }[u R_0'(k) v\, T(k)] \qquad (4.36)$$

$$= 4^{-1}(ik)^{-3} (v, T(k) u) - 2^{-1} k^{-1} \text{Tr}[M'(k) T(k)].$$

For more details we refer to [69] – [70]. Using this we derive the low-energy sum rules by contour integration starting from the function

$$F_N(k) = k^{-2N}\left\{\text{Tr}[R(k) - R_0(k)] - \sum_{n=q-2}^{2N-2} (ik)^n \Delta_n\right\}, \quad N = 0,1,\ldots \quad (4.37)$$

There are no high-energy corrections necessary since, using estimates of the type (3.3)-(3.5) for the free Green function (4.23) into Eq. (4.36) shows that [70]

$$|\text{Tr}[R(k) - R_0(k)]| \le c|k|^{-3}, \quad |k| \ge k_0 > 0, \quad \text{Im }k > 0. \qquad (4.38)$$

So we have for V satisfying (4.25) and $(v,u) \ne 0$

$$\int_0^\infty dk^2\, k^{-2N} \left\{\text{Im Tr}[R(k) - R_0(k)] - \sum_{n=-2}^{N-2} (-1)^n k^{2n+1} \Delta_{2n+1}\right\}$$

$$= -\pi \sum_{j=1}^{N_b} (\lambda_j)^{-N} + \pi(-1)^{N-1} \Delta_{2N-2}, \quad N = 0,1\ldots \qquad (4.39)$$

Similar results can be derived for N complex. Let us look here only at the $N = 0$ relation in more detail. Introducing the phase shift $\delta(k)$ by

$$\det S(k) = e^{2i\delta(k)}, \quad k > 0, \qquad (4.40)$$

with $\lim_{k \to \infty} \delta(k) = 0$, we can show that $\delta(k)$ is continuously differentiable in $k > 0$ and

$$\text{Im Tr}[R(k) - R_0(k)] = -\frac{i}{4k} \text{Tr}[S^*(k) S'(k)] = \frac{1}{2k} \delta'(k), \quad k > 0. \quad (4.41)$$

Using this, we get the N = 0 rule

$$\delta(0+) = \pi(N_b + \Delta_{-2}), \quad \delta(\infty) = 0, \qquad (4.42)$$

where

$$\Delta_{-2} = -\tfrac{1}{2} \text{ in Case I },$$
$$\Delta_{-2} = 0 \text{ in Cases IIa)-c) }.$$

So the structure of Levinson's theorem for scattering on the line is completely different from its analogue in three dimensions. Indeed, when there is no zero-energy resonance present, we not only get on the r.h.s. of (4.42) the term proportional to the number of negative energy bound states, πN_b, but an additional factor $-\pi/2$ appears. In the case of a zero-energy resonance, we simply obtain the term πN_b. The difference is of course due to the additional Dirichlet boundary conditions at the origin when considering Schrödinger operators on the half line.

We end this one dimensional discussion with two remarks. First, the case $(v,u) = 0$ has been considered recently in full detail [71]. The case distinctions (4.34) are still valid but Q has to be replaced by $\tilde{Q} = 1-\tilde{P}$, with $\tilde{P} = c^{-1} P M_0$, $c = (v, M_0 u)$. Furthermore, c_1 and c_2 now read $c_1 = c^{-1}(v, M_0^2 \phi_0)$, $c_2 = 2^{-1}((\cdot)v, \phi_0)$. In general, the Δ_n will be different since they depend on the details of the interaction V but, as one expects, the form of Levinson's theorem (4.42) stays the same. Secondly, just as in three dimensions, the stringent conditions (4.25) can be relaxed to finiteness of a number of moments of V if one is only interested in asymptotic expansions and a finite number of low-energy sum rules.

C. Two-Dimensional Results

Recently there has been a lot of interest in two-dimensional scattering systems from different theoretical as well as experimental points of view (see references in [77] - [78]). A discussion of low-energy sum rules, which has only been started recently [72], parallels the one-dimensional treatment since also here the free Green function

$$R_0(k,\underline{x},\underline{y}) = \frac{i}{4} H_0^{(1)}(k|\underline{x}-\underline{y}|), \quad \underline{x} \neq \underline{y} \qquad (4.43)$$

has a singularity as $k \to 0$. However, this singularity is of a logarithmic type such that the technical details are much more involved. We again decompose [82]

$$u R_0(k) v = (2\pi)^{-1}[-\ln k - (i\pi/2) + \ln 2 + \Psi(1)](v,\cdot)u + M(k) \qquad (4.44)$$

$$\text{Im } k \geq 0, \quad k \neq 0,$$

where $M(k) \in B_2(L^2(\mathbb{R}^2))$ for all $\text{Im } k \geq 0$ and Ψ is the digamma function. In particular, the kernel of $M(0) \equiv M_{00}$ reads

$$M_{00}(\underline{x},\underline{y}) = -(2\pi)^{-1} u(\underline{x}) \ln |\underline{x}-\underline{y}| v(\underline{y}), \quad \underline{x} \neq \underline{y}. \qquad (4.45)$$

Introducing the projector P (4.28) we can show the following.

Let V satisfy

$$\int d^2x \, (1 + |\underline{x}|^{2+\delta})|V(\underline{x})| < \infty, \quad \int d^2x \, |V(\underline{x})|^{1+\delta} < \infty, \quad \delta > 0 \qquad (4.46)$$

$$\int d^2x \, V(\underline{x}) \neq 0.$$

Assume that $QM_0 Q\phi = -\phi$ for some $\phi \in L^2(\mathbb{R}^2)$ and define the zero-energy resonance wave function ψ by

$$\psi(\underline{x}) = -(v,u)^{-1} (v,M_0\phi) - (2\pi)^{-1} \int d^2y \, \ln |\underline{x}-\underline{y}| v(\underline{x}) \phi(\underline{y}). \qquad (4.47)$$

Then

(i) $\psi \in L^2_{loc}(\mathbb{R}^2)$ and $H\psi = 0$ in the sense of distributions

(ii) $u(\underline{x})\, \psi(\underline{x}) = -\phi(\underline{x})$ a.e.

(iii) $[\psi + (v,u)^{-1}(v,M_0\phi) - (2\pi)^{-1}|x|^{-2}\underline{x}\cdot((\underline{\cdot})v,\phi)] \in L^2(\mathbb{R}^2)$, in particular $\psi \in L^2(\mathbb{R}^2) \Leftrightarrow (v,M_0\phi) = ((\underline{\cdot})v,\phi) = 0$.

We can then introduce the following case distinctions concerning the zero-energy behavior of H [72]

<u>Case I</u>: -1 is not an eigenvalue of QM_0Q.

<u>Case II</u>: -1 is an eigenvalue of QM_0Q of multiplicity $M \leq 3$, $QM_0Q\phi_j = -\phi_j$, $\phi_j \in L^2(\mathbb{R}^2)$, $0 \leq j \leq 2$

and

a) $M = 1$, $c_1^{(0)} \neq 0$, $\underline{c}_2^{(0)} = 0$ or $\underline{c}_2^{(0)} \neq 0$

or

b) $M \leq 2$, $c_1^{(j)} = 0$, $\underline{c}_2^{(j)} \neq 0$, $1 \leq j \leq 2$

if $M = 2$ then $\underline{c}_2^{(1)}$, $\underline{c}_2^{(2)}$ are linearly independent

or

c) $2 \leq M \leq 3$, $c_1^{(0)} \neq 0$, $\underline{c}_2^{(0)} = 0$ or $\underline{c}_2^{(0)} \neq 0$,

$c_1^{(j)} = 0$, $\underline{c}_2^{(j)} \neq 0$, $1 \leq j \leq 2$

if $M = 3$ then $\underline{c}_2^{(1)}$, $\underline{c}_2^{(2)}$ are linearly independent.

<u>Case III</u>: -1 is an eigenvalue of QM_0Q, $QM_0Q\phi_j = -\phi_j$,

$\phi_j \in L^2(\mathbb{R}^2)$, $3 \leq j \leq 3+N$, $N \in \mathbb{N}$ and

$c_1^{(j)} = \underline{c}_2^{(j)} = 0$.

<u>Case IV</u>: -1 is an eigenvalue of QM_0Q, $QM_0Q\phi_j = -\phi_j$,

$\phi_j \in L^2(\mathbb{R}^2)$, $0 \leq j \leq 3+N$, $N \in \mathbb{N}$ and

a) $c_1^{(0)} \neq 0$, $\underline{c}_2^{(0)} = 0$ or $\underline{c}_2^{(0)} \neq 0$,

$c_1^{(j)} = \underline{c}_2^{(j)} = 0$, $3 \leq j \leq 3+N$

or

b) $c_1^{(j)} = 0$, $\underline{c}_2^{(j)} \neq 0$, $1 \leq j \leq 2$,

$c_1^{(j)} = \underline{c}_2^{(j)} = 0$, $3 \leq j \leq 3+N$,

or

c) $c_1^{(0)} \neq 0$, $\underline{c}_2^{(0)} = 0$ or $\underline{c}_2^{(0)} \neq 0$,

$c_1^{(j)} = 0$, $\underline{c}_2^{(j)} \neq 0$, $1 \leq j \leq 2$,

$c_1^{(j)} = \underline{c}_2^{(j)} = 0$, $3 \leq j \leq 3+N$.

Here the coefficients $c_1^{(j)}$, $\underline{c}_2^{(j)}$ are defined by

$$c_1^{(j)} = (v,u)^{-1}(v,M_0\phi_j), \quad \underline{c}_2^{(j)} = (2\pi)^{-1}((\underline{\cdot})v,\phi_j) \tag{4.48}$$

where j takes on the values described above. In all cases we have $(v,\phi_j) = 0$.

We are presently studying the low-energy expansions of $T(k)$ and $\text{Tr}[R(k)-R_0(k)]$ in all cases in order to find the correction terms necessary to derive the low-energy sum rules. It is interesting to remark already that in Case IIa, which corresponds to a s-wave zero-energy resonance when V is spherically symmetric, $T(k)$ behaves like $\ln k$. Case IIb, corresponding to a p-wave zero-energy resonance for spherically symmetric V, gives a $(k^2 \ln k)^{-1}$ behavior of $T(k)$. Of course, in the zero-energy bound state case, Case III, $T(k)$ behaves like k^{-2}.

V. RESULTS FOR OTHER SCATTERING SITUATIONS

In this section, we present a short overview of the study of some of these spectral sum rules in other, sometimes more complicated, scattering situations. We start with

A. Non-Local Interactions

As we have already mentioned in Section IV, both sets of sum rules can be written down for non-local interactions ([6], [30], [31], [79],

[80]). These interactions usually support positive energy bound states (i.e. eigenvalues embedded in the continuous spectrum). The manner in which these bound states of positive energy enter into the generalized Levinson theorem has been a matter of some confusion. (There is quite an extensive literature on this point. We refer to e.g. [6], [30], [31], [79], [80], [87] and references cited therein.)

Here we follow [6] and show that there is certainly no contradiction between the two formulations of Levinson's theorem that show up in the literature, viz.

$$\delta_1(k^2) \text{ is continuous and } \delta_1(0) = \pi(N_{B<} + N_{B>}), \tag{5.1}$$

$$\delta_2(k^2) \text{ jumps at } k_B^2 \text{ and } \delta_2(0) = \pi N_{B<}, \tag{5.2}$$

where the notation is obvious. (We use the energy as variable.)

Under reasonable assumptions on the scattering system, $\det S(k^2)$ has the following properties

$$(\text{i}) \quad \det S(k^2) = \frac{D(k^2)^*}{D(k^2)} \quad \text{for all } k^2 \notin E_B, \tag{5.3}$$

where $D(k^2)$ is holomorphic and non-zero in the open upper half energy plane, with boundary value $D(k^2) = \lim_{\varepsilon \to 0} D(k^2 + i\varepsilon)$ and zero-set $E_B = \{k_B^2 \in \mathbb{R} | D(k_B^2) = 0\}$. (Compare Eq. (2.65).)

(ii) E_B consists of $N_{B<}$ negative bound states ($0 \leq N_{B<} < \infty$) and $N_{B>}$ continuum, positive energy bound states ($0 \leq N_{B>} < \infty$).

(iii) $D(k^2)$ is continuous. Moreover, for all $k_B^2 \in E_B$

$$\lim_{k^2 \downarrow k_B^2} \det S(k^2) = \lim_{k^2 \uparrow k_B^2} \det S(k^2) = 1. \tag{5.4}$$

Consequently, there are two obvious ways to define the phase shift. The first makes use of the continuity property (5.4) of $\det S(k^2)$ and defines the phase shift as a continuous logarithm

$$\delta_1(k^2) = \frac{1}{2i} \ln \det S(k^2) , \quad \delta_1(\infty) = 0 . \tag{5.5}$$

The second makes use of the existence of analytic extensions and defines the phase shift as a boundary value

$$\delta_2(k^2) = \lim_{\varepsilon \downarrow 0} [-\arg D(k^2 + i\varepsilon)], \quad \delta_2(\infty) = 0 . \tag{5.6}$$

$\delta_1(k^2)$ is continuous and satisfies the modified form of Levinson's theorem (5.1). δ_2 is not continuous, at $k_B^2 \in E_B$ it has a jump

$$\lim_{k^2 \downarrow k_B^2} \delta_2(k^2) - \lim_{k^2 \uparrow k_B^2} \delta_2(k^2) = \pi , \tag{5.7}$$

elsewhere it is continuous. It satisfies Levinson's theorem (5.2) in its original form. It is easy to see that (5.1) and (5.2) follow from each other when taking Eq. (5.7) into account. So both results are correct and there is no contradiction at all.

A discussion that was still going on quite recently [87] is then which phase shift one should choose as the "more physical" one. Should one insist on continuity and take δ_1? Or should one take δ_2 because it has the following advantage: Assume there is a positive energy bound state pole for a certain interaction. If a small perturbation causes this pole to move down onto the second sheet of the Riemann surface, this pole disappears and a scattering resonance at a nearby energy replaces it. The phase shift then increases continuously but rapidly in this region by approximately π. So δ_2 has a kind of stability with respect to certain small perturbations.

We remark that a similar discussion is going on in the physics literature for absorptive interactions. For more details we refer e.g. to [88] - [89] and the references therein.

B. Coulomb Plus Short-Range Interactions

For three-dimensional attractive Coulomb plus spherically symmetric short-range interactions Yafaev [90] has studied trace relations of the type (3.24).

Considering first the radial scattering problem (see also [4], [91], [92]) with the short-range potentials satisfying $\int_0^1 dr\, r|V(r)| + \int dr\, r^{2/3} |V(r)| < \infty$, he proves that the quantum defect of the continuous spectrum, the phase shift $\eta_\ell(k^2)$ equals the quantum defect of the discrete spectrum δ_ℓ, viz.

$$\eta_\ell(\infty) - \eta_\ell(0) = \pi\, \delta_\ell, \tag{5.8}$$

where δ_ℓ is a real number depending on $V(r)$ that can be understood as follows. When we add a short-range potential to the Coulomb potential, the Balmer levels are changed according to

$$E_{n'\ell} = -\frac{e^4}{4(n'+\ell+1)^2}, \tag{5.9}$$

where n' differs from the quantum number n in the pure Coulomb case by

$$n' = n - \delta_\ell(n), \quad n = 0, 1 \dots . \tag{5.10}$$

The zero's of the Jost function for the total problem in the upper half k-plane in the limit $n \to \infty$ are given by

$$k_{n,\ell} = i[n + \ell - \delta_\ell + O(1)]^{-1}, \tag{5.11}$$

with $\delta_\ell = \delta_\ell(n \to \infty)$.

For the three-dimensional problem with spherically symmetric short-range potentials satisfying $V \in L^1(\mathbb{R}^3) \cap L^2(\mathbb{R}^3)$ Yafaev [90] proves a set of higher order Levinson theorems of the type (3.24). Because of the Coulomb potential, two changes have to be made. First, the contribution from the discrete spectrum has to be replaced by

$$\sum_{j=1}^{N_b} \lambda_j^N \to \sum_{n=1}^{\infty} |k_{n,\ell}|^{2N} - \sum_{n=1}^{\infty} \left(\frac{1}{n+\ell}\right)^{2N}, \tag{5.12}$$

where $k_{n,\ell}$ is given by (5.11). Secondly, for $N \geq 2$, correction terms of the form $(\ln k)^j$, j dependent on N, appear. For the exact expressions of the relations we refer to [90].

Here, we add the remark that Levinson's theorem for systems in which both the unperturbed and the total Hamiltonian have a finite number of discrete eigenvalues has been considered e.g. in [93].

C. Impurity Scattering, Friedel's Sum Rule

In the course of a recent study of inverse scattering in one dimension for Hill's equation modified by a nonperiodic potential that tends to zero as $|x| \to \infty$, a Levinson theorem for such a system, representing local impurity scattering in a crystal, has been derived [94], [95]. For basic results on periodic Hamiltonians we refer to [84], [94], [96] and references therein. Here we restrict ourselves to stating that in this type of problem, the spectrum displays a band structure with bound states in the gaps. When the energy is in the allowed band, one can define an S-matrix, a transmission amplitude T and a phase shift δ according to the following formulas

$$\det S(k^2) = T(k^2)/T^*(k^2), \quad T = |T|e^{-\bar{\delta}}, \quad (5.13)$$

and

$$\det S(k^2) = e^{2i\delta(k^2)}, \delta = \bar{\delta}(\bmod \pi), \delta(-k^2) = -\delta(k^2). \quad (5.14)$$

(Compare Eq. (4.40).) In the band gaps T is real and hence $\bar{\delta} = 0(\bmod \pi)$. It is important to remark (see VA) that δ in [94]-[95] is defined to be continuous in the allowed bands for $k^2 > 0$. In the generic case (see also IVB) T has simple zeros at the periodic spectrum as a function of k. As one circumscribes such a pole clockwise, its phase decreases by $\pi/2$. Therefore δ will be defined so that at each band gap the difference between its left-hand limit at the left gap end and its right-hand limit at the right gap end is π. As $k^2 \to \infty$, δ goes to zero. This defines δ uniquely in all the allowed regions. Then one can prove a Levinson theorem for every gap connecting the phase shift at the endpoints of the gap with the bound-states in between [94]-[95]. For results on the distribution of bound-states among the band gaps see [83], [97].

When talking about impurity scattering, one certainly has to mention a standard result in solid state physics that can be considered an equivalent of Levinson's theorem, i.e. Friedel's sum rule [98]-[100]. Here we build up this rule by using physical arguments [100]. Suppose an impurity atom is introduced into a metal. Let the nucleus of the impurity atom have a charge Z units bigger than that of the metal. The charge distribution ρ of the latter will distort so as to screen the long-range Coulomb potential at large distances. Charge will accumulate around the impurity in a sufficient amount to balance the excess ionic charge and to produce a system which is electrically neutral. It must therefore be required that the total displaced charged in $\Delta\rho$ equals Z, viz.

$$\int d^3r\, \Delta\rho(\underline{r}) = Z \ . \tag{5.15}$$

From the meaning of the quantities involved it is clear that this is equivalent to

$$\int_0^{E_F} dE\, \Delta N(E) = Z \ , \tag{5.16}$$

where ΔN is the change in density of states between the perfect crystal and the real, imperfect one and E_F is the Fermi energy. The possible change in the Fermi energy of the system is of order $1/N$ for a single impurity. It is negligible in this case and can still be neglected when the impurity concentration is small. The representation of ΔN in terms of scattering parameters is then

$$\Delta N(E) = \frac{2}{\pi} \sum_s g_s \frac{d\delta_s(E)}{dE} \ , \tag{5.17}$$

(compare Eqs. (2.52)-(2.56)), where $\delta_s(E)$ is the phase shift for impurity scattering in representation s (the irreducible representation of the symmetry group under which the potential is invariant; this is usually not spherical symmetry but e.g. the crystal point group or some subgroup of it), and g_s is the degeneracy of this representation. A factor of 2 counts both directions of electron spin. Assuming $\delta_s(0) = 0$ we then get

$$\sum_s g_s \, \delta_s(E_F) = \frac{\pi Z}{2} \, . \qquad (5.18)$$

This is Friedel's sum rule. It may be imposed as a self-consistency requirement on impurity potentials. Recently, this rule has been applied to magnetic monopoles in field theory, showing e.g. its equivalence with Levinson's theorem ([101], [102] and Section VI).

D. Classical Scattering

Consider the problem of a point mass moving in a classical potential field, once the center-of-mass motion has been removed. Its position will be given by a vector $\underline{r}$ in a d-dimensional Euclidean space, its momentum is denoted by $\underline{p}$. The classical Hamiltonian is defined as

$$H(\underline{r},\underline{p}) = \sum_{i=1}^{n} p_i^2 + V(\underline{r}) \, . \qquad (5.19)$$

Under the following conditions on the potentials

(a) $V(\underline{r})$ is bounded from below and continuous with bounded derivatives up to order 2 on $\{\underline{r}: |V(\underline{r})| < M, M < \infty\}$

(b) $|\underline{\nabla} V(\underline{r})| < |\underline{r}|^{-2-\varepsilon}$, $\varepsilon > 0$,

Newton's equations of motion

$$(\dot{\underline{r}},\dot{\underline{p}}) = (2\underline{p}, -\underline{\nabla} V) \qquad (5.20)$$

have unique solutions for initial conditions $(\underline{r}_o,\underline{p}_o)$. The map

$$S_t : (\underline{r}_o,\underline{p}_o) \to (\underline{r}_t,\underline{p}_t) \qquad (5.21)$$

defines a one-parameter group of canonical transformations from the phase space $\Gamma = \{(\underline{r},\underline{p}) : |H(\underline{r},\underline{p})| < \infty\}$ onto itself [103]. This phase space can be decomposed into the regions of bound orbits and scattering trajectories.

Next we consider the restricted phase space determined by the conditions

$$H(\underline{r},\underline{p}) \leq \varepsilon, \quad \underline{r} \in \Sigma, \quad \Sigma \subseteq \mathbb{R}^n, \mu(\Sigma) < \infty \quad (5.22)$$

with ε the energy and μ the Lebesgue measure. Then a set of local sum rules (cf. Eq. (3.29)) can be derived using phase space methods [59]. Here we present the results for $d=3$ and $d=2$, viz.

$$\int_0^\infty d\varepsilon \, \varepsilon^N \left\{ \mathrm{Tr}\, \tau_\Sigma(\varepsilon) - \frac{1}{3\pi} \sum_{j=1}^{N+1} \binom{3/2}{j} (3/2-j)^{1/2-j} \int_\Sigma d^3r (-V(\underline{r}))^j \right\}$$
$$= -2\pi \int_{-\infty}^{+\infty} d\varepsilon \, \varepsilon^N \, n_\Sigma(\varepsilon) , \quad (5.23)$$

$$\int_0^{a_\Sigma} d\varepsilon \, \varepsilon^N \, \mathrm{Tr}\, \tau_\Sigma(\varepsilon)$$
$$= -2\pi \int_{-\infty}^{a_\Sigma} d\varepsilon \, \varepsilon^N \, n_\Sigma(\varepsilon) - \frac{1}{2(N+1)} \int_\Sigma d^2r \, (V(\underline{r}))^{N+1} , \quad (5.24)$$

with

$$a_\Sigma = \sup_{\underline{r} \in \Sigma} \Theta(V(\underline{r})) V(\underline{r}) . \quad (5.25)$$

In these formulas $\tau_\Sigma(\varepsilon)$ is the time delay for the region Σ or, in other words, the difference between the transit times through Σ for the interacting and non-interacting system, $n_\Sigma(\varepsilon)$ is the density of bound orbits in Σ.

Let us immediately remark that the rules (5.23)-(5.24) stay valid for $\Sigma = \mathbb{R}^d$. In that case it is easy to show that they correspond to the classical limit of the relations (3.24) respectively (3.26) (where we first have to reintroduce the $\hbar$ factors). The correction terms $P_n(\underline{x},\underline{x})$ in the latter then become exactly $(V(\underline{x}))^n$, the first terms on the r.h.s. of Eqs. (3.19). Like in the quantum mechanical rules, the difference in structure between even and odd dimensions is apparent. Furthermore, in classical mechanics it is even sufficient for even dimensions to integrate up to a_Σ when $a_\Sigma < \infty$. This has to do with the fact that in that case $(\varepsilon-V(r))^{d/2}$ is a polynomial. The most dramatic example of this difference occurs for potentials that are everywhere attractive. Then (5.24) collapses to a statement predicting the integral over the energy moments of the bound state density in terms of an integral over the potential. In these derivations there is no requirement that $V(\underline{r})$ conserves angular momentum. The lowest order (N = 0,1) global rules for d = 3 have also been derived in [104].

In Refs. [105], [106] a simple alternative derivation of Levinson's theorem for $d \leq 3$ (N = 0 in (5.23), (5.24)) has been given, using the fact that the Möller and S-transformations preserve the volume in phase space as canonical transformations. For angular momentum conserving potentials, these authors show that the quasiclassical approximation for the quantum mechanical phase shift is the generator of the classical canonical S-transformation, viz.

$$S(d,\chi,\xi,k,L,L_3) = (d - 2(\tfrac{\partial}{\partial k}\delta)(k,L), \chi - 2(\tfrac{\partial}{\partial L}\delta)(k,L), \xi, k, L, L_3), \quad (5.26)$$

with

$$\delta(k,L) = \lim_{R \to \infty} \left\{ \int_{r'}^{R} dr [k^2 - \frac{L^2}{r^2} - V(r)]^{1/2} - \int_{r'_0}^{R} dr [k^2 - \frac{L^2}{r^2}]^{1/2} \right\}, \quad (5.27)$$

where the canonical coordinate system has been chosen as the momentum k, the angular momentum and its third component $\underline{L}$ and L_3, $d = (\underline{r}\cdot\underline{k})/|\underline{k}|$, the angle χ of the momentum in the plane of motion and the angle ξ of the projection of L in the $(\underline{x},\underline{y})$-plane, and where r' and r'_0 are the classical turning points. This discussion has been given also for the Coulomb modified phase shift in [15]. Of course, $\delta(k,L)$ shows up in Levinson's theorem ((5.23), (5.24)) for

N = 0) through the relation [104], [15]

$$\tau(k^2,L) = \frac{1}{k} \frac{\partial}{\partial k} \delta(k,L), \quad \varepsilon = k^2 \; . \tag{5.28}$$

E. Remarks on Relativistic and Many-Particle Scattering

For these scattering situations, we restrict ourselves to merely presenting a list of references and a few remarks. Levinson's theorem for the Dirac Hamiltonian has been discussed by Barthélémy [107] and by Ni Guang-jiong [108]. This author also considers the Klein-Gordon Hamiltonian. There exists also a new preprint on the Dirac case by E.J. and Th.V. Kanellopoulos [109].

Concerning many-channel and many-particle scattering, we refer to [4], [47], [48], [110] - [112]. In Ref. [110] a formal generalization has been given of Jauch's proof for Levinson's theorem (see Section II) to three-body systems, in terms of eigenphase shifts. Refs. [47], [48] present a generalization of the higher-order Levinson theorems (3.24) and (3.26), in their local and global form, to three-particle systems with general, not necessarily spherically symmetric, interactions. They use contour integration around the spectrum of the three-particle Hamiltonian and discuss the correction terms in detail.

VI APPLICATIONS

A. Statistical Mechanics

The first application of the sum rules derived in the previous sections concerns the general area of statistical mechanics and plasma physics. In particular, they are connected with the S-matrix approach [113] - [115] to statistical mechanics, which formulates the statistical behavior of a gas in terms of the collision processes of the constituent particles. For an extensive list of references in this context, we refer to [116].

Consider a gas of N distinguishable particles having identical mass. The strength of the interaction is not restricted so that eventually stable n-particle bound states ($n \leq N$) can be formed. Therefore, the equation of state will involve many types of collisions like coalescence, rearrangement, breakup and elastic scattering of both free particles and clusters so that a description in terms of appropriate multichannel scattering theory is needed. For the moment we assume Boltzmann statistics for the motion of stable clusters on particles.

As is well-known [117], the grand-canonical partition function of a Boltzmann gas, that depends on the volume, the temperature and the fugacity, can be expanded in a series in the fugacity using the cluster expansion. The coefficients in this expansion are the cluster integrals b_n which can be written, assuming they exist in the thermodynamic limit [118]

$$b_1 = \lambda^{-3}$$
$$b_2 = \frac{2^{3/2}}{2!\lambda^3} \, \text{Tr} \, [e^{-\beta H} - e^{-\beta H_0}] \,, \qquad (6.1)$$
$$b_3 = \frac{3^{3/2}}{3!\lambda^3} \, \text{Tr}\left[e^{-\beta H^{(3)}} - e^{-\beta H_0^{(3)}} - \sum_{i<j=1}^{3} (e^{-\beta(H_0^{(3)}+V_{ij})} - e^{-\beta H_0^{(3)}}) \right].$$
$$\ldots \ldots \,.$$

Here λ is the thermal wavelength given by

$$\lambda = (4\pi\beta)^{1/2} \,, \quad \beta = 1/kT \,, \qquad (6.2)$$

H and H_0 are the two-particle Hamiltonians (cf. Eq. (2.2)), $H^{(3)}$ and $H_0^{(3)}$ are the three-particle Hamiltonians (see e.g. [47]) and V_{ij} the interactions between particles i and j. The factors in front of the trace come from the center-of-mass motion. The operator structures are shown to be trace-class e.g. in [119] - [121].

Once we have an explicit form for the b_n, the grand canonical partition function and all other thermodynamic properties of the system are determined. The equation of state e.g. can be expanded in the density of particles ρ, viz.

$$P = kT \sum_{n=1}^{\infty} a_n \rho^n \qquad (6.3)$$

and its coefficient a_n, the virial coefficients are given by

$$a_1 = 1, \quad a_2 = -b_2 b_1^{-2},$$
$$a_3 = 4a_2^2 - 2b_3 b_1^{-3}. \qquad (6.4)$$

......

We remark that the nth virial coefficient and cluster integral only involve n and fewer particle effects. For results on the convergence of these type of expansions, see e.g. [118], [122].

We now want to evaluate the cluster integrals (6.1) in terms of scattering quantities. The standard method is to use the Watson transform and write the operators $\exp(-\beta H)$ as

$$e^{-\beta H} = -\frac{1}{2\pi i} \oint_C dk^2 \, e^{-\beta k^2} R(k). \qquad (6.5)$$

where C is a contour integral around the spectrum of H in the complex energy plane (see Section III). This leads to (for the moment we assume that there are no zero-energy bound states and/or zero-energy resonances)

$$b_2(\beta) = \frac{2^{1/2}}{\lambda^3} \left\{ \sum_{j=1}^{N_b} e^{-\beta \lambda_j} + \frac{1}{2\pi} \int_0^\infty dk^2 \, e^{-\beta k^2} \, 2\mathrm{Im} \, \mathrm{Tr}[R(k)-R_0(k)] \right\} \qquad (6.6)$$

and similarly for $b_3\ldots$ [116], [119]–[121], [123], [124]. This result can also be written down in terms of time delay or in terms of the S-matrix (cf. Eqs. (2.49) and (2.51)). It generalizes the well-known results of Beth and Uhlenbeck ([125]–[127]) e.g. to non-spherically symmetric interactions. For alternative methods of deriving these results and further discussion see [116].

Using the sum rules developed in Section III and IV, we now derive high-temperature and low-temperature expansions for b_2. We restrict ourselves to the second cluster coefficient in three dimensions. We start with the high-temperature results.

Let us look at the continuum part of (6.6) and write it as

$$b_2^c(\beta) = \frac{2^{1/2}}{2\pi\lambda^3} \int_0^\infty dk^2 \, e^{-\beta k^2} \left\{ 2 \, \text{Im} \, \text{Tr}[R(k)-R_0(k)] + \frac{1}{4\pi k} \int d^3x \, V(\underline{x}) \right\}$$

$$- \frac{2^{1/2}}{2\pi\lambda^3} \int_0^\infty dk^2 \, e^{-\beta k^2} \frac{1}{4\pi k} \int d^3x \, V(\underline{x}) \, . \quad (6.7)$$

The second integral can be worked out explicitly. The first one can be written by making a total differential out of its second factor, viz.

$$- \frac{2^{1/2}}{2\pi\lambda^3} \int_0^\infty dk_2^2 \, e^{-\beta k_2^2} \, d \int_{k_2^2}^\infty dk_1^2 \left\{ 2 \, \text{Im} \, \text{Tr}[R(k_1)-R_0(k_1)] + \frac{1}{4\pi k_1} \int d^3x \, V(\underline{x}) \right\}.$$

$$(6.8)$$

Partial integration with respect to k_2^2 and use of the higher-order Levinson theorem Eq. (3.24) for $N=0$ to evaluate the surface term give then the following expression

$$b_2^c(\beta) = \frac{2^{1/2}}{2\pi\lambda^3} \left\{ \frac{-1}{4\pi^{1/2}\beta^{1/2}} \int d^3x \, V(\underline{x}) - 2\pi \, N_b \right.$$

$$\left. -\beta \int_0^\infty dk_2^2 \, e^{-\beta k_2^2} \int_{k_2^2}^\infty dk_1^2 \left\{ 2 \, \text{Im} \, \text{Tr}[R(k_1)-R_0(k_1)] + \frac{1}{4\pi k_1} \int d^3x \, V(\underline{x}) \right\} \right. .$$

$$(6.9)$$

It is clear that this process may be repeated. For example, we now add $16\pi k_1^{-3} \int d^3x \, P_2(\underline{x},\underline{x})$ (see Eq. (3.24)) to the last integral in (6.9) and calculate this integral with the help of an integration by parts and the use of the $N=1$ higher-order Levinson theorem to determine the nonvanishing part of the surface term. In this way, we finally arrive at [40], [37]

$$b_2^c(\beta) = \frac{-2^{1/2}}{\lambda^3} \sum_{j=1}^{N_b} e^{-\beta\lambda_j} + \sum_{n=1}^{\infty} \frac{(-1)^n}{(4\pi)^3 n!} \int d^3x \, P_n(\underline{x},\underline{x}) \beta^{n-3} \,, \quad (6.10)$$

or

$$a_2(\beta) = -2^{1/2} \sum_{n=1}^{\infty} \frac{(-1)^n}{n!} \int d^3x \, P_n(\underline{x},\underline{x}) \, \beta^n \,. \quad (6.11)$$

This result is nothing but the Wigner-Kirkwood expansion, which can be derived in many other ways. The unknown functions P_n entering in (6.11) are the Korteweg-de Vries type invariants discussed in Section III (Eq. (3.28)). This has been noticed first in [128], [129], [50] in a 1-dimensional context and in [37] for three dimensions. (See also [51], [130], [41].) As we have seen, these functions also appear as high-energy corrections in the sum rules (3.24). We recall that in deriving (3.24) we have assumed smooth potentials (cf. Section III). For Yukawa or Coulomb potentials e.g. non-analytic terms of order $\beta^{5/2}$... appear in (6.11) ([130] - [131]). Finally, we mention that an analogous expansion has been derived for classical scattering (e.g. [59]). It is interesting to note that the quantum effects in (6.11) appear from $n = 3$ onwards, which is clear when looking at the expressions for the P_n (Eq. (3.19)).

The advantage of the method of derivation of (6.11) presented here in some detail is apparent in Eqs. (6.9) and (6.10). There we see an explicit cancellation between the bound state and continuum contributions to $a_2(\beta)$. It is important to note that this cancellation is rigorously valid on a fully quantum mechanical level. It is essential in the applications of the theory of virial coefficients, as we will shortly illustrate in the following two examples.

In chemical equilibrium calculations in partial ionized plasmas, consisting out of atoms and dimers, the bound state contributions are taken as the appropriate low-order term to calculate the internal partition function. However, to avoid unphysical discontinuous changes in this partition function when new bound states are formed, it was realized in the literature that some compensation had to be included with part of the scattering contributions, whose effects only appear in the higher-order terms. This has been studied in the framework of partial wave scattering, first numerically, then in WKB approximation. For the history and more details on this we refer to [37], [132] and the references cited therein. A rigorous full quantum mechanical proof of the total cancellation of the bound state contributions, valid for non-spherically symmetric interaction, has only been achieved through the use of the higher-order Levinson theorems, as we have reported above.

The second example concerns the study of the thermodynamic properties of a quantum Lorentz gas model and a Born-Oppenheimer model of a gas mixture, using the virial expansion [133]. The question comes up whether the Efimov effect, known to occur in these models, does not cause a divergence in the third virial coefficient. Indeed, this effect generates an increasing number of low-energy three-body bound states when a zero-energy two-body bound state is created, because the effective interaction then becomes long range, such that the bound state contribution to the third virial diverges. For a binary gas mixture of light and heavy particles interacting with a separable s-wave Yamaguchi potential, this divergence is explicitized but shown to be cancelled by continuum state contributions. For more details see [133].

We end this subsection on applications to statistical mechanics with a description of the results on the low-temperature behavior of the virial coefficients. For the second virial in three dimensions, we start again from Eq. (6.6) but we now allow zero-energy bound states and/or zero-energy resonances such that an additional contribution Δ_{-2} occurs ($\Delta_{-2}^{I} = 0$, $\Delta_{-2}^{II} = 1/2$, $\Delta_{-2}^{III} = N_0$, $\Delta_{-2}^{IV} = (N_0-1)+1/2$). Following a similar procedure as before, now using the low-energy results (4.14) and (4.17), we arrive at [68]

$$a_2(\beta) = -\sqrt{2}\,\lambda^3 \left\{ \sum_{j=1}^{N_b} e^{-\beta\lambda_j} + \Delta_{-2} + \sum_{n=0}^{N} 2(-1)^{n+1} \frac{\Gamma(n+\frac{1}{2})}{\beta^{n+1/2}} \Delta_{2n-1} + O(\beta^{-N-1/2}) \right\}. \quad (6.12)$$

A similar result has been obtained (for $\beta = it$) using different methods in [63]. From (6.12) we infer that the negative energy bound states form the most important contribution at low temperatures. The next order contribution is given by Δ_{-2}. Finally, the leading scattering contribution is always given by Δ_{-1} (see also [134]) that is related with the generalized scattering length (Cases I and III) or effective range (Cases II and IV) as we have seen in Section IV. Analogous results are valid in other dimensions [64], [66], [67].

B. Other Fields

We briefly overview the applications of the sum rules in nuclear physics and field theory. We restrict ourselves to giving some references and occasional remarks.

1. Nuclear physics

Here the result that is most interesting is the low-energy sum rule (4.19). The reason is that it connects different scattering observables, namely the S-matrix, the generalized scattering length and the bound state energies with the potential. So it can be used e.g. as a working tool to eliminate ambiguities in phase shift analysis and to determine uncertain low-energy parameters. In its partial wave form, this sum rule has already been used in that sense in discussions of nucleon-nucleon and neutron-deuteron scattering ([46], [135] - [139]).

2. Field theory

For earlier discussions of Levinson's theorem in the context of field theory, we refer to [4]. More recently, applications of Levinson's theorem and more general the one-dimensional relations (3.27) have appeared in the context of the massive Thirring model, the two-dimensional σ-model, solitons in polyacetylene and scattering of fermions from a magnetic monopole [101], [102], [140] - [147]. We also recall that in [54], some aspects of these relations (3.27) and their applications are treated. Finally, the use of Levinson's theorem in studying the effect of random potentials on the Hall current has been discussed in Ref. [148].

Notes Added in Proof

New proofs of the three-dimensional Levinson theorem have appeared, using the Sturm-Liouville theorem [149] - [151]. In Ref. [149] a potential with a tail $V(r) \sim br^{-2}$, $r \to \infty$, has been included. In the other works, it has been shown that Levinson's theorem, extended to include repulsive Coulomb tails [151], provides information on the nodal structure of the zero-energy wave function of the problem [150], [151].

Concerning the two-dimensional low-energy sum rules, briefly discussed in Section IVC, there have been some recent, surprising developments. One can indeed show that Levinson's theorem in the Cases I-III in that section reads [152]

$$\int_0^\infty dk^2 \, 2 \, \text{Im Tr} \, [R(k)-R_0(k)] = -2\pi N_b - 2\pi D - (v,u)/2 \, ,$$

where N_b is the number of negative energy bound states and where

$$D^I = 0, \quad D^{IIa} = 0, \quad D^{IIb} = M, \quad D^{IIc} = M-1, \quad D^{III} = N.$$

(One assumes that there are no positive energy bound states embedded in the continuum.) This means that a possible s-wave type zero-energy resonance does not contribute at all to this theorem, while there possibly exist also p-wave type zero-energy resonances that contribute each a term $(-\pi)$, exactly like (zero-energy) bound states.

In connection with Section VE we mention that new discussions of Levinson's theorem have appeared for Dirac particles interacting with finite-range potentials [153], interacting with long-range potentials [154], moving in a background magnetic monopole field [155].

Finally for one-dimensional supersymmetric models, Levinson's theorem in the presence of a zero-energy resonance has been used to explain the (possible) fractional value of the regularized Witten index [156], whenever the continuum in the spectrum of the model extends down to zero-energy [157].

ACKNOWLEDGMENTS

It is a pleasure to thank Professors H. Ezawa, A. Fonseca, F. Gesztesy, H. Grosse and L. Streit for discussions, In particular, I would like to thank Professor F. Gesztesy for some remarks improving the manuscript. I am indebted to Professor L. Streit for inviting me to take part in Project Nr. 2: Mathematics + Physics, and for his warm hospitality at the Zentrum für interdisziplinäre Forschung der Universität Bielefeld, FRG.

APPENDIX

In this Appendix we list the expressions for Δ_{-1} and Δ_0 occurring in the low-energy sum rule (4.19).

Case I:

$$\Delta_{-1} = \frac{1}{8\pi}(v, T_0 u), \quad \Delta_0 = -2\Delta_{-1}^2 - \text{Tr}(r_2 T_0 r_0).$$

Case II:

$$\Delta_{-1} = 2\pi \frac{(\tilde{\phi}, r_2 \phi)}{|(v,\phi)|^2}$$

$$\Delta_0 = -2\Delta_{-1}^2 + \frac{(\tilde{\phi}, r_2 \phi)}{|(v,\phi)|^2}(v, T_0 u) + 4\pi \frac{(\tilde{\phi}, r_3 \phi)}{|(v,\phi)|^2} - \frac{(\tilde{\phi}, r_2 \phi)}{(\tilde{\phi},\phi)}$$

$$- |(v,\phi)|^{-2} \left[(v,\phi)(\tilde{\phi}, r_2 T_0 u) + (\tilde{\phi}, u)(v, T_0 r_2 \phi) \right] - \text{Tr}(r_2 T_0 r_0).$$

Introducing the shorthand notation $(O)_{ij} \equiv (\tilde{\phi}, O\phi)_{ij}$ for any operator O, we have

Case III:

$$\Delta_{-1} = \frac{1}{8\pi}(v, T_0 u) + \frac{1}{2} \sum_{i,j=1}^{N_0} (r_2)_{ij}^{-1}(r_3)_{ji}$$

$$\Delta_0 = -\frac{1}{32\pi^2}(v, T_0 u)^2 - \text{Tr}(r_2 T_0 r_0) - \sum_{i,j}(r_2)_{ij}^{-1}(r_2 T_0 r_2)_{ji}$$

$$- \frac{1}{2} \sum_{i,j,k,\ell}(r_2)_{ij}^{-1}(r_3)_{jk}(r_2)_{k\ell}^{-1}(r_3)_{\ell i} + \sum_{i,j}(r_2)_{ij}^{-1}(r_4)_{ji}$$

$$- \sum_j (\tilde{\phi}_j, \phi_j)^{-1}(r_2)_{jj}.$$

Case IV:

$$\Delta_{-1} = -2\pi[(v,\phi_1)(r_2)^{-1}_{11}(\tilde{\phi}_1,u)]^{-1} \left\{ (4\pi)^{-1} \sum_{j,k} (v,\phi_1)(r_2)^{-1}_{1j}(r_3)_{jk}(r_2)^{-1}_{k1}(\tilde{\phi}_1,u) - 1 \right\}$$

$$+ \frac{1}{2} \sum_{i,j} (r_2)^{-1}_{ij} (r_3)_{ji} .$$

In these formulas T_0 denotes the reduced resolvent defined by Eq. (4.9) and the r_n are the coefficients in the expansion of $uR_0(k)v$ given by Eqs. (4.11) and (4.12).

REFERENCES

[1] N. Levinson, Kgl. Danske Videnskab. Selskab, Mat.-fys. Medd. $\underline{25}$ (9) (1949)

[2] J.M. Jauch, Helv. Phys. Acta $\underline{30}$. 143 (1957)

[3] R.G. Newton, Scattering Theory of Waves and Particles, 2nd ed., (Springer, N.Y. 1982)

[4] P. Beregi, B.N. Zakhar'ev, and S.A. Niyazgulav, Sov. J. Particles Nucl. $\underline{4}$, 217 (1973)

[5] V.S. Buslaev, Topics in Math. Phys., Ed. M.Sh. Birman, Vol. 1, p. 69 (1967)

[6] T. Dreyfus, J. Phys. $\underline{A9}$, L187 (1976)

[7] B. Simon, Quantum Mechanics for Hamiltonians defined as Quadratic Forms (Princeton University Press, Princeton, N.J. 1971)

[8] M. Reed and B. Simon, Methods of Modern Mathematical Physics, Vol. III, Scattering Theory (Academic, N.Y. 1971)

[9] W.O. Amrein, J.M. Jauch and K.B. Sinha, Scattering Theory in Quantum Mechanics (Benjamin, Reading, Mass. 1977)

[10] V.S. Buslaev, Sov. Phys. Dokl. $\underline{7}$, 295 (1962)

[11] L.D. Faddeev, Mathematical Aspects of the Three-Body Problem in the Quantum Scattering Theory (Sivan, Jerusalem 1965), Eq. (8.16)

[12] T.A. Osborn and D. Bollé, J. Math. Phys. $\underline{18}$, 432 (1977)

[13] H.M. Nussenzveig, Causality and Dispersion Relations (Academic, New York, 1972), Chapt. 6

[14] Ph.A. Martin, Acta Physica Austriaca, Suppl. $\underline{23}$, 157 (1981)

[15] D. Bollé, F. Gesztesy, and H. Grosse, J. Math. Phys. $\underline{24}$, 1529 (1983)

[16] H. Narnhofer, J. Math. Phys. $\underline{25}$, 987 (1984)

[17] J.M. Jauch, K.B. Sinha, and B.N. Misra, Helv. Phys. Acta $\underline{45}$, 398 (1972)

[18] See e.g. W.M. Gibson, M. Maruyama, Y. Hashimoto, E.P. Kanter, G.M. Temmer, R.K. Keddy, D.W. Mingay, J.P.F. Sellschop, Nucl. Phys. $\underline{A317}$, 313 (1979)

[19] S. Yoshida, Phys. Lett. $\underline{45B}$, 324 (1973)

[20] K. Yazaki and S. Yoshida, Nucl. Phys. $\underline{A232}$, 249 (1974)

[21] M.G. Krein, Mat. Sb. $\underline{33}$, 75 (1953)

[22] M.G. Krein, Sov. Math. Dokl. $\underline{3}$, 707 (1962)

[23] M.G. Birman and M.G. Krein, Sov. Math. Dokl. $\underline{3}$, 740 (1962)

[24] K. Sinha, On the theorem of M.G. Krein, Univ. of Geneva preprint (1975)

[25] H. Baumgärtel and M. Wollenberg, Mathematical Scattering Theory (Birkhäuser, Basel 1983), Sect. 19.1

[26] R.G. Newton, J. Math. Phys. $\underline{18}$, 1348 (1977)

[27] M. Cheney, J. Math. Phys. $\underline{25}$, 1449 (1984)

[28] T. Dreyfus, Helv. Phys. Acta $\underline{51}$, 321 (1978)

[29] T. Dreyfus, J. Math. Anal. Appl. $\underline{64}$, 114 (1978)

[30] M. Wollenberg, Math. Nachr. $\underline{78}$, 223 (1977)

[31] M. Wollenberg, Math. Nacht. $\underline{78}$, 369 (1977)

[32] I. Waller, Z. Phys. $\underline{79}$, 380 (1932)

[33] S. Albeverio, Ph. Blanchard, and R. Høegh-Krohn, Commun. Math. Phys. $\underline{83}$, 49 (1982)

[34] R.G. Newton, Phys. Rev. $\underline{101}$, 1588 (1956)

[35] I.C. Percival, Proc. Phys. Soc. $\underline{80}$, 1290 (1962)

[36] I.C. Percival and M.J. Roberts, Proc. Phys. Soc. $\underline{82}$, 519 (1963)

[37] D. Bollé, Ann. Phys. $\underline{121}$, 131 (1979)

[38] T.A. Osborn and Y. Fujiwara, J. Math. Phys. $\underline{24}$, 1093 (1983)

[39] P. Greiner, Arch. Ration. Mech. Anal. $\underline{41}$, 163 (1971)

[40] D. Bollé and H. Smeesters, Phys. Lett. $\underline{A62}$, 290 (1977)

[41] S.F.J. Wilk, Y. Fujiwara, and T.A. Osborn, Phys. Rev. $\underline{A24}$, 2187 (1981)

[42] B. Simon, Bull. Am. Math. Soc. $\underline{7}$, 447 (1982)

[43] M.J. Roberts, Proc. Phys. Soc. $\underline{84}$, 825 (1964)

[44] R.D. Puff, Phys. Rev. $\underline{A11}$, 154 (1975)

[45] V.S. Buslaev and L.D. Faddeev, Sov. Math. Dokl. $\underline{1}$, 451 (1960)

[46] M.J. Roberts, Proc. Phys. Soc. $\underline{83}$, 503 (1964)

[47] D. Bollé and T.A. Osborn, Phys. Rev. $\underline{A26}$, 3062 (1982)

[48] D. Bollé, C. Danneels and T.A. Osborn, in Few Body Problems in Physics: Vol. II, ed. B. Zeitnitz, p. 349 (Elsevier Science 1984) and in preparation

[49] V.E. Zakharov and L.D. Faddeev, Funct. Anal. Appl. $\underline{5}$, 280 (1971)

[50] I.M. Gelfand and L.A. Dikii, Russian Math. Surveys $\underline{30}$. 77 (1975)

[51] A.M. Perelomov, Ann. Inst. Henri Poincaré, Sect. A, $\underline{24}$, 161 (1976)

[52] C.S. Gardner, J.M. Greene, M.D. Kruskal and R.M. Miura, Phys. Rev. Lett. $\underline{19}$, 1095 (1967)

[53] M.D. Kruskal, R.M. Miura, C.S. Gardner, and N.J. Zabusky, J. Math. Phys. $\underline{11}$, 952 (1970)

[54] H. Grosse, The inverse method in quantum mechanics and field theory, this volume

[55] D. Bollé and D. Roekaerts, Phys. Rev. $\underline{A30}$, 2024 (1984)

[56] D. Bollé and D. Roekaerts, Acta Physica Austriaca, Suppl. $\underline{26}$. 371 (1984)

[57] T.A. Osborn, K.B. Sinha, D. Bollé and C. Danneels, J. Math. Phys. $\underline{26}$, 2796 (1985)

[58] L.W. MacMillan and T.A. Osborn, Ann. Phys. $\underline{126}$, 1 (1980)

[59] D. Bollé and T.A. Osborn, J. Math. Phys. $\underline{22}$, 883 (1981)

[60] T.A. Osborn, R.G. Froese and S.F. Howes, Phys. Rev. Lett. $\underline{45}$, 1987 (1980)

[61] F. Gesztesy, in Lecture Notes in Physics $\underline{211}$, ed. by S. Albeverio, L. Ferreira and L. Streit (Springer,Verlag 1984) 78

[62] D. Bollé, Phys. Lett $\underline{A74}$, 26 (1979)

[63] A. Jensen and T. Kato, Duke Math. J. $\underline{46}$, 583 (1979)

[64] A. Jensen, Duke Math. J. 47, 57 (1980)

[65] S. Albeverio, F. Gesztesy and R. Høegh-Krohn, Ann. Inst. Henri Poincaré, Sect. A, 37, 1 (1982)

[66] M. Murata, J. Funct. Anal. 49, 10 (1982)

[67] A. Jensen, J. Math. Anal. Appl. 101, 397 (1984)

[68] D. Bollé and S.F.J. Wilk, J. Math. Phys. 24, 1555 (1983)

[69] D. Bollé, F. Gesztesy and S.F.J. Wilk, Phys. Lett. A97, 30 (1983)

[70] D. Bollé, F. Gesztesy and S.F.J. Wilk, J. Operator Theory 13, 1 (1985)

[71] D. Bollé, F. Gesztesy and M. Klaus, "Scattering theory for one-dimensional systems with $\int dx\, V(x) = 0$", J. Math. Anal. Appl. (to appear)

[72] D. Bollé, F. Gesztesy, C. Danneels, and S.F.J. Wilk, in preparation

[73] M. Klaus and B. Simon, Ann. Phys. 130, 251 (1981)

[74] A.M. Badalyan, L.P. Kok, M.I. Polykarpov, and Yu.A. Simonov, Phys. Rep. 82, 31 (1982)

[75] S. Albeverio, D. Bollé, F. Gesztesy, R. Høegh-Krohn, and L. Streit, Ann. Phys. 148, 308 (1983)

[76] D. Bollé, F. Gesztesy, W. Schweiger, J. Math. Phys. 26, 1661 (1985)

[77] D. Bollé, F. Gesztesy, Phys. Rev. Lett. 52, 1469 (1984)

[78] D. Bollé, F. Gesztesy, Phys. Rev. A30, 1279 (1984)

[79] R.G. Newton, J. Math. Phys. 18, 1582 (1977)

[80] D. Bollé, F. Gesztesy, C. Nessmann and L. Streit, "Scattering theory for general, non-local interactions: threshold behavior and sum rules, preprint KUL-TF-85/14

[81] S. Albeverio, F. Gesztesy, R. Høegh-Krohn, W. Kirsch, J. Operator Theory 12. 101 (1984)

[82] B. Simon, Ann. Phys. 97, 279 (1976)

[83] M. Klaus, Helv. Phys. Acta 55, 49 (1982). See also F.S. Rofe-Bikitov, Sov. Math. Dokl. 5, 689 (1964)

[84]   P. Deift, E. Trubowitz, Commun. Pure Appl. Math. $\underline{32}$, 121 (1979)

[85]   R.G. Newton, J. Math. Phys. $\underline{21}$, 493 (1980)

[86]   K. Chadan, P.C. Sabatier, Inverse problems in quantum scattering theory (Springer, New York 1977)

[87]   L.L. Foldy, Phys. Rev. $\underline{C26}$, 1818 (1982)

[88]   W. Cassing, M. Stingl, A. Weiguny, Phys. Rev. $\underline{C26}$, 22 (1982)

[89]   S. Joffily, Phys. Rev. $\underline{C27}$, 1830 (1983)

[90]   D.R. Yafaev, Theor. Math. Phys. $\underline{11}$, 358 (1972)

[91]   P. Swan, Nucl. Phys. $\underline{A119}$, 40 (1968)

[92]   L.A. Sakhnovich, Izv. Akad. Nauk. SSSR, Ser. Matem., $\underline{30}$, 1297 (1966)

[93]   C.M. Becchi and R. Collina, J. Math. Phys. $\underline{13}$, 1499 (1972)

[94]   R.G. Newton, J. Math. Phys. $\underline{24}$, 2152 (1983)

[95]   R.G. Newton, J. Math. Phys. $\underline{26}$, 311 (1985)

[96]   M. Reed and B. Simon, Methods of Modern Mathematical Physics, Vol. IV: Analysis of Operators (Academic, N.Y. 1978)

[97]   V.A. Zheludev, Topics in Mathematical Physics, Vol. $\underline{2}$, 87 (1968); Vol. $\underline{4}$, 55 (1971)

[98]   J. Friedel, Philos. Mag. $\underline{43}$, 153 (1952)

[99]   J. Friedel, Nuovo Cimento Suppl. $\underline{7}$, 287 (1958)

[100] J. Callaway, "Quantum Theory of the Solid State" (Academic, N.Y. 1974)

[101] B. Grossman, Phys. Rev. Lett. $\underline{50}$, 464 (1983)

[102] B. Grossman, Phys. Rev. Lett. $\underline{51}$, 959 (1983)

[103] W. Thirring, Classical Dynamical Systems (Springer, N.Y. 1978)

[104] T.A. Osborn, R.G. Froese, and S.F. Howes, Phys. Rev. $\underline{A22}$, 101 (1980)

[105] H. Narnhofer and W. Thirring, Phys. Rev. $\underline{A23}$, 1688 (1981)

[106] W. Thirring, Acta Physica Austriaca, Suppl. $\underline{23}$, 3 (1981)

[107] M.C. Barthélémy, Ann. Inst. Henri Poincaré $\underline{7}$, 115 (1967)

[108] Ni Guany-jiong, Phys. Energ. Fort. Nucl. $\underline{3}$, 432 (1979)

[109] E.J. Kanellopoulos and Th.V. Kanellopoulos, Levinson's theorem for Schrödinger and Dirac case, Universität Tübingen, FRG

[110] J.A. Wright, Phys. Rev. $\underline{139B}$, 137 (1965)

[111] R.G. Newton, Trudy. Mat. Inst. Steklov, $\underline{136}$, 224 (1975)

[112] W. Glöckle and J. Le Tourneux, Nucl. Phys. $\underline{A269}$, 16 (1976)

[113] R. Dashen, S. Ma and H. Bernstein, Phys. Rev. $\underline{187}$, 345 (1969)

[114] R. Dashen and S. Ma, J. Math. Phys. $\underline{11}$, 136 (1970)

[115] R. Dashen and S. Ma, J. Math. Phys. $\underline{12}$, 689 (1971)

[116] D. Bollé, Nucl. Phys. $\underline{A353}$, 377C (1981)

[117] K. Huang, Statistical mechanics (Wiley N.Y. 1964)

[118] D. Ruelle, Statistical mechanics (Benjamin, N.Y. 1969)

[119] V.S. Buslaev and S.P. Mercuriev, Proc. Steklov Inst. Math. $\underline{110}$. 28 (1970)

[120] V.S. Buslaev and S.P. Mercuriev, Theor. Math. Phys. $\underline{5}$, 1216 (1970)

[121] V.S. Buslaev and S.P. Mercuriev, Sov. Phys. Dokl. $\underline{14}$, 1055 (1970)

[122] G. Mack and M. Göpfert, Commun. Math. Phys. $\underline{81}$, 97 (1982)

[123] T.A. Osborn and T.Y. Tsang, Ann. Phys. $\underline{101}$, 119 (1976)

[124] F.A. Berezin, Sov. Phys. Dokl. $\underline{9}$, 641 (1965)

[125] G.E. Uhlenbeck and E. Beth, Physica $\underline{3}$, 729 (1936)

[126] E. Beth and G.E. Uhlenbeck, Physica $\underline{4}$, 915 (1937)

[127] L. Gropper, Phys. Rev. $\underline{50}$, 963 (1936)

[128] M. Kac and P. Van Moerbeke, Proc. Nat. Acad. Sci. USA $\underline{71}$, 2350 (1974)

[129] H.P. McKean and P. Van Moerbeke, Invent. Math. $\underline{30}$. 217 (1975)

[130] J.E. Avron, Ann. Phys. $\underline{108}$, 448 (1977)

[131] H.E. DeWitt, J. Math. Phys. $\underline{3}$, 1003 (1962)

[132] D. Bollé, in Mathematical methods and applications of scattering theory, eds. J.A. De Santo, A.W. Saenz, W.W. Zachary (Springer, Berlin 1980) 211

[133] W. Hoogeveen and J.A. Tjon, Physica $\underline{108A}$, 77 (1981)

[134] S.K. Adhikari and R.D. Amado, Phys. Rev. Lett. $\underline{27}$, 485 (1971)

[135] K. Chadan and A. Motes Lozano, Phys. Rev. $\underline{164}$, 1762 (1967)

[136] D.A. Kirhznits and N.Zh. Takibaev, Proc. Europ. Symp. on Few-particle problems (Potsdam 1977) p 98

[137] D.A. Kirhznits and N.Zh. Takibaev, Sov. J. Nucl. Phys. $\underline{16}$, 242 (1973)

[138] D.A. Kirhznits and N.Zh. Takibaev, Sov. J. Nucl. Phys. $\underline{25}$, 370 (1977)

[139] N.Zh. Takibaev, Sov. J. Nucl. Phys. $\underline{26}$, 42 (1977)

[140] R. Dashen and G.L. Kane, Phys. Rev. $\underline{D11}$, 136 (1975)

[141] R. Jackiw and G. Woo, Phys. Rev. $\underline{D12}$, 1643 (1975)

[142] S.S. Shu, Phys. Rev. $\underline{D14}$, 535 (1976)

[143] L.F. Abbott and H.J. Schnitzer, Phys. Rev. $\underline{D14}$, 1977 (1976)

[144] D.K. Campbell and Y.T. Liao, Phys. Rev. $\underline{D14}$, 2093 (1976)

[145] B. Berg, M. Karowski, W.R. Theis, and H.J. Thun, Phys. Rev. $\underline{D17}$, 1172 (1978)

[146] H. Takayama, Y.R. Lin-Liu and K. Maki, Phys. Rev. $\underline{B21}$, 2388 (1980)

[147] B. Berg, Lecture Notes in Physics $\underline{126}$, 317 (1980)

[148] W. Brenig, Z. Phys. $\underline{B50}$, 305 (1983)

[149] Zhong-Qi Ma, J. Math. Phys. $\underline{26}$, 1995 (1985)

[150] Z.R. Iwinski, Leonard Rosenberg and Larry Spruch, Phys. Rev. $\underline{A31}$, 1229 (1985)

[151] Z.R. Iwinski, Leonard Rosenberg and Larry Spruch, Phys. Rev. Lett. $\underline{54}$, 1602 (1985)

[152] D. Bollé, F. Gesztesy, C. Danneels and S.F.J. Wilk, "Threshold

behavior and Levinson's theorem for two-dimensional scattering systems: a surprise", preprint KUL-TF-85/15

[153] Zhong-Qi Ma and Guang-Jiong Ni, Phys. Rev. $\underline{D31}$, 1482 (1985)

[154] Zhong-Qi Ma, Phys. Rev. $\underline{D32}$, 2213 (1985)

[155] Zhong-Qi Ma, Phys. Rev. $\underline{D32}$, 2203 (1985)

[156] E. Witten, Nucl. Phys. $\underline{B202}$, 253 (1982)

[157] D. Boyanowski and R. Blankenbecler, Phys. Rev. $\underline{D30}$, 1821 (1985); see also R. Blankenbecler and D. Boyanowski, Phys. Rev. $\underline{D31}$, 2089 (1985)

NON LINEAR EVOLUTION EQUATIONS.
CAUCHY PROBLEM AND SCATTERING THEORY

J. GINIBRE
Laboratoire de Physique Théorique et Hautes Energies*
Université de Paris-Sud, F-91404 Orsay Cedex, France

G. VELO
Dipartimento di Fisica, Università di Bologna
and INFN, Sezione di Bologna, Italy

(0) Introduction

In these lectures I should like to present a general and unified framework to study the Cauchy problem and the theory of scattering for a class of non linear evolution equations which occurs in physics. There are two main ingredients. The first one is a non linear perturbation or contraction method used to solve the local Cauchy problem both at finite and infinite initial time. The second one is the use of a priori estimates associated with conservation laws of the equations. Energy estimates are used at finite time to prove the existence of global solutions. Other estimates, associated for instance with dilatational or conformal invariance, are used at infinite time to prove asymptotic completeness.

The general framework applies in particular to the following equations:

The non linear Schrödinger (NLS) equation

$$i \frac{d\varphi}{dt} = - \frac{1}{2} \Delta\varphi + f_0(\varphi) \ . \qquad (0.1)$$

---

*Laboratoire associé au Centre National de la Recherche Scientifique

The Hartree equation

$$i \frac{d\varphi}{dt} = -\frac{1}{2}\Delta\varphi + \varphi(V*|\varphi|^2) \ . \tag{0.2}$$

The non linear Klein-Gordon (NLKG) equation

$$\Box\varphi + m^2\varphi + f_0(\varphi) = 0 \ , \tag{0.3}$$

where φ is a complex-valued function defined in $n+1$ dimensional space time $\mathbb{R}^{n+1}$, Δ is the Laplace operator in $\mathbb{R}^n$, $\Box = \frac{\partial^2}{\partial t^2} - \Delta$, f_0 is a non linear function of φ, a typical example being

$$f_0(\varphi) = \lambda_1 |\varphi|^{p_1-1}\varphi + \lambda_2 |\varphi|^{p_2-1}\varphi \ , \tag{0.4}$$

and V is a real even function of space.

The general framework will also apply, at least as far as the Cauchy problem is concerned, to the Yang-Mills equations, and to various systems of equations of the previous type coupled between themselves or with otherwise linear equations such as the Maxwell and Dirac equation, with reasonable (but in general non linear) couplings. Rather than giving a list of equations and systems and of the available results of them, I shall concentrate on the abstract theory, and illustrate each of its steps with the example of the NLS equation and to a lesser extent of the NLKG equation. The theory has reached roughly the same stage of development for these two equations as regards both the Cauchy problem and the theory of scattering. The case of the NLS equation is slightly more difficult (although less complicated) than that of the NLKG equation, essentially because there exists a larger wealth of results and especially estimates for the corresponding free equation in the latter than in the former case. Also the result for the NLS equation are more recent and less well-known. The Hartree equation can be treated exactly in the same way as the NLS equation, with similar results. Other equations are in a much less advanced stage of development. Systems including the Dirac equation are plagued by the non positivity of the energy, already at the stage of the Cauchy problem, while systems involving the Maxwell and a fortiori the Yang-Mills equations exhibit long range effects which make the theory of scattering much more difficult.

The Cauchy problem for some of the previous equations, in particular the NLS and the NLKG equations, can also be treated by a different method. One first proves existence of global weak solutions (solutions in the distribution sense) by compactness methods using the energy estimates. One then proves uniqueness by a partial contraction method on bounded sets of the energy space. We refer to [17] for general information and to [13], [14] for recent results and references to previous work.

The general framework was originally developed by Segal [24] and in a less general form by Browder [6] following earlier work by Jörgens [15] on the NLKG equation. There is a vast amount of literature on the NLKG equation. The main results concerning scattering were obtained by Strauss [27] and Moravetz and Strauss [26], following earlier work of Segal, and with later contributions of several authors. The corresponding results for the NLS equation were obtained in [1], [8], [9] and [16]. The Cauchy problem for the Yang-Mills equation was solved locally in [25] and globally in [10] in dimensions 1+1 and 2+1 and in [7] in dimension 3+1. There is no complete theory of scattering for that case.

The bibliography given at the end is far from being complete and is mostly restricted to the papers directly relevant for the topics covered in the lectures. Previous exposition of similar topics include [23] and [27].

(1) The Cauchy Problem at Finite Times

The equations considered here can be written in general as

$$\frac{du}{dt} = Ku + f(t,u) \qquad (1.1)$$

where u is a function from space time to a finite dimensional vector space, K a linear differential operator which we assume for simplicity to be time independent and f a non linear interaction term, in general of lower degree than K in the space derivatives. For instance, for the NLS equation, one can take $u = \varphi$, $K = \frac{i}{2}\Delta$ and $f(t,u) = -if_0(\varphi)$. For the NLKG equation, one can take $u = \binom{\varphi}{\psi}$, $K = \begin{pmatrix} 0 & 1 \\ \Delta-m^2 & 0 \end{pmatrix}$ and $f(u) = \binom{0}{-f_0(\varphi)}$ so that ψ is simply $\frac{d\varphi}{dt}$ as a consequence of (1.1). The Cauchy problem consists in solving the equation (1.1) with initial condition $u(t_0,x) = u_0(x)$ for some $t_0 \in \mathbb{R}$ and some prescribed u_0. It is convenient and traditional to split that problem into two separate ones. The first problem is the <u>local</u> Cauchy problem and consists in solving (1.1) in some small interval around t_0. The second problem is the global Cauchy problem and consists in extending the solutions thereby obtained to all times. We shall consider these two problems successively.

(1.a) <u>The local Cauchy problem</u>

A general abstract method for studying the Cauchy problem was proposed by Segal in 1963 [24]. It consists in recasting the Cauchy problem in the form of an integral equation and solving that equation by a fixed point method. For that purpose, one introduces the one parameter group of operators $U(t) = \exp(tK)$ formally generated by K (to be called henceforth the free evolution group) and rewrites the Cauchy problem in the form

$$u(t) = U(t-t_0)u_0 + \int_{t_0}^{t} d\tau U(t-\tau)f(\tau,u(\tau)) \equiv A(t_0,u_0)u(t) . \qquad (1.2)$$

One then looks for solutions of (1.2) in the space $C(I,X)$ of continuous function of the time t in some interval $I \subset \mathbb{R}$ to a suitable Banach space X. For compact I, $C(I,X)$ is itself a Banach space with norm

$$|u|_I \equiv \sup_{t \in I} \|u(t)\|_X ,$$

where $\|\cdot\|_X$ denotes the norm in X. We denote by $B(I,\rho)$ the ball of radius ρ in $C(I,X)$. If one can choose X in such a way that
(1) $U(\cdot)$ is a continuous one parameter group of bounded operators in X, with

$$\|U(t)\|_{X \leftarrow X} \leq \mu(|t|), \qquad (1.3)$$

(2) f is a continuous function from $\mathbb{R} \times X$ to X satisfying a Lipschitz condition

$$\|f(t,u_1) - f(t,u_2)\|_X \leq C(I,\rho) \|u_1 - u_2\|_X \qquad (1.4)$$

for all $u_i \in X$, $\|u_i\|_X \leq \rho$ and all $t \in I$, then one sees easily that for $u_0 \in X$ and for $I = [t_0 - T, t_0 + T]$ with T sufficiently small (depending on u_0), the operator $A(t_0, u_0)$ leaves invariant a ball $B(I,\rho)$ in $C(I,X)$ containing the free term $U(\cdot - t_0)u_0$ and is a contraction in that ball, so that the equation (1.2) has a unique solution in $B(I,\rho)$ by the contraction mapping principle, thereby providing a solution of the local Cauchy problem. The method extends in a straightforward way to the case where $U(\cdot)$ is only a semi-group, or where $U(\cdot)$ is replaced by a two parameter group or semi-group of bounded propagators $U(t,s)$, corresponding to the case where the operator K is time dependent.

The theory just described applies to the equations mentioned in the introduction for suitable choices of the space X. However, it imposes restrictions on that choice that are both unnecessarily strong and inconvenient for various purposes. This is especially true for the NLS equation, as we shall see below. It is therefore useful, and actually possible, to develop a more general formalism, where we relax the assumption that the free group $U(\cdot)$ be bounded in X. For that purpose, we assume that there exists a "large" space Z (in practice one can often take $Z = \mathcal{D}'(\mathbb{R}^n)$) such that f be a continuous map from $\mathbb{R} \times X$ to Z and $U(\cdot)$ a one parameter continuous group in Z, so that the integrand $U(t-\tau)f(\tau,u(\tau))$ in (1.2) is well-defined (in Z)

for $u \in C(I,X)$ and τ in I. We then replace the separate assumptions (1.3) and (1.4) on U and f by a joint assumption on the pair (U,f). Since in addition we are interested in the integral in (1.2) rather than in the integrand, we formulate this assumption on the integral itself. There is, however, one complication coming from the fact that U is no longer assumed to be bounded in X, namely the fact that $v \in C(I,X)$ no longer implies that the function $(t,s) \to U(t-s)v(s)$ belongs to $C(\mathbb{R} \times I, X)$. We take that fact into account by stating the basic assumption in terms of integrals more general than that occurring in (1.2), and formally defined by

$$G([s_1,s_2],u)(t) \equiv \int_{s_1}^{s_2} d\tau\, U(t-\tau)f(\tau,u(\tau)) \ . \qquad (1.5)$$

It then turns out that the main results relative to the local Cauchy problem can be derived from the following assumption on G.

Assumption 1.1. For any interval I, any $t \in \mathbb{R}$ and any $u \in C(I,X)$, the function $\tau \to U(t-\tau)f(\tau,u(\tau))$ is locally Bochner integrable from I to X, and the function $(s_1,s_2,t) \to G([s_1,s_2],u)(t)$ is continuous from $I \times I \times \mathbb{R}$ to X. For any bounded intervals I and J, for any $\rho > 0$, for any u_1 and $u_2 \in B(I,\rho)$, G satisfies the following Lipschitz condition

$$|G(I,u_1) - G(I,u_2)|_J \leq C(I,J,\rho)|u_1 - u_2|_I \ , \qquad (1.6)$$

where $C(I,J,\rho)$ is separately non decreasing in I, J and ρ, and tends to zero when I tends to zero for fixed J and ρ.

Note that the monotonicity of $C(I,J,\rho)$ in each of its arguments is quite natural, from the very nature of the estimate (1.6). The purpose of introducing the interval J is to control the t-dependence of G, which is not controlled otherwise since we no longer assume U to be bounded in X. Assumption 1.1 is easily seen to be satisfied under the separate assumptions made in Segal's theory, and in particular (1.6) follows easily from (1.3) and (1.4). The next level of generality, which is sufficient to cover all the applications considered later, is the situation where there exists in addition to X an auxiliary Banach space $\bar{X}$ such that for each $t \neq 0$, $U(t)$ is a bounded operator from $\bar{X}$ to X, continuous in t for $t \neq 0$, and satisfying an estimate

$$\|U(t)\|_{X \leftarrow \bar{X}} \leq \mu(|t|) \tag{1.3}$$

for some function μ continuous for $t \neq 0$ and integrable at zero, and where f is a continuous function from $\mathbb{R} \times X$ to $\bar{X}$ satisfying a Lipschitz condition

$$\|f(t,u_1) - f(t,u_2)\|_{\bar{X}} \leq C(I,\rho) \|u_1 - u_2\|_X \tag{1.4}$$

for $u_i \in X$, $\|u_i\|_X \leq \rho$ and $t \in I$.

The reasons for stating Assumption 1.1 in the form given rather than to assume the above separate properties on U and f are twofold. On the one hand, that form might be actually weaker than the separate assumptions on (U,f) in more complicated situations, and in any case it is exactly what is needed for the local Cauchy problem. On the other hand, it anticipates on the form of the assumption needed to study the local problem at infinity, which is the basic step in the theory of scattering.

Under Assumption 1.1, one derives by various arguments revolving around the contraction mapping principle the basic results concerning the local Cauchy problem: existence of solutions in a small time interval, uniqueness of solutions in an arbitrary interval and continuity of the solutions with respect to the initial data in the neighborhood of a given solution.

<u>Proposition 1.1</u>. Let Assumption 1.1 hold. Let $t_o \in \mathbb{R}$. Assume (for simplicity) that the interaction is source-free, namely $f(t,0) = 0$. Then

(1) For any $\rho > 0$, there exists $T \equiv T(t_0,\rho)$ such that (with $I \equiv [t_0-T, t_0+T]$) for any $u_0 \in X$ such that $U(\cdot-t_0)u_0 \in B(I,\rho)$, the Equation (1.2) has a unique solution in $B(I,2\rho)$.

(2) For any interval $I \ni t_0$ and any $u_0 \in X$ such that $U(\cdot-t_0)u_0 \in C(I,X)$, the Equation (1.2) has at most one solution in $C(I,X)$.

(3) Let I and J be compact intervals, $I \subset J$, $t_0 \in I$, $u_0 \in X$ such that $U(\cdot-t_0)u_0 \in C(J,X)$, and $u \in C(I,X)$ a solution of (1.2). Then there exists a neighborhood $\mathcal{U}$ of (t_0,u_0) such that for any (t_0',u_0') in $\mathcal{U}$, the Equation (1.2) with (t_0,u_0) replaced by (t_0',u_0') has a unique solution u' in $C(I,X)$, and that solution depends continuously on (t_0',u_0'). (The relevant topology on (t_0,u_0) is that induced by the topology of $\mathbb{R}$ for t_0 and by the norm $|U(\cdot-t_0)u_0|_J$ on u_0. The relevant topology on u is that induced by the norm $\sup_{s\in I} |U(\cdot-s)u(s)|_J$).

The basic problem one encounters when trying to apply the abstract theory to a specific equation is of course to choose the space X. For that purpose, it is useful to look ahead and anticipate on the treatment of the global Cauchy problem. There, an essential role is played by the conservation laws associated with the Equation (1.1). The net effect of these conservation laws is the existence of a space X_2 such that any solution with initial data in X_2 is a priori controlled in X_2. In general, the dominant part of the conserved quantities comes from the free part of the equation, and therefore is quadratic. As a consequence, in general X_2 is a Hilbert space. Furthermore, the free group U is in general bounded in X_2. X_2 is the natural space where to look for global solutions, and therefore also the natural candidate for X. Failure to choose $X = X_2$ will be a source of difficulties at the stage of the global Cauchy problem.

In the subsequent applications we shall use mainly the usual L^q spaces with norm $\|\cdot\|_q$, ($1 \le q \le \infty$) and the Sobolev space $W^{k,q}$ (k integer, $k \ge 0$, $1 \le q \le \infty$) defined by

$$W^{k,q} = \{u : \sum_{\alpha : 0 \leq |\alpha| \leq k} \|D^\alpha u\|_q \equiv \|u\|_{kq} < \infty\}$$

where $\alpha = (\alpha_1, \ldots \alpha_n)$ denotes a multiindex of space derivatives, and $|\alpha| = \alpha_1 + \ldots + \alpha_n$.

We shall also use the notation H^k for $W^{k,2}$. Sobolev spaces satisfy various embedding relations. In particular $H^1 \subset L^q$ for $2 \leq q \leq \frac{2n}{n-2}$ ($n \geq 3$), and $H^k \subset L^\infty$ for $k > n/2$.

Furthermore, H^k is an algebra (for the pointwise operations) for $k > n/2$.

We now show how the abstract theory can be applied to the NLS equation. The free group is $U(t) = \exp(i\frac{t}{2}\Delta)$ and satisfies the following two properties, the only ones to be used at this stage:

• $U(t)$ is unitary in H^k for any $k \geq 0$.

• $U(t)$ is bounded and strongly continuous in $t \neq 0$ from $L^{\bar{q}}$ to L^q (more generally, from $W^{k,\bar{q}}$ to $W^{k,q}$), where $2 \leq q \leq \infty$ and, for any such $q, \bar{q}$ denotes the dual exponent, $1/q + 1/\bar{q} = 1$, with an estimate

$$\|U(t)\|_{L^q \leftarrow L^{\bar{q}}} \leq (2\pi|t|)^{-\delta(q)} \tag{1.7}$$

with

$$\delta(q) = \frac{n}{2} - \frac{n}{q}. \tag{1.8}$$

Under suitable assumptions the space X_2 suggested by the conservation laws turns out to be H^1, resulting from the conservation of the L^2-norm and the energy. Unfortunately, there does not seem to be any simple way to take $X = H^1$, except for $n = 1$, and we must make another choice for X. There is a large amount of freedom in that choice. "Large" spaces (typically such that $X \supset H^1$) have the obvious advantage that one can handle a larger set of initial data, while "small" spaces (typically such that $X \subset H^1$) provide additional smoothness of the solutions. Additional differences will appear at the stage of globalization. We give below for illustration three examples where Assumption 1.1 is satisfied and for which the local theory applies. For each of them, we indicate the choice of the space and the relevant assumptions on the interaction f_0.

<u>Case 1</u> (small space) [1], [11], [22], [28], n arbitrary, $X = H^k$, $k > n/2$, $f_0 \in C^{k+1}$ with $f_0(0) = 0$. This case is covered by the theory in Segal's original form. Note that there is no restriction on the behaviour of f_0 at infinity, as is typical of theories where X consists of bounded functions.

<u>Case 2</u> (large space) [8], $n \geq 2$, $X = L^{r_1} \cap L^{r_2}$ with

$$\frac{1}{2} - \frac{1}{n} < \frac{1}{r_2} < \frac{1}{r_1} \leq \frac{1}{2} \tag{1.9}$$

so that $X \supset H^1$ by the Sobolev embedding theorem; $f_0 \in C^1$ with $f_0(0) = 0$, and in addition

$$|f_0'(z)| \leq C(|z|^{p_1-1} + |z|^{p_2-1}) \tag{1.10}$$

with

$$\frac{r_1}{r_2} \leq p_1 \leq p_2 \leq \frac{r_2}{r_1} . \tag{1.11}$$

The conditions (1.9) and (1.11) imply

$$0 \leq p_1 - 1 \leq p_2 - 1 < \frac{4}{n-2} \tag{1.12}$$

and an additional coupled restriction between p_1 and p_2 to the effect that they are not too far from each other. In particular for a single power p, one can take any $p \in [1, 4/(n-2))$ for a suitable choice of X. That case requires the full generality of the abstract theory. Assumption 1.1 is satisfied via (1.3) and (1.4) with the choice $\bar{X} = L^{r_1} \cap L^{r_2}$. In particular (1.3) follows from (1.7).

A similar theory can be made for $n = 1$, but there is hardly any point in doing so since for $n = 1$ one can work directly with $X = X_2 = H^1$.

<u>Case 3</u> (intermediate space) [9], $n = 3$, $X = H^1 \cap L^\infty$, $f_0 \in C^2$ with $f_0(0) = f_0'(0) = 0$. That case has been studied in some detail because of the special interest in dimension 3 and in working with the largest convenient space of bounded functions. That theory falls again in the general case, (1.3) and (1.4) now hold with $X = H^1 \cap W^{1,q}$ for an arbitrary $q \in (3,6)$.

We next show how the abstract theory can be applied to the NLKG equation. In this case the free group is

$$U(t) = \begin{pmatrix} \cos\omega t & \omega^{-1}\sin\omega t \\ -\omega\sin\omega t & \cos\omega t \end{pmatrix} \qquad (1.13)$$

where $\omega = (-\Delta)^{1/2}$ and where the mass term has been incorporated into f_0. Conservation of the energy for the free equation implies that $U(t)$ is a one parameter continuous group in $H^k \oplus H^{k-1}$ with

$$\|U(t)\|_{H^k \oplus H^{k-1}} \leq C(1 + |t|) \qquad (1.14)$$

for any $k \geq 1$. For the NLKG equation one obtains already an interesting theory with the simplest choice $X = X_2$. We give that example first followed by three other ones patterned after those given for the NLS equation.

<u>Case 0</u> [6], [15], [24], n arbitrary, $X = H^1 \oplus L^2$. $f_0 \in C^1$, $f_0(0) = 0$ and in addition, for $n \geq 2$,

$$|f_0'(z)| \leq C(1 + |z|^{p-1}) \qquad (1.15)$$

with p arbitrary for $n = 2$ and

$$0 \leq p - 1 \leq \frac{2}{n-2} \qquad \text{for } n \geq 3. \qquad (1.16)$$

This case is covered by the original version of the theory, but the upper bound (1.16) on $p-1$ is lower than the expected one, namely $4/(n-2)$.

<u>Case 1</u> (small space) [2], [5], [20], [21], n arbitrary, $X = H^k \oplus H^{k-1}$, $k > n/2$, $f_0 \in C^k$, $f_0(0) = 0$. This case can also be covered by the original version of the theory. Note that there is no restriction on the behaviour of f_0 at infinity.

<u>Case 2</u> (large space), $n \geq 3$. One uses the estimate [18]

$$\|\omega^{-1}\sin\omega t\,\psi\|_r \leq C|t|^{1+n/r-n/s}\|\psi\|_s \tag{1.17}$$

with the restrictions

$$\begin{cases} \frac{1}{2} - \frac{1}{n+1} \leq \frac{1}{r} \leq \frac{1}{2} \\ \frac{n}{r} \leq \frac{n}{s} \leq \frac{1}{r} + \frac{n+1}{2} \end{cases} \tag{1.18}$$

One takes $X = (L^2 \cap L^r) \oplus H^{-1}$, $f_0 \in C^1$, $f_0(0) = 0$ and f_0 satisfying in addition (1.15) and

$$0 \leq p - 1 \leq \frac{4}{n-1}. \tag{1.19}$$

This case requires the full generality of the abstract theory. The upper bound (1.19) coincides with (1.16) for $n = 3$ and becomes better for $n > 3$. It is always smaller than the expected value $4/(n-2)$, but comes within $O(n^{-2})$ from it for larger values of n.

<u>Case 3</u> (intermediate space) [15], [26], $n = 3$, $X = (H^1 \cap L^\infty) \oplus L^2$, $f_0 \in C^2$, $f_0(0) = 0$. This example falls again in the general case and no restriction on f_0 at infinity is required. However, if f_0 satisfies (1.15) with p not satisfying (1.16) one can replace L^∞ by L^r with any $r \geq 2p$ in the definition of X.

We conclude this section with some general comments on smoothness. The question is to ascertain whether solutions of (1.2) in $C(I,X)$

have additional smoothness properties in space and time if the initial data are smooth. At the abstract level, where no mention is made of space variables, smoothness in space is naturally replaced by the property that the solution remains in the domain of some power of the free generator, while smoothness in time is expressed in terms of differentiability. A general study along these lines has been made by Segal in the original version of the theory. It does not extend in any obvious way to the more general framework given here. Actually, we shall see below on the examples of the NLS and NLKG equations that serious difficulties are likely to occur. The difficulty is already apparent when considering the differential equation (1.1), where smoothness would be expected to mean for instance that $\frac{du}{dt}$ and Ku belong to X, while the interaction term is a priori expected to belong to the auxiliary space $\bar{X}$ (if any), which may be very different from X.

At the concrete level of a given PDE, smoothness in space can be analyzed in a much more flexible way by asking whether a given solution $u \in C(I,X)$ of (1.2) with suitable initial data is such that Du remains in X for some class of differential operators D in the space variables, not restricted to the single operator K.

In any case, at the abstract as well as at the concrete level, the proof of smoothness often boils down to studying a linearized version of the Equation (1.2) in a neighborhood of a given solution. Typically, for a concrete PDE, if D is a differential operator commuting with $U(\cdot)$ and f a local function of u, one has to study the linear inhomogeneous equation

$$Du(t) = U(t-t_0)Du_0 + \int_{t_0}^{t} d\tau \, U(t-\tau) \left\{ \frac{\partial f}{\partial u}(\tau, u(\tau)) Du(\tau) + \text{lower order terms} \right\} \quad (1.20)$$

for $v = Du$ considered as an unknown function, and prove that it has a solution for v with more smoothness than was anticipated for Du. As we shall see below, this technical step, and thereby the smoothness problem, has a close connection with the problem of globalization in small spaces.

(1.b) The global Cauchy problem

The natural question that arises next is whether the solutions obtained at the previous step can be continued for all times, namely the problem of globalization. It appears already on elementary examples of ODE that global solutions may fail to exist. For instance, the differential equation

$$\frac{dy}{dt} = y^2$$

with initial condition $y(0) = y_0 > 0$ is solved by $y(t) = y_0(1-ty_0)^{-1}$, which blows up at time $t = y_0^{-1}$. When the local existence result of Section (1.a) is available, one is tempted to construct global solutions by iterating the local construction: one solves successively the Cauchy problem with initial times $t_j (j = 0,1,2,...)$ and initial data $u(t_j)$ in time intervals $[t_j, t_j+1]$ with $t_{j+1} = t_j + T_j$. The reason why that method fails to yield global solutions is simple: at each step of the resolution, the norm of $u(t_j)$ can increase by some factor $\lambda > 1$. The time T_j of local resolution, on the other hand, in general decreases as a negative power of that norm (typically a power of $-(p-1)$ if f_0 has degree p). As a consequence the T_j form a convergent geometric series, and the solution cannot be continued beyond $\bar{t} = t_0 + \sum_{j=0}^{\infty} T_j$. In addition, this argument shows that the solution ceases to exist because its norm tends to infinity. It also points out to a possible way to circumvent the difficulty, namely the method of a priori estimates: that method consists in proving that any solution of the Equation (1.2) is a priori bounded for each time, in terms of the initial data, in the norms which are relevant to solve the local problem. If this is the case, the T_j that occur in the previous iteration, which are expressed in terms of the appropriate norm of $u(t_j)$, are controlled by the a priori estimate and therefore do not tend to zero for any finite t_j, thereby preventing the convergence of the previous series.

In the original version of the theory (with U a bounded group in X) it is sufficient to obtain an a priori estimate of the norm of $u(t)$ in X, namely to prove that there exists a function $M(t_0,u_0,t)$ such that any solution of (1.2) satisfies

$$\|u(t)\|_X \leq M(t_0,u_0,t) \tag{1.21}$$

for all t for which it is defined. In the generalized framework, the previous condition has to be replaced by the following condition. For any compact interval J containing t_0, there exists $M(t_0,u_0,J)$ such that any solution $u \in C(I,X)$ of (1.2) in an interval I, $t_0 \in I \subset J$, satisfies

$$\sup_{s \in I} |U(\cdot-s)u(s)|_J \equiv \sup_{\substack{s \in I \\ t \in J}} \|U(t-s)\|_X \leq M(t_0,u_0,J) . \tag{1.22}$$

The basic tool that enters in the derivation of the a priori estimates (1.21) or (1.22) consists of the conservation laws associated with the equation. In a number of interesting cases, these conservation laws follow by a straightforward application of Noether's theorem from the fact that (1) the equation under consideration is the Euler-Lagrange equation associated with some Lagrangian and (2) the Lagrangian is invariant under some transformation group. This is the case in particular for the NLS equation (0.1) and for the NLKG equation (0.3) provided f_0 is of the form

$$f_0(z) = \frac{\partial V}{\partial \bar{z}} \tag{1.23}$$

for some real function $V(z) = V(|z|)$ depending only on $|z|$ (derivatives being taken with respect to z and $\bar{z}$, regarded as independent variables). The Lagrangian densities are then

$$L = \frac{i}{2}(\bar{\varphi}\partial_t\varphi - \varphi\partial_t\bar{\varphi}) - \frac{1}{2}|\nabla\varphi|^2 - V(\varphi) \tag{1.24}$$

for the NLS equation, and

$$L = \partial_\mu \bar{\varphi} \partial^\mu \varphi - m^2 |\varphi|^2 - V(\varphi) \tag{1.25}$$

for the NLKG equation.

These Lagrangians are invariant under gauge transformation of the first kind $\varphi \to e^{i\Theta}\varphi$, and in addition under transformations of the Galilei group for the NLS equation and of the Poincaré group for the NLKG equation.

Of special interest are the conservation laws that give rise to quantities having some positivity properties, so that they can be used to control some norm of the solutions. For the NLS equation, gauge invariance yields the conservation of the L^2-norm and time translation invariance yields the conservation of the energy:

$$\|\varphi(t)\|_2 = \|\varphi_0\|_2, \quad E(\varphi(t)) = E(\varphi_0) \tag{1.26}$$

where

$$E(\varphi) = \frac{1}{2} \|\nabla\varphi\|_2^2 + \int dx \, V(\varphi) . \tag{1.27}$$

These two conservation laws together imply at least formally, that the solutions are uniformly bounded in $X_2 = H^1$ provided $\varphi_0 \in L^2$ and $E(\varphi_0)$ is finite, and provided V satisfies some lower boundedness condition that prevents the kinetic and potential energies to become separately infinite for fixed total energy $E(\varphi)$. A sufficient condition to that effect is that V satisfies the condition

$$V(\rho) \geq -C(\rho^2 + \rho^{1+p_3}) \tag{1.28}$$

with $p_3 < 1 + 4/n$.

Similarly, for the NLKG equation, gauge invariance yields the conservation of the charge, which is not positive, while the energy is given by

$$E(\varphi) = \|\dot{\varphi}\|_2^2 + \|\nabla\varphi\|_2^2 + m^2 \|\varphi\|_2^2 + \int dx\, V(\varphi). \tag{1.29}$$

Energy conservation then implies formally that for any solution φ of the NLKG, $u \equiv (\varphi, \dot{\varphi})$ is uniformly bounded in $X_2 = H^1 \oplus L^2$ under a suitable lower boundedness condition on V. The relevant condition in that case is

$$V(\rho) \geq -C\rho^2. \tag{1.30}$$

One is then faced with the task of proving the (so far formal) conservation laws in the functional framework where one solves the local Cauchy problem. Two difficulties may occur, depending on the relation of X with X_2: The first one is that the conservation laws are easily derived in differential form from the Equation (1.1), whereas solutions of (1.2) in $C(I,X)$ may not be sufficiently smooth for (1.1) to hold in a sufficiently strong sense. That difficulty can be circumvented by a cut-off and limiting procedure. One regularizes both the equation and the initial data by introducing suitable cut-offs, one proves that the solutions of the regularized equation satisfy a regularized form of the conservation law, and one removes the cut-off by a limiting procedure. A more serious difficulty occurs if X is a "large" space, in the sense that $X \not\subset X_2$. In that case, one does not even know in advance that a solution in $C(I,X)$ with initial data in $X \cap X_2$ remains in X_2 for all times where it is defined, and one must prove this fact together with the conservation law. For that purpose, the cut-off and limiting procedure has to be supplemented with a weak compactness argument in X_2, in which the conservation law for the regularized solutions is used to prove their boundedness in X_2 uniformly with respect to the cut-off.

Once it is proved that the solutions in $C(I,X)$ with initial data in $X \cap X_2$ satisfy the relevant conservation laws, and are estimated in X_2 in terms of the initial data by virtue of these conservation laws, the last step in the globalization problem is to derive the a priori estimates (1.21) or (1.22) from the conservation laws. Here again the situation depends critically on the relation between X and X_2, but now in contrast with the previous step, the trouble comes from "small" spaces. For "large" spaces $X \supset X_2$, boundedness in X_2 immediately implies the required estimates in X. For small spaces $X \subset X_2$ on the contrary, additional estimates are needed. In the typical case when X is a Sobolev space one generally needs to control suitable L^q-norms of derivatives D of higher order than occurs in the conservation laws. This can (or cannot) be done by starting from equations of the type (1.20) for such derivatives and using the estimates provided by the conservation laws to obtain sublinear integral inequalities for their relevant norms, from which the required a priori estimates follow by Gronwall's inequality. This step is technically very similar to that required for the proof of smoothness properties of the solutions. As mentioned earlier, smoothness is closely connected with globalization in small spaces.

The NLS and the NLKG equations provide enlightening examples of the various possibilities described above, depending on the choice of X. We discuss first the case of the NLS equation. We consider again the three examples given above and indicate briefly the corresponding results for the global Cauchy problem. In all cases we make again the assumptions needed for the local Cauchy problem and we assume (1.10) and (1.28) with $p_3 < 1 + 4/n$. We present the three cases in the order of increasing difficulty, which is not the same as before, as just explained. In all three cases $X_2 = H^1$.

Case 2 (large space), $n \geq 2$, $X = L^{r_1} \cap L^{r_2}$. Then for all $\varphi_0 \in X_2 (\subset X)$ the NLS equation has a unique solution in $C(\mathbb{R}, X_2)$, and that solution is uniformly bounded in X_2 (and therefore in X).

Case 3 (intermediate space), $n = 3$, $X = H^1 \cap L^\infty$. Assume in addition that f_0 satisfies the condition

$$|f_0'(z)| \leq C|z|^{p_2 - 1} \quad \text{for} \quad |z| \geq 1 \tag{1.31}$$

with

$$0 \leq p_2 - 1 < \frac{4}{n-2} . \qquad (1.32)$$

Then the NLS equation has a unique solution in $C(\mathbb{R}, X)$ and that solution is uniformly bounded in X_2. Actually it turns out that the solution is also uniformly bounded in L^∞ and therefore in X provided the free term $U(\cdot - t_0)u_0$ satisfies that property.

Note that the condition (1.32) which was needed in Case 2 already at the local stage is needed also here, but only at the global stage. This indicates that the splitting in two stages is somewhat artificial.

<u>Case 1</u> (small space), $X = H^k$ with $k > \frac{n}{2}$. The globalization proof just sketched breaks down for large n. For $n \leq 9$, one can prove that the NLS equation has a unique solution in $C(\mathbb{R}, X)$ by adding the assumptions $f_0'(0) = 0$ and (1.31), (1.32) for $2 \leq n \leq 9$, plus additional mild restrictions on the behaviour of f_0'', f_0''' and f_0'''' for $4 \leq n \leq 9$.

Comparing the situation in Cases 1 and 2 leads to suspect that for high dimensions, the global solutions obtained in Case 1 may fail to remain smooth even for smooth initial data, and therefore that the smoothness problem is much more complicated in the general framework of Section (1.a) than in Segal's original theory.

We next consider the case of the NLKG equation. We make again the Assumption (1.15) and the Assumption (1.30). We treat again the various examples in the order of increasing complexity. In all cases $X_2 = H^1 \oplus L^2$.

<u>Cases 0 and 2</u>, $X = H^1 \oplus L^2$ or $X = (L^2 \cap L^r) \oplus H^{-1}$ with the corresponding assumption of f_0 (see above). Then, in both cases, for any $(\varphi_0, \psi_0) \in X_2$, the NLKG equation has a unique solution in $C(\mathbb{R}, X)$.

<u>Case 3</u> (intermediate space), $n = 3$, $X = (H^1 \cap L^\infty) \oplus L^2$. Assume in addition that f_0 satisfies the condition (1.15) with

$$0 \leq p - 1 < \frac{4}{n-2} . \qquad (1.33)$$

Then the NLKG equation has a unique solution in $C(\mathbb{R}, X)$

<u>Case 1</u> (small space), $X = H^k \oplus H^{k-1}$, $k > n/2$. The globalization proof breaks down for $n \geq 10$. For $n \leq 9$ the NLKG equation has a unique solution in $C(\mathbb{R}, X)$ under the assumptions (1.15), (1.33) and additional mild restrictions on the higher derivatives of f_0.

We conclude this section with some brief comments on the necessity of the lower bound (1.28) or (1.30) on V for the existence of global solutions. A condition of this type is indeed necessary. In fact, if

$$V(\rho) = -C\rho^{p_3+1} \tag{1.34}$$

with $C > 0$ and $p_3 \geq 1 + 4/n$ for the NLS equation, $p_3 > 1$ for the NLKG equation, then one can show that solutions of (1.1) blow up in a finite time if the initial data are large. The general method to prove such a result consists in choosing cleverly a suitable norm of the solution and deriving for that norm, from the differential equation (1.1), a differential inequality which implies some form of blow up. In the case of the NLS equation, a suitable norm is the moment of inertia of solution $\frac{1}{2} \|x\varphi\|_2^2$. For interactions of the type (1.34) with $p_3 > 1 + 4/n$,

$$\frac{d^2}{dt^2} \frac{1}{2} \|x\varphi\|_2^2 \leq E(\varphi) \tag{1.35}$$

so that $\|x\varphi\|_2$ vanishes for some finite time if $E(\varphi) < 0$. (Remember that $E(\varphi)$ is constant), a condition which for any (suitably regular) φ_0 is always fulfilled by $\lambda\varphi_0$ for sufficiently large λ, since the negative potential energy increases faster than quadratically in λ. The fact that $\|x\varphi\|_2 \to 0$ implies that the kinetic energy tends to infinity, since

$$\|x\varphi\|_2 \|\nabla\varphi\|_2 \geq \frac{n}{2} \|\varphi\|_2^2 \tag{1.36}$$

and $\|\varphi\|_2$ is constant. It also suggests that the solution actually "blows in" by concentrating at a single point. By a slightly more refined argument, one can see that concentration is most likely to occur at the center of mass $\bar{x}$ of the solution, defined by $\bar{x} \|\varphi\|_2^2 = <\varphi, x\varphi>$.

In the case of the NLKG equation, a suitable norm to consider is simply $\|\varphi\|_2$. For interactions of the type (1.34) with $p_3 = 1+2\delta > 1$ one can show that $F \equiv \|\varphi\|_2^{-\delta}$ satisfies the inequality

$$\frac{d^2 F}{dt^2} \leq \delta(\delta+1) \, E \, F^{(1+2/\delta)} \tag{1.37}$$

which for $E < 0$ implies that $\|\varphi(t)\|_2$ becomes infinite in a finite time.

(2) <u>Scattering theory</u>

We now turn to the problem of the asymptotic behaviour in time of the solutions of the Equation (1.1). Such problems are in general difficult, but some progress can be made in situations where in addition to the given evolution equation, there exists another, simpler, evolution equation to the solutions of which one can compare those of the original equation. That comparison is the basic purpose of scattering theory. In the present case, the simpler equation to which (1.1) will be compared is the free equation

$$\frac{du}{dt} = Ku \, . \tag{2.1}$$

The first step consists in looking for dispersive solutions of (1.1), namely for solutions that behave asymptotically in time like solutions of (2.1). If u is such a solution, one expects that there exists $u_\pm \in X$ so that $u(t) \sim U(t)u_\pm$ as $t \to \pm\infty$ or, more precisely, that $\tilde{u}(t) \to u_\pm$ as $t \to \pm\infty$, where $\tilde{u}$ is defined by

$$\tilde{u}(t) = U(-t)u(t) \, . \tag{2.2}$$

If this is the case, one obtains formally from (1.2)

$$u_\pm = \tilde{u}(t_0) + \int_{t_0}^{\pm\infty} d\tau\, U(-\tau) f(\tau, u(\tau)) \tag{2.3}$$

and therefore

$$u(t) = U(t) u_\pm + \int_{\pm\infty}^{t} d\tau\, U(t-\tau) f(\tau, u(\tau)) \ . \tag{2.4}$$

Dispersive solutions are obtained by solving (2.4) for u, namely by solving the Cauchy problem with infinite initial time. More generally, it is useful to study the equation

$$u(t) = U(t) \tilde{u}_0 + \int_{t_0}^{t} d\tau\, U(t-\tau) f(\tau, u(\tau)) \equiv A_0(t_0, \tilde{u}_0) u(t) \tag{2.5}$$

for given $\tilde{u}_0$ and t_0 in a neighborhood of $\pm\infty$, and to derive results on that equation that are uniform in t_0 and have some continuity in t_0 in the neighborhood of $\pm\infty$.

In terms of scattering theory, the maps $u_\pm \to u(0)$, where u is solution of (2.4), are simply the wave operators, the construction of which is therefore a by-product of the solution of the Cauchy problem at infinity. That problem will be considered in Section (2a) at a level of abstraction comparable with that of the Cauchy problem at finite times.

The second and more difficult step in the theory of scattering is to ascertain whether all solutions of the Equation (1.1) or rather (1.2) are actually dispersive. This is the problem of asymptotic completeness. It is much more dependent on the specific equation under consideration, and the available treatment both for the NLS and NLKG equations are heavily based on the conservation laws associated with those equations. That problem will be considered in Section (2b).

(2a) The Cauchy problem at infinity

We want to solve the Equation (2.5) with t_0 at or in a neighborhood of infinity (we restrict our attention to $t \to -\infty$ for definiteness). For that purpose, we split again the problem in two steps: the first one is the local Cauchy problem at infinity and consists in solving (2.5) in some interval $[T,\infty)$ with T sufficiently large. The second one consists in extending the solutions thereby obtained to all values of time and is therefore covered by the results of Section (1b). We therefore concentrate on the first step, which we treat again by a contraction method similar to that of Section (1a). We now look for solutions of (2.5) in a space $X_0(I) \subset C(I,X)$, which for unbounded I includes some time decay in its definition, so that the integrals that occur in (2.4), (2.5) are convergent at infinity for $u \in X_0(\cdot)$ and exhibits some uniformity properties in t and t_0. There are various ways to formulate that time decay. The simplest one is to take a family of semi-norms $\|\cdot\|_\alpha$, $\alpha \in A$, on X, all bounded by the norm in X and including the norm in X itself, and a family of continuous functions m_α from $\mathbb{R}$ to $[1,\infty)$ such that for any compact interval I, $\underset{\alpha}{\mathrm{Sup}}\ \underset{t \in I}{\mathrm{Sup}}\ m_\alpha(t) < \infty$, and to define for any interval I

$$X_0(I) = \{u \in C(I,X) : |u|_{0I} < \infty\} \qquad (2.6)$$

where

$$|u|_{0I} = \underset{\alpha}{\mathrm{Sup}}\ \underset{t \in I}{\mathrm{Sup}}\ m_\alpha(t) \|u(t)\|_\alpha . \qquad (2.7)$$

Such spaces will be referred to as uniform spaces. For compact I, $X_0(I) = C(I,X)$ with $|u|_{0I} \geq |u|_I$, while the inclusion $X_0(I) \subset C(I,X)$ is strict for unbounded I.

A more complicated family of spaces is obtained by taking, in addition to some family of semi-norms of the previous type, another family of semi-norms $\|\cdot\|_\beta$, $\beta \in B$, all bounded by the norm in X, and for each β a real number $q(\beta) \geq 1$, and defining $X_0(I)$ by (2.6) with

$$|u|_{0I} = \text{Max } \{\sup_\alpha \sup_{t \in I} m_\alpha(t) \| u(t) \|_\alpha, \sup_\beta \| \|u(\cdot)\|_\beta \|_{q(\beta),I}\},$$
(2.8)

where $\| \cdot \|_{q,I}$ denotes the norm in $L^q(I,dt)$. Such spaces will be referred to as integral spaces. One could of course define more general spaces with weighted L^q-norms in the time variable.

It then turns out that the main results relative to the Cauchy problem at infinity can be derived from a set of abstract assumptions that are very similar to Assumption 1.1. We restrict our attention to the case of uniform spaces, where the assumptions can be stated entirely in terms of the integrals $G([s_1,s_2],u)$ defined by (1.5). For any interval $I \subset \mathbb{R}$ we denote by $\bar{I}$ its closure in $\bar{\mathbb{R}} = \mathbb{R} \cup \{+\infty, -\infty\}$ with the obvious topology. In particular, if $I = [T,\infty)$, then $\bar{I} = [T,\infty]$. We denote by $B_0(I,\rho)$ the ball of radius ρ in $X_0(I)$.

Assumption 2.1. For any closed interval $I \subset \mathbb{R}$, any $t \in \mathbb{R}$ and any $u \in X_0(I)$, the function $\tau \to U(t-\tau)f(\tau,u(\tau))$ is Bochner integrable from I to X and the function $(s_1,s_2) \to G([s_1,s_2],u)$ is continuous from $\bar{I} \times \bar{I}$ to $X_0(\mathbb{R})$. For any closed interval I, for any $\rho > 0$, for any u_1 and u_2 in $B_0(I,\rho)$, G satisfies the following Lipschitz condition

$$|G(I,u_1) - G(I,u_2)|_{0\mathbb{R}} \leq C_0(I,\rho)|u_1 - u_2|_{0I} \qquad (2.9)$$

where $C_0(I,\rho)$ is separately non-increasing in I and ρ and tends to zero when I tends to zero for fixed ρ. (By "I tends to zero" we mean that both ends of I tend to a common value, possibly $+\infty$ or $-\infty$.)

Under Assumption 2.1, one derives the basic results concerning the Cauchy problem at infinity by arguments similar to those used in the study of the local Cauchy problem at finite times, and based again on the contraction mapping principle. These results include the existence of asymptotic states (i.e. of the limits $u_\pm = \lim \tilde{u}(t)$ as $t \to \pm\infty$) for dispersive solutions (now technically defined as solutions in $X_0(\mathbb{R})$), the existence an uniqueness of the solutions of (2.5) in a neighborhood of infinity, and continuity of the solutions with respect to the initial time and initial data. In order to state the results, it is convenient to define the space

$$X_0 = \{u \in X : U(\cdot)u \in X_0(\mathbb{R})\} \tag{2.10}$$

with norm

$$\|u\|_0 = |U(\cdot)u|_{0\mathbb{R}} . \tag{2.11}$$

Clearly X_0 is a Banach space continuously embedded in X.

Proposition 2.1. Let Assumption 2.1 hold. Assume (for simplicity) that the interaction is source-free, namely $f(t,0) = 0$. Then

(1) Let $I = [T,\infty)$, $t_0 \in \bar{I}$, $\tilde{u}_0 \in X_0$ and let $u \in X_0(I)$ be a solution of (2.5). Then $\tilde{u} \in C(\bar{I}, X_0)$ (see (2.2)). In particular, there exists u_+ in X_0 such that $\tilde{u}(t) \to u_+$ in X_0 when $t \to \infty$.

(2) For any $\rho > 0$, there exists $T_0(\rho) > \infty$ such that for any $t_0 \in \bar{I}$ where $I = [T_0(\rho),\infty)$ and for any $\tilde{u}_0 \in X_0$ with $\|\tilde{u}_0\|_0 \leq \rho$, equation (2.5) has a unique solution in $B_0(I, 2\rho)$. That solution is actually unique in $X_0(I)$.

(3) In the situation of (2), the maps $(t_0, \tilde{u}_0) \to u$ and $(t_0, \tilde{u}_0) \to u$ are continuous, with the topology of $I \times X_0$ on $(t_0, \tilde{u}_0)$, of $X_0(I)$ on u and of $C(\bar{I}, X_0)$ on u.

Similar results can be obtained with the integral spaces of the type (2.6) - (2.8), although the abstract formulation is slightly more complicated.

As a by-product of the local theory at infinity, one obtains in general a proof of existence of global solutions and of asymptotic completeness (namely all solutions are dispersive) for small initial data. Contrary to the global existence proof of Section (1b), that proof does not depend on a priori estimates of the solutions. In fact, the applicability of the contraction argument leading to Proposition 2.1, part

(2), relies on the possiblity of making $C_0(I,\rho)$ small. That can be achieved in general not only ba taking I small for given ρ, but also by taking $I = \mathbb{R}$ and ρ small.

Proposition 2.2. Let Assumption 2.1 hold and assume in addition that $C_0(\mathbb{R},\rho) \to 0$ when $\rho \to 0$. Assume that $f(t,0) = 0$. Then there exists $\rho > 0$ such that for any $\tilde{u}_0 \in X_0$ with $\|u_0\|_0 \leq \rho$ and any $t_0 \in \bar{\mathbb{R}}$, the Equation (2.5) has a unique solution in $X_0(R)$ (if t_0 is finite, the solution is actually unique in $C(\mathbb{R},X)$).

When trying to apply the abstract theory to a specific equation, the basic problem one encounters is to choose the spaces $X_0(\cdot)$. That choice is dictated by the available decay estimates on the free group $U(\cdot)$, and the need to prove the basic estimate (2.9). In particular, the time decay to be included in the definition of $X_0(\cdot)$ can be at most that of the solutions of the free equation (2.1). On the other hand, as in linear scattering theory, some decay of the interaction in space is needed. For interactions of the typical form (0.4), that condition takes the form of lower bounds on the allowed values of p_1, p_2.

We now show how the abstract theory can be applied to the NLS equation. The basic decay estimate on the free group is (1.7). We consider only the cases 2 and 3 of Section (1a).

Case 2. $X = L^{r_1} \cap L^{r_2}$. For the norms $\|\cdot\|_\alpha$ and the estimating functions $m_\alpha(t)$ of the abstract theory, we take the norms in L^r, $r_1 \leq r \leq r_2$, with $m_r(t) = (1+|t|)^{\bar{\delta}(r)}$ and $\bar{\delta}(r) = \text{Min}(\delta(r),\delta)$ for some δ, $0 < \delta < 1$, and with $\delta(r)$ defined by (1.8), so that for any interval I

$$|u|_{0I} = \sup_{t \in I} \sup_{r_1 \leq r \leq r_2} (1+|t|)^{\bar{\delta}(r)} \|u(t)\|_r . \qquad (2.12)$$

One can then show that Assumption 2.1 holds provided f_0 satisfies the assumptions of Section (1a) (namely $f_0 \in C^1$, $f_0(0) = 0$ and (1.10), (1.11)) and in addition

$$p_1 > \text{Max}(\delta^{-1}, \ 1 + \frac{2}{n}(1+\delta)). \tag{2.13}$$

The condition on p_1 resulting from the best choice of δ can be rewritten as

$$p_1 \delta(p_1 + 1) > 1 \tag{2.14}$$

or, equivalently,

$$p_1 > \frac{1}{2n}\{n+2 + ((n+2)^2 + 8n)^{1/2}\}, \tag{2.15}$$

namely $p_1 > 1 + \sqrt{2}$ for $n = 2$, $p_2 > 2$ for $n = 3$, etc.

The same equation can be treated in integral spaces of the type (2.6), (2.8) under exactly the same assumptions on f_0 and with essentially the same results. For that purpose, one takes for the norms $\|\cdot\|_\alpha$ and associated function m_α again the L^r norms for $r_1 \leq r \leq r_2$, with $m_r(t) = 1$, and for the norms $\|\cdot\|_\beta$ and associated exponents $q(\beta)$ again the same norms with $q(r) = (1+\varepsilon)\bar{\delta}(r)^{-1}$ for some ε sufficiently small, such that

$$p_1 \geq \text{Max}\{\delta^{-1}(1+\varepsilon(1-\delta(r_1))), \ 1 + \frac{2}{n}(1+\delta+\varepsilon(1-\delta))\}. \tag{2.13}_\varepsilon$$

A nice application of the previous integral spaces can be made by using an extension of an inequality of Strichartz, which states that for all $\varphi \in L^2$, $U(\cdot)\varphi \in L^q(\mathbb{R}, L^r(\mathbb{R}^n))$ for $0 \leq 2/q = \delta(r) < 1$. Together with elementary arguments, this implies that the previous spaces for any $\delta \leq 1$ and $\varepsilon = 1$ are such that $X_0 \supset H^1$. In particular, if one is in a situation where energy conservation holds and where the H^1 norm of solutions is bounded, namely if f_0 satisfies (1.23) and (1.28), then one obtains from Proposition 2.2 a proof of asymptotic completeness for small data in H^1. For interactions of the type (0.4) with a single power p and positive λ, this covers the case when $4/n \leq p - 1 < 4/(n-2)$.

Case 3. $n = 3$, $X = H^1 \cap L^\infty$. In that case one can define $X_0(\cdot)$ is a way similar to that of Case 2, namely

$$|u|_{0I} = \underset{t}{\text{Sup Max}} \{\|u(t)\|_{1,2}, \underset{2\leq r\leq \infty}{\text{Sup}} (1+|t|)^{\bar{\delta}(r)} \|u(t)\|_r\} \quad (2.16)$$

with again $\bar{\delta}(r) = \text{Min}(\delta(r), \delta)$ for some δ, $0 < \delta \leq \frac{n}{2}$.

One can then show that Assumption 2.1 holds if in addition to the assumptions made in Section (1a) for the local Cauchy problem, f_0 satisfies the condition

$$p_1 > \text{Max} \{1 + \delta^{-1}, 1 + \frac{2}{n}(1 + \delta)\} . \quad (2.17)$$

That condition is stronger than (2.13), indicating that the present choice of X is less natural than the previous one.

When combined with the assumptions and results of Section (2a), the results of this section provide some additional information. For instance, starting with initial data $\tilde{u}_0 \in X_0$ at some sufficiently large positive time t_0, one can construct a solution of the Cauchy problem that is dispersive at $+\infty$ and continue that solution to all times. In particular, if $t_0 = +\infty$, $\tilde{u}_0 = u_+$, the wave operator Ω_+: $u_+ \to u(0)$ is well-defined. Such solutions need not be dispersive at $t \to \infty$ however, i.e. asymptotic completeness may fail to hold. Another result of interest may be the extension of conservation laws to infinite times. For instance, for the NLS equation in Case 2, one can show under the assumptions made in Sections (2a) and (1b) that the dispersive solutions obtained in Proposition 2.1 satisfy $\tilde{u} \in C(I, H^1)$ and satisfy the identities

$$\|\varphi\|_2 = \|\varphi_+\|_2, \quad E(\varphi) = \frac{1}{2}\|\nabla\varphi_+\|_2^2$$

the second of which is strongly reminiscent of the interwining property of the wave operators familiar in linear scattering theory.

Similar results hold for the NLKG equation.

(2b) Asymptotic completeness

Once one knows how to solve the Cauchy problem at infinity (for initial data in X_0) and in particular how to construct the wave operators $\Omega_\pm : u_\pm \to u(0)$ (as operators from X_0 to X_0), the next question to ask is whether all solutions of the Equation (2.5) with finite t_0 are dispersive, namely lie in $X_0(\mathbb{R})$, for all initial data in X_0 or at least in some space Σ densely and continuously embedded in X_0. If in addition Σ is stable under $\Omega_\pm$, this property implies asymptotic completeness in Σ.

Such a property holds only in special cases and is strongly dependent on the equation under consideration. It has been established only for the NLS and NLKG equation with interaction satisfying a suitable repulsivity condition. The proofs are based on a priori estimates on the solutions, derived from conservation laws satisfied by the equations. There are basically two methods available. The first one is based on conformal invariance for the NLKG equation (in which case it works only in the massless case [26]) and its Galilean analogue, hereafter called pseudoconformal invariance, for the NLS equation [8]. The second method, which is more complicated, is based on a modified form of dilational invariance and applies to the massive NLKG equation [4], [19] and to the NLS equation [12], [16]. Here, as an illustration, we briefly sketch the pseudo-conformal invariance method for the NLS equation.

One first notices that the free Schrödinger equation, and more remarkably, the NLS equation, in the case of a single power interaction (cf. (0.4)) with $p = 1 + 4/n$, is invariant under a projective representation of a group G_S which is **larger than the** Galilei group and is generated by the Galilei group, the dilations $(t,x) \to (t\, e^{2\theta}, xe^\theta)$ and a one parameter group of transformations, hereafter called pseudoconformal, $(t,x) \to (t(1+at)^{-1}, x(1+at)^{-1})$ or, in projective coordinates $(t,x,1) \to (t,x,1+at)$. That group G_S is obtained by enlarging the Galilei group by its transform under the external automorphism generated by the inversion $(t,x) \to t^{-1}, t^{-1}x)$ (in projective coordinates $(t,x,1) \to (1,x,t)$. Under that automorphism space translations are exchanged with pure Galilei transformations, and time translations with pseudo-conformal transformations. G_S is sometimes called the Schrödinger group. The projective representation under which the free Schrödinger equation is invariant has the pseudo-conformal transformations represented by

$$\varphi(t,x) \to (1-at)^{-n/2} \exp[-\frac{i}{2} a x^2 (1-at)^{-1}]$$

$$\varphi(t(1-at)^{-1}, x(1-at)^{-1}) \ . \quad (2.18)$$

For a general interaction f_0 (satisfying (1.23)), the NLS equation is no longer invariant. However by a standard application of Noether's theorem, one derives easily the following approximate conservation law

$$\frac{1}{2} \|x \tilde{\varphi}(t)\|_2^2 + t^2 \int dx \, V(\varphi(t)) = \text{idem} \, (t \to s)$$

$$+ \int_s^t d\tau \, \tau \int dx \, W(\varphi(\tau)) \quad (2.19)$$

where $W(z) = W(|z|)$ is defined by

$$W(\rho) = (n+2) V(\rho) - \frac{n}{2} V'(\rho) \quad (2.20)$$

and $\tilde{\varphi}$ is defined in analogy with (2.2).

Define now Σ as the Hilbert space with norm

$$\|\varphi\|_\Sigma^2 = \|\varphi\|_2^2 + \|\nabla \varphi\|_2^2 + \|x\varphi\|_2^2 \ . \quad (2.21)$$

Let the interaction f_0 satisfy (1.23) and be repulsive in the sense that $V \geq 0$ and $W \leq 0$, and let φ be a solution of (2.5) with initial time $t_0 = 0$ (for simplicity) and with $\tilde{\varphi}_0 (= \varphi_0) \in \Sigma$. It follows then from L^2-norm and energy conservation and from (2.19) that for all t

$$\|\tilde{\varphi}(t)\|_\Sigma^2 \leq \|\varphi_0\|_2^2 + \|x\varphi_0\|_2^2 + 2E(\varphi_0) \qquad (2.22)$$

so that $\tilde{\varphi}$ is bounded in Σ uniformly in time. By a straightforward application of (1.7) and the Sobolev inequalities, this implies that φ itself satisfies the time decay

$$\|\varphi(t)\|_r \leq C(1 + |t|)^{-\delta(r)} \qquad (2.23)$$

for all r satisfying $\frac{1}{2} - \frac{1}{n} < \frac{1}{r} \leq \frac{1}{2}$.

The previous argument, which is partly formal at the moment, has to be combined with the functional framework developed in Section (2.a) to study the Cauchy problem at infinity. The results are most satisfactory in case 2, with $X = L^{r_1} \cap L^{r_2}$. In that case $\Sigma \subset X_0$, the preceding arguments boil down to a proof of asymptotic completeness in Σ for repulsive interactions in the previous sense. In particular (2.22), (2.23) imply that all solutions with initial data $\tilde{\varphi}_0 \in \Sigma$ belong to $X_0(\mathbb{R})$.

Similar results hold for the massless NLKG equation.

References

[1] J.B. Baillon, T. Cazenave, M. Figuera, C.R. Acad. Sci. Paris 284, 869-872 (1977).

[2] P. Brenner, Math. Z. 145, 251-254 (1975).

[3] P. Brenner, Math. Z. 167, 99-135 (1979).

[4] P. Brenner, Math. Z. 186, 383-391 (1984).

[5] P. Brenner, W. von Wahl, Math. Z. 176, 87-121 (1981).

[6] F.E. Browder, Math. Z. 80, 249-264 (1962).

[7] D. Eardley, V. Moncrief, Commun. Math. Phys. 83, 171-212 (1982).

[8] J. Ginibre, G. Velo, J. Funct. Anal. 32, 1-71 (1979).

[9] J. Ginibre, G. Velo, Ann. IHP (Phys. Théor.), 28, 287-316 (1978).

[10] J. Ginibre, G. Velo, Commun. Math. Phys. 82. 1-28 (1981); Ann. IHP (Phys. Théor.), 36, 59-78 (1982).

[11] J. Ginibre, G. Velo, Ann. IHP (Anal. non lin.), 1, 309-323 (1984).

[12] J. Ginibre, G. Velo, C.R. Acad. Sci. Paris, 298, 137-140 (1984) and J. Math. Pur. Appl., in press.

[13] J. Ginibre, G. Velo, Ann. IHP (Anal. non lin.), 2, 309-327 (1985).

[14] J. Ginibre, G. Velo, Math. Z. 189, 487-505 (1985).

[15] K. Jörgens, Math. Z. 77, 295-307 (1961).

[16] J.E. Lin, W. Strauss, J. Funct. Anal. 30, 245-263 (1978).

[17] J.L. Lions, Quelques méthodes de résolution des problèmes aux limites non linéaires, Dunod Gauthier-Villars, Paris, 1969.

[18] B. Marshall, W. Strauss, S. Wainger, J. Math. Pur. Appl. 59, 417-440 (1980).

[19] C. Morawetz, W. Strauss, Comm. Pure Appl. Math. 25, 1-31 (1972).

[20] H. Pecher, Math. Z. 150, 159-183 (1976).

[21] H. Pecher, Math. Z. 161, 9-40 (1978).

[22] H. Pecher, W. von Wahl, Manuscripta Math. 27, 125-157 (1979).

[23] M. Reed, Abstract non-linear wave equations, Springer, Berlin, 1976.

[24] I.E. Segal, Ann. Math. $\underline{78}$, 339-364 (1963).

[25] I.E. Segal, J. Funct. Anal. $\underline{33}$, 175-194 (1978).

[26] W. Strauss, J. Funct. Anal. $\underline{2}$, 409-457 (1968).

[27] W. Strauss, in "Invariant Wave Equations" (Erice, 1977), Lecture notes in Physics, Vol. 73, 197-249, Springer, Berlin, 1978.

[28] M. Tsutsumi, N. Hayashi, Math. Z. $\underline{177}$, 217-234 (1981).

THE INVERSE METHOD IN QUANTUM MECHANICS AND FIELD THEORY

H. GROSSE
Institut für Theoretische Physik
der Universität Wien
A-1090 Wien, Austria

PREFACE

Although the inverse problem in quantum mechanics has been developed and solved in the sixties, it is of great importance even nowadays. The reason is, that it has been realized, that this method allows to find soliton solutions of certain nonlinear field equations, it identifies certain models as being completely integrable and it allows for a quantum generalization.

In these lectures, we follow the historical development and divide into four parts:

I) In the first Part, the direct and inverse problem for the one dimensional Schrödinger equation is discussed. In addition, we remark the deep connection to the Korteweg-de Vries (KdV) equation which was the starting point for developing the inverse scattering transform method (ISTM). All reflexionless potentials for the Schrödinger equation are explicitly given.

II) In order to get all soliton solutions of KdV-equation, one has to find the time evolution of scattering data. We sketch first an explicit way of obtaining it, but give also the more general schemes, the Lax and the AKNS-Zhakarov-Shabat scheme. This explains the reason of the wide applicability of the method.

III) It furthermore points to the fact that certain partial differential equations, which are solvable by that method, can be considered as completely integrable hamiltonian dynamical systems. We illustrate this notion first for finite dimensional systems and sketch afterwards the proof of integrability for the nonlinear Schrödinger equation.

IV) Recently, the quantization of the full direct and inverse problem has led to new insights into models for which Bethe states were known. The algebraic structure which is behind it is indicated. The difficult problem of changing the representation of the canonical commutation relations is mentioned at the end.

Clearly, we attempt to give only a rough overview on that subject. We next list books and reviews which are of general interest and we will refer to them as [B Nr.]. At the end, we give only a few additional references, which are explicitly mentioned in the text.

ACKNOWLEDGEMENT

It is a great pleasure for me to thank Professor Ludwig Streit for the kind invitation to the Project Nr. 2 on "Mathematics + Physics" and for the possibility to lecture in Bielefeld. I also would like to thank the theory group of Brookhaven National Laboratory for the kind hospitality in Summer 1984, where the final version of these lectures was worked out.

REVIEWS AND BOOKS

[B1] ABLOWITZ, M.J. and SEGUR, H., Solitons and the inverse scattering transform, SIAM, Philadelphia, 1981

[B2] BULLOUGH, R. and CAUDRY, P.J. (eds.), Solitons, Springer, Berlin 1977

[B3] CALOGERO, F. and DEGASPERIS, A., Spectral Transform and Solitons I, North-Holland, Amsterdam, 1982

[B4] CHADAN, K. and SABATIER, P.C., Inverse problems in quantum scattering theory, Springer, New York, 1977

[B5] DODD, R.K., EILBECK, J.C., GIBBON, J.D. and MORRIS, H.C., Solitons and nonlinear wave equations, Academic Press, New York, 1982

[B6] ECKHAUS, W. and Van HARTEN, A., The inverse scattering transformation and the theory of solitons, North-Holland, Amsterdam, New York, 1981

[B7] EILENBERGER, G., Solitons, Series in Solid State Science, Nr. 29, Springer, Berlin 1981

[B8] FADDEEV, L.D., SKLYANIN, E.K. and TAKHTAJAN, L.A., Theor. mat. fiz. 40, (1979) 194

FADDEEV, L.D., TAKHTAJAN, L.A., Uspekti mat. nauk 34, (1978) 13

FADDEEV, L.D., Soviet Sci. Reviews, Cont. Math. Phys. C1 (1980) 107

FADDEEV, L.D., Physica Scripta 23 (1981)

FADDEEV, L.D., in "Structural Elements in Particle Physics and Statistical Mechanics", Honerkamp J., Pohlmeyer, K. and Römer, H., Nato advanced Study Inst. Scies. B, Vol. 82 (1983) 93

[B9] KULISH, P.P. and SKLYANIN, E.K. in "Proceedings of the Int. Symposium on Integrable Quantum Fields", Hietarinta J. and Montonen, C. (eds.) (Springer, Berlin 1981) Lecture Notes in Physics 151

[B10] THACKER, H.B., Review Mod. Phys. 53 (1981) 253, in "Proceedings of the Int. Symposium on Integrable Quantum Fields" Hietarinta J. and Montonen, C. (eds.) (Springer, Berlin 1981) Lecture Notes in Physics 151

[B11] REBBI, C. and SOLIANI, G. Solitons and Particle Physics, World Scientific Publishing Co., Singapore 1984

[B12] SCOTT, A.C., CHU, F.Y.E. and McLAUGHLIN, D.W. Proc. of the IEEE 61 (1973) 1443

CONTENTS

I. DIRECT AND INVERSE PROBLEM FOR THE SCHRÖDINGER EQUATION

   1. Introduction
   2. Direct Problem for the Schrödinger Equation
   3. Inverse Problem for the Schrödinger Equation
   4. Reflexionless Potentials

II. LAX SCHEME AND AKNS-ZHAKAROV-SHABAT METHOD

   1. Remarks on Solitons
   2. Time Evolution of Scattering Data
   3. Lax Method
   4. AKNS-Zhakarov-Shabat Method

III. COMPLETELY INTEGRABLE HAMILTONIAN DYNAMICAL SYSTEMS

   1. Finite Dimensional Systems
   2. Nonlinear Schrödinger Equation
   3. Derivation of Sum Rules
   4. Derivation of Poisson Brackets

IV. QUANTUM INVERSE SCATTERING TRANSFORM METHOD

   1. r-Matrix Approach
   2. Bethe States
   3. Quantize Direct Step
   4. Quantize Inverse Step

I. DIRECT AND INVERSE PROBLEM FOR THE SCHRÖDINGER EQUATION

1. Introduction

To start with, let us give one example of an inverse problem which is closely related to what follows. One of the most famous inverse problems concerns the drum problem [1]: Consider the d-dimensional Laplacian on functions which are restricted to a certain volume Ω with Dirichlet boundary conditions. The question is whether the domain is uniquely determined by the given discrete spectrum. An extremely simple argument shows uniqueness in case of a spherical drum. Expand for small β

$$\mathrm{Tr}\, e^{-\beta H} \underset{\beta \to 0}{\cong} c_0 \frac{|\Omega|}{\beta} + c_1 \frac{L}{\beta^{1/2}} + c_2(1-n) + \ldots \qquad (I.1)$$

where c_i are constants, $|\Omega|$ denotes the area, L the length of the boundary and n the number of holes. Since there is a general isoperimetric inequality

$$|\Omega| \leq L^2/4\pi \qquad (I.2)$$

which becomes an equality iff Ω is a sphere, the above stated result follows: Given all energies means that all expansion coefficients on the right of Eq. (I.1) are determined, given these, Inequality (I.2) can be checked.

Remarks:
- α) Since n is determined by the energy levels as well, one can hear the number of holes in the drum.
- β) There is obviously a connection of the above problem to index theorems.
- γ) To the best of our knowledge, the general inverse problem for the drum is not yet solved.
- δ) There is an example of drums in $d = 24$ due to Milnor showing nonuniqueness of the inverse problem.

The analog of the drum problem for the one dimensional Schrödinger operator has been worked out as well [2]. Start with

$$H = -\frac{d^2}{dx^2} + V(x) \quad \text{on} \quad L^2([0,L],dx) \tag{I.3}$$

and expand

$$\text{Tr } e^{-\beta H} \underset{\beta \to 0}{\cong} \frac{1}{\sqrt{\beta}} \sum_{n=0}^{\infty} \beta^n I_n(V) . \tag{I.4}$$

The I_n's are generalized moments; the first few are given by

$$I_0 = \text{const.} = k_0, I_1 = k_1 \int_0^L dx \, V(x), \quad I_2 = k_2 \int_0^L dx \, V^2(x)$$

$$I_3 = k_3 \int_0^L dx \left[V_x^2(x) + 4V^3(x) \right] . \tag{I.5}$$

It is possible to show that I_3 has a unique infimum for given I_1 and I_2; therefore a situation similar to the drum case occurs.

Remarks:

α) The above mentioned infimum is given by a solution of Lamê's equation, in the limit $L \to \infty$ it becomes the one soliton potential.

β) Assume V depends on a parameter such that all eigenvalues are independent of it; this shows that all I_n's are also independent of that parameter.

γ) The I_n's are indeed the infinite number of conserved quantities of the KdV equation.

δ) There exists a second representation of the l.h.s. of (I.4) in terms of scattering data; expanding in β gives an infinite number of sum rules which will be discussed later on.

We next sketch the way which allows to solve the initial value problem for certain partial differential equations with the help of the inverse problem. We shall do it for the KdV-equation but the scheme works for a large class of equations. The KdV-equation for $v(t,x)$:

$$v_t = 6vv_x - v_{xxx} \qquad (I.6)$$

describes waves propagating in a narrow channel; that phenomenon was described more than 150 years ago by J.S. Russell, Eq. (I.6) was written down in 1895, but after the numerical work of Zabusky and Kruskal in 1965 it was realized in 1967 by Gardner, Green, Kruskal and Miura [3] that a complete characterization of the initial value problem can be obtained. The combination of a weak nonlinearity, which would give rise to shock waves, with a dispersion, which would lead to spreading of wave packets, leads to soliton-like behavior.

Instead of determining the mapping from initial data $v(0,x)$ to $v(t,x)$ immediately one replaces that problem by a three step procedure [B5]. Through the direct problem (step A) one takes $v(0,x)$ as potential in a one dimensional Schrödinger equation and goes over to scattering data at time zero $\{R(0,k), \varepsilon_\ell(0), c_\ell(0)\}$, where R denotes the reflection coefficient, ε_ℓ are energy eigenvalues and c_ℓ are normalization constants. It turns out that the time evolution of scattering data can be found in a simple way (step B). It is especially remarkable that the energy eigenvalues ε_ℓ do not depend on time. This runs under the name of isospectral invariance of the KdV flow. If $v(t,x)$ evolves according to the KdV equation, the energy levels remain invariant. Finally, in step C one recovers $v(t,x)$ by solving a Gelfand-Levitan-Marchenko (GLM)-type equation.

2. Direct Problem for the Schrödinger Equation

Starting from the Schrödinger equation for the full line

$$\left[-\frac{d^2}{dx^2} + V(x)\right] \psi(k,x) = \varepsilon \psi(k,x), \quad \varepsilon = k^2 \qquad (I.7)$$

and imposing conditions on V (real, locally integrable, V and $xV \in L^1(\mathbb{R})$), one defines first different solutions of (I.7) by imposing certain boundary conditions [B4]. Jost solutions are given by

$$\lim_{x \to \infty} f_1(k,x) e^{-ikx} = 1, \quad \lim_{x \to -\infty} f_2(k,x) e^{ikx} = 1 . \quad (I.8)$$

Since $f_1(-k,x)$ is a solution to (I.7), too, and linear independent of $f_1(k,x)$ and $f_2(k,x)$ for $k \neq 0$, one may write

$$f_2(k,x) = a(k) f_1(-k,x) + b(k) f_1(k,x). \quad (I.9)$$

The physical solution is then given by

$$\varphi(k,x) = T(k) f_2(k,x) = R(k) f_1(k,x) + f_1(-k,x) \quad (I.10)$$

where the reflexion and transmission coefficients are related to a and b by

$$a(k) = \frac{1}{T(k)}, \quad b(k) = \frac{R(k)}{T(k)} . \quad (I.11)$$

In a standard way one can obtain properties of these solutions: Going over to the Volterra integral equation and incorporating the boundary conditions, one deduces analyticity of $f_{1,2}$ in the open upper half plane in k and using properties of V one gets the bound

$$|f_1(k,x) - e^{ikx}| < c \frac{e^{-x \, \mathrm{Im} \, k}}{1 + |k|} , \quad \mathrm{Im} \, k \geq 0, \, x > 0 \quad (I.12)$$

implying

$$|f_1(k,x) - e^{ikx}| \in L^2((-\infty,\infty),dk), \quad k \text{ real} . \quad (I.13)$$

An analog of the Paley-Wiener theorem implies the so-called Levin representation for f_1:

$$K(x,y) = \int_{-\infty}^{\infty} \frac{dk}{2\pi} e^{iky} (f_1(k,x) - e^{iky}) \quad \text{if } y \geq x$$
$$K(x,y) = 0 \quad \text{if } y < x . \quad (I.14)$$

Taking the inverse Fourier transform of (I.14) gives a representation of the Möller operator in terms of a kernel K which is <u>independent of momentum</u> k :

$$f_1(k,x) = e^{ikx} + \int_x^\infty dy \, e^{iky} K(x,y) \; . \qquad (I.15)$$

Remarks:

α) Eq. (I.15) is the key for a solution of the inverse problem. It follows from the fact that the Fourier transform of the regular solution has compact support (by the Paley-Wiener theorem).

β) There exists a similar representation for f_2

$$f_2(k,x) = e^{-ikx} + \int_{-\infty}^x dy \, e^{-iky} \bar{K}(x,y) \; . \qquad (I.16)$$

3. Inverse Problem for the Schrödinger Equation

Up to now we did not impose the condition that f_1 has to be a solution of the Schrödinger Equation (I.7). Doing this relates the kernel K to the potential [B12]. Apply the Schrödinger operator to f_1-exp(ikx) and take the Fourier transform in k. This gives

$$\left(\frac{\partial^2}{\partial x^2} - \frac{\partial^2}{\partial y^2} - V(x)\right) K(x,y) = V(x) \, \delta(x-y) \; . \qquad (I.17)$$

Introduce light cone variables ξ = y+x, η = y-x; (I.17) turns into

$$-4 \frac{\partial^2}{\partial \xi \partial \eta} K(x,y) - V(\tfrac{\xi-\eta}{2}) K(x,y) = V(\tfrac{\xi-\eta}{2}) \delta(\eta) \; . \qquad (I.18)$$

Integrating Eq. (I.18) from -ε to ε, and observing the fact that $K(x,y) = 0$ for $x > y$ gives finally

$$-2 \frac{\partial}{\partial x} K(x,x) = V(x) \; . \qquad (I.19)$$

It is a little bit more complicated to relate the kernel K to the scattering data. From the fact that $\lim_{k \to \infty} T(k) = 1$ and $\lim_{k \to \infty} R(k) = 0$ we may rewrite (I.10) in such a way, so that all terms have Fourier transforms. This gives

$$\int_{-\infty}^{\infty} \frac{dk}{2\pi} (T(k)-1) f_2(k,x) e^{iky} = \int_{x}^{\infty} dz\, \tilde{R}(y+z) K(x,z) + \tilde{R}(x+y) + K(x,y) \quad (I.20)$$

where $\tilde{R}$ denotes the F.T. of $R(k)$. The l.h.s. of (I.20) can be evaluated using Wronski identities and Cauchy's integration theorem; one obtains contributions from the poles of $T(k)$ which correspond to bound states (Im $k_\ell > 0$)

$$(\text{l.h.s. of (I.20)}) = \sum_{\ell=1}^{N} c_\ell^2 e^{-\kappa_\ell y} f_1(i\kappa_\ell, x), \quad k_\ell = i\kappa_\ell \quad (I.21)$$

where the c_ℓ^2 are normalization constants of the bound state wave function:

$$c_\ell^{-2} = \int_{-\infty}^{\infty} dx\, f_1^2(i\kappa_\ell, x) \quad . \quad (I.22)$$

Use next the representation of the wave function (I.15) with $k = i\kappa_\ell$ and define a kernel G by

$$G(x,y) = \tilde{R}(x+y) + \sum_{\ell=1}^{N} c_\ell^2 e^{-\kappa_\ell(x+y)} \quad (I.23)$$

which is determined by the scattering data. From Eqs. (I.20) to (I.23) one finally obtains the GLM equation

$$K(x,y) + G(x,y) + \int_{x}^{\infty} dz\, K(x,z) G(z,y) = 0 \quad x < y \quad . \quad (I.24)$$

Together with Eqs. (I.15) and (I.19) it gives the solution to the inverse problem.

<u>Remarks</u>:

α) (I.24) is a Fredholm integral equation (for fixed x); since it is easy to see that the homogeneous equation has no trivial solution, its solution is unique.

β) Although not many solutions are known with $R \neq 0$, the case of $R(k) \equiv 0$ can be solved completely.

4. Reflexionless Potentials

For a large class of smooth potentials an incoming wave goes through the potential suffering only a phase shift but nothing is reflected. The S-matrix is therefore nontrivial. The occurrence of such a phenomenon may be surprising (it gives the chance to construct glasses which do not reflect at all for all frequencies!), the explicit construction of <u>all</u> such potentials may surprise even more [3]:

<u>Proposition</u>: Given $\{\varepsilon_\ell, c_\ell\}_{\ell=1}^N$; all reflexionless potentials with eigenvalues ε_ℓ and wave function normalization constants c_ℓ are given by

$$V(x) = -2 \frac{d^2}{dx^2} \ln \det(1 + C)$$
$$C_{\ell m} = \frac{c_\ell c_m}{\kappa_\ell + \kappa_m} e^{-(\kappa_\ell + \kappa_m)x} , \quad \kappa_\ell^2 = -\varepsilon_\ell . \quad (I.25)$$

Remarks:

α) These potentials are rational functions of exponentials.

β) They form all the pure soliton solutions of the KdV equation.

γ) The well-known special case with $V(x) = -n(n+1)/\text{ch}^2 x$ (potential has n bound states) is included in (I.25).

<u>Proof</u>: Let $\kappa_\ell \neq \kappa_m$ for $\ell \neq m$ and insert the kernel G into the GLM equation; since G is separable, one makes the ansatz for K:

$$K(x,y) = - \sum_{\ell=1}^N c_\ell \psi_\ell(x) e^{-\kappa_\ell y} \quad (I.26)$$

where ψ_ℓ turns out to be the normalized bound state wave function. The

GLM equation now becomes

$$(1+C(x))_{m\ell}\psi_\ell(x) = c_m e^{-\kappa_m x}, \quad C_{\ell m}(x) = \frac{c_\ell c_m}{\kappa_\ell + \kappa_m} e^{-(\kappa_\ell + \kappa_m)x} \qquad (I.27)$$

since $C \geq 0$, $(1+C)^{-1}$ exists and ψ can be evaluated from (I.27). Using trivial algebra, one finally finds (I.25).

Remark: In order to obtain solutions of the KdV equation, one has to determine the time evolution of scattering data. This we shall treat in Part II.

II. LAX SCHEME AND AKNS-ZHAKAROV-SHABAT METHOD

1. Remarks on Solitons

The reflexionless potentials are special in several respect. Let us remark that they fulfill a nonlinear autonomous system of equations; or stated differently: they are solutions of selfconsistently determined potentials and solve Hartree-type equations.

Lemma: If $R(k) \equiv 0$ and $\psi_{\ell,xx} = (V + \kappa_\ell^2)\psi_\ell$ one has

$$V(x) = -4 \sum_{\ell=1}^{N} \kappa_\ell \psi_\ell^2 \quad . \qquad (II.1)$$

Remarks:

α) (II.1) can be shown by differentiating (I.27) w.r.t. x, multiplying by ψ_ℓ and summing over ℓ on the one hand; multiplying on the other hand by $2\psi_\ell \kappa_\ell$, summing over ℓ and subtracting both expressions.

β) In Ref. [4] we obtained a nonlinear system of the type (II.1) starting from a quite different problem and we have been able to show that it is a completely integrable Hamiltonian system [5], by interpreting x as "time" and considering the system on a 2N dimensional phase space.

γ) For a one soliton solution V is proportional to ψ^2, therefore $W = \int_{-\infty}^{\infty} dx\, V^2(x)$ can be considered as the potential energy.

δ) From a sum rule (see III.3) one gets the inequality

$$\int_{-\infty}^{\infty} dx\, V^2(x) \geq \frac{16}{3} \sum_{\ell} \kappa_\ell^3 \qquad (II.2)$$

with equality iff $R \equiv 0$. Such bounds on the moments of energy levels have been of great interest in the last few years [4 to 7].

ε) From the last remark one deduces that one may call $\bar{V}(x) = V(x) + 4 \sum_{\ell=1}^{N} \kappa_\ell \psi_\ell^2$ the nonsoliton part of a general solution of the equation.

φ) Since

$$3 \frac{d}{dt} \int_{-\infty}^{\infty} dx\, v^2(t,x) = 0 \Rightarrow \frac{d}{dt} \int_{-\infty}^{\infty} dx\, x\, v(t,x) = \text{const.} \qquad (II.3)$$

one gets conservation of the center of mass of the distribution v, if it evolves according to the KdV equation.

ξ) It is possible to show that, if $v(0,x)$ is smooth

$$\lim_{t \to \infty} \sup_{x} |v(t,x) + \sum_{\ell=1}^{N} 4\kappa_\ell \psi_\ell^2| = 0 \qquad (II.4)$$

which means that in L^∞-norm smooth solutions become asymptotically pure solitons. This is not true in L^2-norm.

In addition to the above stated properties of soliton solutions there are other ones like stability properties under collisions, behavior under perturbations, etc., which are of importance for various applications.

2. Time Evolution of Scattering Data

One of the deep results concerning the connection of the KdV equation and the one-dimensional Schrödinger equation is the fact that the energy levels do not depend on time if $V(t,x)$ evolves according to the KdV equation:

<u>Lemma</u>: Let $V(t,x)$ be a solution of Eq. (I.6) and take it as a potential in the Schrödinger equation (I.7); then $\frac{\partial \varepsilon}{\partial t} = 0$ for all discrete eigenvalues ε.

<u>Proof</u>: According to the Feynman Hellmann Theorem we have

$$\frac{\partial \varepsilon}{\partial t} = (\psi, \frac{\partial V}{\partial t} \psi) = (\psi, (6VV_x - V_{xxx})\psi) . \qquad (II.5)$$

By partial integrations using again the Schrödinger equation, it is trivial to see that the r.h.s. of (II.5) vanishes identically.

<u>Remark</u>: This argument goes through if one takes the correct ratio -6 for the factors in the KdV equation; a scale transformation would clearly change them.

The next question concerns the time evolution of the reflection and transmission coefficient as well as that of the constants c_ℓ. We sketch first the naive method [3]; scattering data fulfill simple equations in t, since they are defined asymptotically as limits $x \to \pm \infty$. The principal idea behind the following is clear: since $\varepsilon_t = 0$ for bound states one forms $\varepsilon_t \psi^2(x)$ and deduces that it has to be a total derivative in x, one finds

$$\varepsilon_t \psi^2 = (\psi_x S - \psi S_x)_x , \quad S = \psi_t + \psi_{xxx} - 3(v+\varepsilon)\psi_x . \qquad (II.6)$$

From (II.6) we deduce for the case of an L^2-function ψ that S obeys the same Schrödinger equation like ψ, therefore $S = C(t)\psi$; multiplying S times ψ and integrating w.r.t. x shows that $C(t) \equiv 0$; introducing the asymptotic form $\psi_\ell \sim c_\ell \cdot \exp(-\kappa_\ell x)$ for $x \to \pm \infty$ we finally obtain

$$\dot{c}_\ell = 4\kappa_\ell^3 c_\ell \Rightarrow c_\ell(t) = e^{4\kappa_\ell^3 t} c_\ell(0) . \qquad (II.7)$$

For the continuous spectrum the last step does not work and $C(t) \neq 0$; from the asymptotic behavior for $x \to \pm\infty$ one gets this time two equations

$$T_t + 4i k^3 T = C(t)T$$

$$(R_t - 4ik^3 R)e^{ikx} + 4ik^3 e^{-ikx} = C \cdot e^{-ikx} + C \cdot R \cdot e^{ikx} \qquad (II.8)$$

from which one deduces the simple evolution equations with solution

$$R(t,k) = e^{8ik^3 t} R(0,k), \quad T(t,k) = T(0,k). \qquad (II.9)$$

Remarks:

α) This completes the explicit construction of solitons of the KdV equation; one takes the time dependence of c_ℓ's of Eq. (I.25) according to (II.7).

β) There are infinitely many nonlinear partial differential equations leaving the spectrum of the Schrödinger equation invariant (higher KdV equations).

γ) There are other linear eigenvalue problems such that the isospectral property gives solutions of nonlinear evolution equations.

δ) One example is the one-dimensional Dirac operator; the spectrum is invariant if the potential evolves according to the modified KdV equation [8]. If one imposes the condition that the potential vanishes at infinity, one gets no soliton solutions; but nontrivial boundary conditions allow for solitons which we have recently written down [9].

ε) The question: given a nonlinear evolution equation, can one find a linear eigenvalue problem such that the isospectral problem allows to determine the time evolution of initial data? is clearly a very difficult and deep one and has not yet been answered in that generality.

3. Lax Idea

The phenomenon that two operators have the same spectrum is well-known from the theory of selfadjoint operators in Hilbert space H: Two unitarily equivalent operators will fulfill it. In the following we shall assume that all operators are defined on a common dense subset of H.

The treatment of the discrete spectrum is easy [B3, B6].

<u>Theorem</u>: Assume that $L(t)$ is a one parameter family of selfadjoint operators in H and continuously differentiable in t. Let

$$L(t)\psi(t) = \lambda(t)\psi(t) , \qquad (II.10)$$

$\lambda(t)$ is an eigenvalue and $\psi(t)$ together with $\frac{\partial \psi(t)}{\partial t} \in H$. Assume there exists a family of operators $B(t)$ such that

$$\frac{\partial L}{\partial t} = BL - LB \quad \text{then} \quad \frac{\partial \lambda}{\partial t} = 0 . \qquad (II.11)$$

<u>Proof</u>: Differentiate (II.10) and use (II.11), this gives

$$(L-\lambda)(\frac{\partial \psi}{\partial t} - B\psi) = \lambda_t \psi . \qquad (II.12)$$

Take the scalar product of (II.12) with ψ and use the fact that L is selfadjoint; this proves that $\lambda_t = 0$.

For the continuous spectrum one has to assume more [B6]:

<u>Theorem</u>: Given a one parameter family of selfadjoint operators which is continuously differentiable; the spectrum will be invariant in time if there exists $B(t)$ such that

$$\text{a)} \qquad \frac{\partial L}{\partial t} = BL - LB \qquad (II.13)$$

the operator equation

$$\text{b)} \quad \frac{\partial U}{\partial t} = BU \quad \text{with} \quad B^+(t) = -B(t) \tag{II.14}$$

has a solution for $U(t)$ $t > 0$ and $U(0) = 1$ and

c) LU is differentiable w.r.t. t.

(In addition, domain questions have to be fulfilled.) If a), b), and c) is true, $\tilde{L}$ given by

$$\tilde{L} = U^{-1}(t) \, L(t) \, U(t) \tag{II.15}$$

is independent of time, and therefore $L(t)$ is a family of operators with the same spectrum.

To precise these remarks we follow [B6] and state first a

Lemma: If $U(t)$ solves (II.14) and $U(0)$ is unitary, $U(t)$ is unitary for all $t > 0$.

Proof: Take $w_i(t) = U(t)w_i(0)$, $i = 1,2$ with $\frac{\partial w_i}{\partial t} = Bw_i$. Compute $(w_1, \frac{\partial w_2}{\partial t}) = (w_1, Bw_2) = (B^+w_1, w_2) = -(\frac{\partial w_1}{\partial t}, w_2)$; it follows that $\frac{\partial}{\partial t}(w_1, w_2) = 0$. Therefore

$$(w_1(t), w_2(t)) = (w_1(0), w_2(0)) = (U(t)w_1(0), U(t)w_2(0))$$

which means that $U(t)$ is unitary.

Proof of Theorem: Consider $\tilde{L}$ from (II.15) and differentiate $LU = U\tilde{L}$; use (II.13) and (II.14). This shows that $\frac{\partial \tilde{L}}{\partial t} = 0$. Therefore, $\tilde{L}$ is time independent.

Examples:

(a)   Take for L the Schrödinger operator, then $\frac{\partial L}{\partial t} = V_t$ is multiplication with V_t. Take for $B = \frac{\partial}{\partial t}$, then one gets the result that if $V_t = V_x$, which means just a shift in x, the spectrum is time independent.

(b) Let L be again the Schrödinger operator, but take B as

$$B = \frac{\partial^3}{\partial x^3} + b \frac{\partial}{\partial x} + \frac{\partial^2}{\partial x^2} b \ . \tag{II.16}$$

A simple calculation of commutators shows that choosing $b = -3V/4$ leads to a cancellation of terms which depend on $\frac{\partial}{\partial x}$ and $\frac{\partial^2}{\partial x^2}$. One obtains finally

$$V_t = [B,L] = \frac{1}{4} V_{xxx} - \frac{3}{2} V V_x$$

which is up to a trivial scale the KdV equation.

<u>Remark</u>: The Lax method not only shows why certain two dimensional field theories can be treated by the ISTM, it also elucidates why certain dynamical systems are completely integrable.

<u>Examples</u>:

(a) Consider the equation of motion of the form

$$\dot{x}_n = P_n = 2b_n, \ \dot{p}_n = e^{-(x_n - x_{n-1})} - e^{-(x_{n+1} - x_n)} = 2(a_n - a_{n+1}). \tag{II.17}$$

Define the operator

$$(L\phi)_n = b_n \phi_n + a_n \phi_{n-1} + a_{n+1} \phi_{n+1}$$

$$(B\phi)_n = a_n \phi_{n-1} - a_{n+1} \phi_{n+1} \ ,$$

then $\dot{L} = [B,L]$ is identical to the Toda lattice equation (II.17).

(b) The Calogero model is defined by

$$\dot{x}_n = p_n, \ \dot{p}_n = \sum_{k \neq n} (x_k - x_n)^{-2}.$$

With the definition of

$$(z^\alpha)_{k\ell} = \begin{cases} (x_k - x_\ell)^{-\alpha} & k \neq \ell \\ 0 & k = \ell \end{cases}, \quad P_{\alpha\beta} = \delta_{\alpha\beta} P_\alpha, \quad D_{\alpha\beta} = \sum_j z^2_{\alpha j} \delta_{\alpha\beta}$$

The Lax pair is $L = p + iz^1$, $B = iD - iz^2$.

Remarks:

α) More general potentials, like the Weierstraß function have been shown to lead to integrable many body problems, too.

β) Expanding the determinant

$$\det(\lambda - L) = \lambda^n + \lambda^{n-1} I_1 + \ldots$$

leads to integrals of motion I_j.

III. COMPLETELY INTEGRABLE HAMILTONIAN DYNAMICAL SYSTEMS

1. Finite Dimensional Systems

The fact that there is a Lax pair and that there are conserved quantities restricts the dynamics and allows to deduce that the system is completely integrable and that there exists a transformation to action-angle variables. For finite dimensional systems these relationships can be made precise [10,11].

A triple (M^{2N}, ω, H) is called a hamiltonian dynamical system, where M^{2N} is a 2N-dimensional differentiable manifold, ω a nondegenerate closed two form and H the Hamiltonian is a function on M^{2N}. Next we discuss these notions in order to make clear the relation between conserved quantities and integrability.

(a) The manifold M^{2n} represents phase space and is the cotangent bundle over a base space M. Let $q \in M$ and p be a point in the fiber over q. On M^{2n} there is a natural one form φ which is given in local coordinates as $\varphi = \sum_{i=1}^{N} p_i dq_i$, where the dq_i are differentiable (and therefore cotangent vectors).

(b) Given a canonical one form φ, one obtains a canonical two form ω:

$$\omega = d\varphi = \sum_i dp_i \wedge dq_i \qquad (III.1)$$

where d denotes exterior differentiation and $\wedge$ denotes the wedge product.

Remarks:

α) Since $d(d\varphi) = 0$ we get $d\omega = 0$ which means that ω is a closed form.

β) Not all closed forms are exact. (A closed form is called exact if there exists a φ such that $\omega = d\varphi$.)

γ) The quotient space (closed k-forms)/(exact k-forms) = $H^k(M,\mathbb{R})$ is called the cohomology group; their dimensions are the Betti-numbers.

With the help of ω one can define Poisson brackets: Given functions f,g on phase space together with their associated vector fields

$$X_f : (q_i, p_i) \longrightarrow \left(q_i, p_i, \frac{\partial f}{\partial q_i}, \frac{\partial f}{\partial p_i}\right) \qquad (III.2)$$

one defines the Poisson bracket as

$$\{f,g\} = \omega(X_f, X_g) = \sum_i \left(\frac{\partial f}{\partial q_i}\frac{\partial g}{\partial p_i} - \frac{\partial f}{\partial p_i}\frac{\partial g}{\partial q_i}\right). \qquad (III.3)$$

Therefore, in the standard chart ω corresponds to the symplectic matrix

$$\omega = \begin{pmatrix} 0 & -\mathbb{1}_n \\ \mathbb{1}_n & 0 \end{pmatrix} \qquad (III.4)$$

where $\mathbb{1}_n$ denotes the $n \times n$ unit matrix. More generally, one calls the pair (M^{2n}, ω) a symplectic manifold.

Definition: A symplectic structure on M^{2n} is given by a closed non-degenerate differentiable two form ω on M^{2n}.

In a local chart ω can be written as

$$\omega = \sum_{i,j=1}^{2N} \omega_{ij}(\xi)\, d\xi_i \wedge d\xi_j, \qquad (III.5)$$

ω nondegenerate means $\det \omega \neq 0$; ω closed means $\partial_i \omega_{jk} + \partial_j \omega_{ki} + \partial_k \omega_{ij} = 0$. Poisson brackets are then given by

$$\{f,g\} = \sum_{i,j} \omega_{ij}(\xi)\, \frac{\partial f}{\partial \xi_i}\, \frac{\partial g}{\partial \xi_j}. \qquad (III.6)$$

(c) The Hamiltonian H is a function on phase space which determines the dynamics of the system. To H there is associated the vector field

$$X_H: (q,p) \to (q,p;\; \frac{\partial H}{\partial q},\, \frac{\partial H}{\partial p})$$

and the equation of motion are given by the Poisson brackets

$$\dot{q} = \{q,H\}, \quad \dot{p} = \{p,H\}.$$

(d) Globally defined constants of motion play an essential role in the theory of integrable systems:

Definition: A globally defined function K on M^{2n} is called constant of motion if $\{H,K\} = 0$.

Let $K_1,\ldots,K_N$ be constants of motion; if $\{K_i,K_j\} = 0$ for all i,j K_i are said to be in involution; if dK_i are linear independent for all i, K_i are said to be linear independent.

Now we can state the most essential

Theorem (Liouville, Arnold) [10]: Given (M^{2n},ω,H) and n linear independent integrals of motion in involution. Then there exists a transformation to action angle variables.

Remarks:

α) The theorem says that if one defines the level set $M_K = \{\xi | K_i(\xi) = \kappa_i, \ i = 1,\ldots,N\}$, M_K is a smooth manifold and invariant under the flow of $H = K_1$.

β) If M_K is compact (and connected), M_K is diffeomorphic to the N-dimensional torus $T^N = \{(\varphi_1,\ldots,\varphi_N) \mod 2\pi\}$.

γ) If β applies, the motion will be periodic $\frac{d\varphi_i}{dt} = \omega_i$, $\varphi_i(t) = \omega_i t + \varphi_i(0)$.

δ) The equation of motion can be integrated by quadratures, that means there are constants $I_i(K_1,\ldots,K_N)$ such that $\frac{dI_i}{dt} = 0$, $\frac{d\varphi_i}{dt} = \omega_i$. $\{I_i, \varphi_i\}$ are the action angle variables. The canonical two form ω is written now as $\omega = \sum_{i=1}^{N} dI_i \wedge d\varphi_i$.

ε) If there are only r-directions in which the motion is restricted to a compact region, M_K will be diffeomorphic to $T^r \times \mathbb{R}^{N-r} = \{(\varphi_1,\ldots,\varphi_r, x_1,\ldots,x_{N-r}), \varphi_i \mod 2\pi, x_i \in \mathbb{R}\}$.

2. Nonlinear Schrödinger Equation

For infinite dimensional hamiltonian systems the general theory is not yet worked out in detail. One therefore looks for explicit transformations to action angle variables and calls such systems integrable. We choose one of the simplest one: The nonlinear Schrödinger equation for $\psi(t,x)$:

$$i\psi_t = -\psi_{xx} + 2c|\psi|^2\psi \qquad (III.7)$$

which has no bound states for positive coupling constant c and the boundary condition $\psi(t,x) \xrightarrow[|x| \to \infty]{} 0$ is assumed. This system can easily be written in hamiltonian form with

$$H = \int_{-\infty}^{\infty} dx \left[|\psi_x|^2 + c|\psi|^4\right] \qquad (III.8)$$

and the canonical two form ω can be written formally as

$$\omega = \text{Im} \int_{-\infty}^{\infty} dx \, \delta\psi^*(x) \wedge \delta\psi(x) \qquad (III.9)$$

such that the Poisson brackets are given by

$$\{f(\psi,\psi^*), g(\psi,\psi^*)\} = i \int dx \left(\frac{\delta f}{\delta\psi} \frac{\delta g}{\delta\psi^*} - \frac{\delta f}{\partial\psi^*} \frac{\delta g}{\delta\psi} \right). \qquad (III.10)$$

The program now concerns the introduction of new variables (scattering data), expressing H in terms of them (thereby one derives an infinite set of sum rules, see Part 3) and expressing ω in terms of the new variables (thereby one derives new Poisson brackets, see Part 4) (see contribution of Faddeev in [B2]).

Similar as for the KdV equation, there exists a Lax pair for our case. Starting from the Lax operator

$$L = \begin{pmatrix} 1 & 0 \\ 0 & -1 \end{pmatrix} \frac{1}{i} \frac{d}{dx} + \begin{pmatrix} 0 & \psi \\ \psi^* & 0 \end{pmatrix} \qquad (III.11)$$

one first does the direct step and goes over from $\psi(x)$ to scattering data $\{R(k) = b(k)/a(k), \varepsilon_\ell, c_\ell\}$. To do that one studies the eigenvalue equation

$$L\phi = k\phi \qquad (III.12)$$

and defines two fundamental solutions behaving asymptotically like plane waves, which one may combine to a two times two matrix $F(k,x)$ with

$$\lim_{x \to \infty} F(k,x) = E(k,x) = \begin{pmatrix} e^{ikx} & 0 \\ 0 & e^{-ikx} \end{pmatrix} . \qquad (III.13)$$

Similarly, let $G(k,x)$ be a two times two matrix combining solutions which behave for $x \to -\infty$ as plane waves

$$\lim_{x \to -\infty} G(k,x) = E(k,x) . \qquad (III.14)$$

The transition matrix T is then defined by

$$F(k,x) = T(k) \cdot G(k,x), \quad T = \begin{pmatrix} a & b \\ b* & a* \end{pmatrix}, \quad |a|^2 - |b|^2 = 1 . \qquad (III.15)$$

From the analyticity of $a(k)$ one deduces the representation

$$a(k) = e^{\frac{i}{\pi} \int \frac{dk' \ln|a(k')|}{k-k'+i\varepsilon}} \left(\prod_\ell \frac{k-\kappa_\ell^*}{k-\kappa_\ell} \right) \qquad (III.16)$$

where the factor in brackets has to be included in case of bound states. In addition, it is possible to show that

$$a(k) \xrightarrow[|k|\to\infty]{} 1 + O(\tfrac{1}{|k|}), \quad b(k) \xrightarrow[|k|\to\infty]{} O(\tfrac{1}{|k|}) . \qquad (III.17)$$

The formulation of the inverse problem runs now along the usual lines. One defines a matrix kernel G in terms of scattering data:

$$G(x) = \begin{pmatrix} 0 & \tilde{G}(x) \\ -\tilde{G}^*(x) & 0 \end{pmatrix}, \quad \tilde{G}(x) = \int_{-\infty}^{\infty} \frac{dk}{2\pi} e^{ik(x+y)} R(k) + \sum_\ell c_\ell e^{-i\kappa_\ell(x+y)}$$

and establishes that a Gelfand-Levitan-Marchenko type equation holds for the kernel $K(x,y)$

$$K(x,y) + G(x,y) + \int_x^\infty dz \, K(x,z) G(z,y) = 0 , \quad x < y . \qquad (III.19)$$

K is a two times two matrix kernel; the potential ψ is related to the one-two component

$$\psi(x) = -2i \, K_{12}(x,x) . \qquad (III.20)$$

3. Derivation of Sum Rules

Next one tries to express the hamiltonian in terms of the new variables; this actually gives sum rules which are connected to an infinite set of conservation laws which exist for these nonlinear equations (see Faddeev's contribution to [B2] or Ref. [B5]).

<u>Proposition</u>: H is a coefficient in the asymptotic expansion of ln a(k) for large k

$$\ln a(k) = \sum_{n=1}^{\infty} \frac{c_n}{k^n} \ . \qquad (III.21)$$

There are two ways of calculating coefficients c_n; setting them equal gives sum rules.

Firstly, use the dispersion relation (III.16). This immediately gives

$$c_n = \frac{i}{n} \int_{-\infty}^{\infty} dk \ k^{n-1} \ \ln|a(k)| - \left(\sum_{=1}^{N} \frac{1}{n} (k^n - k^{*n}) \right) \qquad (III.22)$$

where again the terms in bracket have to be included if bound states exist.

Secondly, express $a(k)$ through $\psi(x)$ and $\psi^*(x)$. Since $f_{11} \sim e^{ikx}(1+o(1))$ for $x \to \infty$ and $f_{11} \sim a(k) e^{ikx}(1+o(1))$ for $x \to \infty$ one may write

$$\ln a(k) = -\int_{-\infty}^{\infty} dx \ \frac{d}{dx} \varphi(x) \ , \quad \varphi(x) = \ln(f_{11} e^{-ikx}) \ . \qquad (III.23)$$

Therefore, the first component ϕ_1 of ϕ is $\exp\varphi(x) = \phi_1$; eliminating the second component ϕ_2 of ϕ from the system (III.12) gives a Riccati equation for φ

$$\psi(\frac{\varphi'}{\psi})' = -|\psi|^2 - 2ik\varphi' - \varphi'^2 \ . \qquad (III.24)$$

Next one makes the Ansatz

$$\varphi'(k,x) = \sum_{n=1}^{\infty} \frac{\chi_n(x)}{k^n} \qquad (III.25)$$

and identifies the constants in (III.21) with the integrals over $-\chi_n$. But χ_n's are determined recursively by (III.24). The first few c_n's

$$c_1 = -\frac{i}{2} \int_{-\infty}^{\infty} dx\ \psi^*(x)\psi(x)\ ,\quad c_2 = \frac{1}{8} \int_{-\infty}^{\infty} dx [\psi_x^*(x)\psi(x) - \psi^*(x)\psi_x(x)]$$

(III.26)

$$c_3 = -\frac{i}{8} \int_{-\infty}^{\infty} dx\ [|\psi_x(x)|^2 + |\psi(x)|^4] \qquad (c = 1)$$

represent particle number, momentum and energy. Comparing to (III.22) yields sum rules (for a more detailed treatment see the article of D. Bollé in this volume). H is proportional to c_3; the appropriate sum rule (for coupling constant $c = -1$) is

$$H = \int_{-\infty}^{\infty} dx [|\psi_x|^2 - |\psi|^4] = -\frac{8i}{\pi} \int_{-\infty}^{\infty} dk\ k^2 \ln |a(k)| - \frac{8i}{3} \sum_{\ell}^{N} (k_\ell^3 - k_\ell^{*3})$$

(III.27)

where the bound state contributions are included. The contributions on the r.h.s. can obviously be interpreted as energies of quasiparticles.

Remarks:

α) The derivation of sum rules for the KdV equation shows the connection to the heat kernel approach and to probabilistic considerations. Starting from the two fundamental solutions of the Schrödinger equation f_1, f_2 (I.8) one constructs the heat kernel $P(x,y;t)$ which solves

$$\frac{\partial P}{\partial t} = \frac{\partial^2 P}{\partial x^2} - V(x)P\ ,\quad P(x,y;0) = \delta(x-y) \qquad \text{(III.28)}$$

explicitly as

$$P(x,y;t) = \int_{-\infty}^{\infty} \frac{dk}{2\pi}\ e^{-k^2 t}\ T(k) f_1(k,x) f_2(k,x)\ . \qquad \text{(III.29)}$$

On the other hand, one writes down a path integral representation for the heat kernel

$$P(x,y;t) = \frac{e^{-(x-y)^2/4t}}{\sqrt{4\pi t}}\ E\left(e^{-\int_0^t d\tau V(x + X(\tau))} \Big|_{x + X(t) = Y} \right)$$

(III.30)

where $X(\tau)$ is Brownian motion and $E(\cdot)$ stands for the expectation value. Take a potential with compact support within $(-a,a)$ and let $y > a$, $x < -a$ and assume for simplicity that there are no bound states. Expanding expression (III.29) and (III.30) for small t gives finally

$$\frac{1}{\pi} \int_{-\infty}^{\infty} dk \ln(1 - |R(k)|^2) = \int_{-a}^{a} dx\, V(x) . \qquad (III.31)$$

β) The general class of sum rules including bound state contributions

$$\sum_{\ell} |\varepsilon_{\ell}|^{(j+1/2)} + (-)^j (2j+1) \int_{-\infty}^{\infty} \frac{dk}{4\pi} k^{2j} \ln(1-|R|^2) = \int_{-a}^{a} dx P_j(V) \qquad (III.32)$$

where $P_j(V)$ is a polynomial in V and derivatives of V, has been used by a number of authors [4 to 7] in order to obtain bounds on the number of bound states and moments of energy levels.

4. Derivation of Poisson Brackets

The most difficult final step consists in expressing ω in terms of scattering data (see contribution of Faddeev to [B2]). This allows to identify certain functions of them as cyclic action angle variables (if their range is restricted) or otherwise as momentum and coordinates of quasiparticles. In the following we give a rough sketch: From (III.9) we see that expressing ω in terms of scattering data means expressing $\delta\psi(x)$ in terms of them. Since ψ is given by $K_{12}(x,x)$ one has to evaluate $\delta K(x,x)$. But K fulfills an equation like

$$K + G + K * G = 0 \qquad (III.33)$$

where we have used a shorthand notation for the kernels; therefore from (III.29) we get

$$\delta K * (1+G) = -(1+K) * \delta G . \qquad (III.34)$$

But δG can be expressed in terms of δR

$$\delta G(x,y) = \int_{-\infty}^{\infty} \frac{dk}{2\pi} \, \delta R(k) \, e^{ik(x+y)} \tag{III.35}$$

where we have treated only the simplest case without bound states. These remarks should make clear that it is possible to express $\delta\psi$ in terms of δR. Explicitly one gets (see contribution of Faddeev in [B2])

$$\delta\psi(x) = \frac{i}{\pi} \int_{-\infty}^{\infty} dk \left(\delta R(k) f_{11}^2(k,x) - \delta R^*(k) f_{12}^2(k,x) \right) . \tag{III.36}$$

Introducing (III.32) into ω shows that one has to evaluate next the integrals over

$$I(k,\ell,x) = f_{11}^2(k,x) f_{12}^2(\ell,x) - f_{12}^2(k,x) f_{11}^2(\ell,x) . \tag{III.37}$$

It turns out that I is a total derivative in x, as one can show by using Wronski identities for four wave functions

$$I(k,\ell,x) = \frac{i}{2(k-\ell)} \frac{d}{dx} \left(f_{11}(k,x) f_{12}(\ell,x) - f_{12}(k,x) f_{11}(\ell,x) \right)^2 . \tag{III.38}$$

This allows to simplify ω to

$$\omega = -\frac{i}{\pi} \int_{-\infty}^{\infty} dk \, \delta R(k) \wedge \delta R^*(k) \, |a(k)|^2 + \frac{2}{\pi} \oint \frac{dk d\ell}{k-\ell} \, \delta \ln|a(k)| \wedge \delta \ln|a(b)| \tag{III.39}$$

which is written already in a very simple way but not yet diagonal. Introducing new variables

$$P(k) = -\frac{2}{\pi} \ln |a(k)| , \quad Q(k) = \arg b(k) \tag{III.40}$$

gives ω in diagonal form

$$\omega = \int_{-\infty}^{\infty} dk \, \delta P(k) \wedge \delta Q(k) \tag{III.41}$$

and shows the main reason why one calls that system completely integrable.

IV. QUANTUM INVERSE SCATTERING TRANSFORM METHOD

1. r-Matrix Approach

There is a different and simpler way to derive the Poisson brackets of Part III, which is more algebraically, allows for a generalization to quantum field theory and has given insight into the structure behind systems for which Bethe states were known. Finally, even exact S-matrices have been derived for certain two dimensional field theories. Here we shall be less explicit and only sketch the main results.

In order to derive the Poisson brackets for scattering data algebraically, one writes the Lax operator as (we follow [B9])

$$\frac{\partial}{\partial x} \phi(x,\lambda) = \ell(x,\lambda)\phi(x,\lambda), \quad \ell(x,\lambda) = i\begin{pmatrix} -\lambda & \sqrt{c}\psi(x) \\ -\sqrt{c}\bar\psi(x) & \lambda \end{pmatrix} \quad (IV.1)$$

which determines the x-dependence of ϕ. (Note the change from ϕ to $\sqrt{c}\psi$). The transition or monodromy matrix $t(x_1,x_2;\lambda)$ is defined by

$$\frac{\partial}{\partial x_2} t(x_1,x_2;\lambda) = \ell(x_2,\lambda)t(x_1,x_2;\lambda) \; ; \quad t(x_1,x_1;\lambda) = 1 \; . \quad (IV.2)$$

For the infinite interval the transition matrix is determined by taking limits

$$t(\lambda) = \lim_{\substack{x_1 \to -\infty \\ x_2 \to +\infty}} e^{i\lambda\sigma_3 x_1} t(x_1,x_2;\lambda) e^{-i\lambda\sigma_3 x_2} \quad (IV.3)$$

where σ_3 denotes the third Pauli matrix. If one writes $t(\lambda)$ as

$$t(\lambda) = \begin{pmatrix} a(\lambda) & b(\lambda) \\ b^*(\lambda) & a^*(\lambda) \end{pmatrix} \quad (IV.4)$$

one has the

Proposition: The Poisson brackets of a, a^*, b and b^* are given by

$$\{a(\lambda),a(\mu)\} = \{a(\lambda),a^*(\mu)\} = \{b(\lambda),b(\mu)\} = 0$$
$$\{b(\lambda),b^*(\mu)\} = 2\pi i \, |a(\lambda)|^2 \, \delta(\lambda-\mu)$$
$$\{a(\lambda),b(\mu)\} = \frac{c}{\lambda-\mu+i\varepsilon} a(\lambda)b(\mu)$$
$$\{a(\lambda),b^*(\mu)\} = \frac{-c}{\lambda-\mu+i\varepsilon} a(\lambda)b^*(\mu) \,. \tag{IV.5}$$

Therefore, $a(\lambda)$ forms a family of commuting operators, but the brackets for a with b and b with b^* are nontrivial

Proof: The simplest trick to derive these relations has been found by the Russian group following ideas of Baxter. One calculates first the brackets for tensor product quantities of ℓ

$$\{\ell(x,\lambda)\otimes\ell(y,\mu)\} = \left[r(\lambda-\mu),\left(\ell(x,\lambda)\otimes\mathbb{1}+\mathbb{1}\otimes\ell(y,\mu)\right)\right]\delta(x-y)$$

$$r(\lambda) = -\frac{c}{\lambda}\begin{pmatrix} 1 & 0 & 0 & 0 \\ 0 & 0 & 1 & 0 \\ 0 & 1 & 0 & 0 \\ 0 & 0 & 0 & 1 \end{pmatrix} \tag{IV.6}$$

which is the easy part. Next one tries to find relations for tensor products of t's, which is the hard part. One finds

$$\{t(x_1,x_2;\lambda)\otimes t(x_1,x_2;\mu)\} = [r(\lambda-\mu),t(x_1,x_2;\lambda)\otimes t(x_1,x_2;\mu)]. \tag{IV.7}$$

Carefully taking limits gives

$$\{t(\lambda)\otimes t(\mu)\} = r_+(\lambda-\mu)t(\lambda)\otimes t(\mu) - t(\lambda)\otimes t(\mu)\, r_-(\lambda-\mu)$$

$$r_\pm(\lambda) = -c\begin{pmatrix} P(\frac{1}{\lambda}) & 0 & 0 & 0 \\ 0 & 0 & \pm i\pi\delta(\lambda) & 0 \\ 0 & \pm i\pi\delta(\lambda) & 0 & 0 \\ 0 & 0 & 0 & P(\frac{1}{\lambda}) \end{pmatrix} \tag{IV.8}$$

where on the l.h.s. the Poisson brackets of the elements of the tensor product is meant.

Remark: It is interesting to note that the same procedure works for the Sine-Gordon equation as well as for the Heisenberg Ferromagnet; the KdV equation has a more complicated structure.

2. Bethe States

The quantization of the nonlinear Schrödinger equation is easy; the Hamilton operator of the model is given by

$$H = \int_{-\infty}^{\infty} dx \; [\psi_x^+ \psi_x + c \, \psi^+ \psi^+ \psi \psi] \tag{IV.9}$$

where ψ is a boson field in one space dimension:

$$[\psi(x), \psi^+(y)] = \delta(x-y), \quad [\psi(x), \psi(y)] = 0 . \tag{IV.10}$$

One can consistently work (for this simple model) in the Fock space of the free field with vacuum state $|0>$, which is annihilated by ψ: $\psi(x)|0> = 0$; $\psi^+(x)$ creates particles. Note that H commutes with the particle operator $N = \int dx \, \psi^+(x)\psi(x)$, which implies that one does not have particle production. Going into the subspace of fixed particle number one gets exact eigenstates of H of the Bethe type:

$$|k_1,\ldots,k_N> = \int \prod_{i=1}^{N} \left(dx_i \; e^{ik_i x_i} \right) \prod_{1 \leq \ell < m \leq N}$$

$$\left(1 - \frac{ic}{k_m - k_\ell} \, \epsilon \, (x_m - x_\ell) \right) \psi^+(x_1),\ldots,\psi^+(x_N)|0> . \tag{IV.11}$$

We quote without proof that

$$H|k_1,\ldots,k_N> = \sum_{i=1}^{N} k_i^2 \, |k_1,\ldots,k_N> . \tag{IV.12}$$

This states have been first written down by Lieb and Liniger [12]; they are actually not normalized to δ-functions. Such normalized states exist too; they are the in-states for given momenta $q_1 > \ldots > q_N$

$$|q_1,\ldots,q_N,\text{in}\rangle = \int \prod_{i=1}^{N} (dx_i e^{iq_i x_i}) \prod_{1\leq \ell < m \leq N} \left(1 + \Theta(x_m - x_\ell)(S_{q_m q_\ell} - 1)\right) \times$$

$$\times \psi^+(x_1),\ldots,\psi^+(x_N)|0\rangle \qquad (IV.13)$$

where $S_{q_m q_\ell}$ denotes the two-to-two particle S-matrix

$$S_{q_m q_\ell} = \frac{q_m - q_\ell - ic}{q_m - q_\ell + ic} \ . \qquad (IV.14)$$

Remarks:

α) This "product" wave function implies factorization of S-matrix. Using such a property and no particle production assumption, it has been possible to write down exact S-matrices for certain models.

β) Using the identity

$$q_m - q_\ell - ic(2\Theta(x_m - x_\ell) - 1) = q_m - q_\ell + ic - 2ic\,\Theta(x_m - x_\ell)$$

one finds the connection between the states of Equation (IV.11) and in-states.

γ) (IV.9) and (IV.10) is actually identical to the N-body point interaction

$$\left[-\sum_{i=1}^{N} \frac{\partial^2}{\partial x_i^2} + c \sum_{i<j} \delta(x_j - x_i)\right] \chi(x,k) = \sum_{i=1}^{N} k_i^2 \chi(x,k). \qquad (IV.15)$$

δ) For $c < 0$, there exist bound states for any N of the form

$$\chi(x,k) \propto e^{\frac{c}{2} \sum_{1\leq \ell < m \leq N} |x_\ell - x_m|} \qquad (IV.16)$$

with center of mass momentum separated out.

3. Quantize Direct Step:

It has been observed by three groups at the same time [13,14,15, B8,B9] that a quantization of the direct step of the ISTM is possible and clarifies the existence of Bethe states. Here we sketch the R-matrix approach to calculate the commutation relations between the operators representing the scattering data. One starts again from the Lax operator $\frac{\partial}{\partial x}\phi = :L\phi:$, but normal orders ψ and ψ^+'s :

$$\frac{1}{i}\frac{\partial}{\partial x}\phi_1 = \lambda\phi_1 + \sqrt{c}\phi_2\psi$$

$$\frac{1}{i}\frac{\partial}{\partial x}\phi_2 = -\lambda\phi_2 - \sqrt{c}\psi^+\phi_1 \quad . \tag{IV.17}$$

If one goes over to an integral equation and expands the expressions for $\phi_{1,2}$ in terms of ψ and ψ^+, they are both normal ordered in the sense that ψ^+'s are to the left of all ψ's.

Next one defines the quantum transition operator by taking $t(x_1,x_2;\lambda)$ from Equation (IV.2) as the Wick symbol for the operator T

$$T(x_1,x_2;\lambda) = :t(x_1,x_2;\lambda): \quad , \quad T(x_1,x_2;\lambda) = 1 \quad . \tag{IV.18}$$

T therefore solves the operator equation

$$\frac{\partial}{\partial x_2} T(x_1,x_2;\lambda) = :L(x_2,\lambda)T(x_1,x_2;\lambda): \quad . \tag{IV.19}$$

The scattering data become operators, too; they are defined as weak limits

$$T(\lambda) = \lim_{\substack{x_1 \to \infty \\ x_2 \to \infty}} e^{i\lambda\sigma_3 x_1} T(x_1,x_2;\lambda) e^{-i\lambda\sigma_3 x_2} = \begin{pmatrix} A(\lambda) & B(\lambda) \\ B^+(\lambda) & A^+(\lambda) \end{pmatrix}. \tag{IV.20}$$

Although the product of two weakly convergent operators need not be convergent, it seems to be that formal operations are justified here and give the following commutation relations for A's and B's :

Proposition: The algebra of A's and B's is given by

$$[A(\lambda),A(\mu)] = [A(\lambda),A^+(\mu)] = [B(\lambda),B^+(\mu)] = 0$$
$$[B(\lambda),B^+(\mu)] = \delta(\lambda-\mu)A(\lambda)A^+(\mu)+B^+(\mu)B(\lambda)\frac{c^2}{(\lambda-\mu+i\epsilon)^2}$$
$$[A(\lambda),B(\mu)] = \frac{-ic}{\lambda-\mu+i\epsilon} A(\lambda)B(\mu)$$
$$[A(\lambda),B^+(\mu)] = \frac{ic}{\lambda-\mu+i\epsilon} B^+(\mu)A(\alpha) \quad . \tag{IV.21}$$

The family $A(\lambda)$ forms again commuting operators. The commutator relations between A's and B's and between the B's are nontrivial. Note the difference between Poisson brackets and commutation relations.

Proof: One may follow the classical case, but has to respect the non-commutativity. For infinitesimal generators there exists a matrix $\bar{R}(\lambda)$ such that

$$\bar{R}(\lambda-\mu)\left[L(x_2,\lambda)\otimes \mathbb{1} + \mathbb{1}\otimes L(x_2,\mu) + c(\sigma_+\otimes \sigma_-)\right] =$$
$$= \left[L(x_2,\lambda)\otimes \mathbb{1} + \mathbb{1}\otimes L(x_2,\mu) + c(\sigma_-\otimes \sigma_+)\right]\bar{R}(\lambda-\mu)$$

$$\bar{R}(\lambda) = 1 - i\frac{c}{\lambda}\begin{pmatrix} 1 & 0 & 0 & 0 \\ 0 & 0 & 1 & 0 \\ 0 & 1 & 0 & 0 \\ 0 & 0 & 0 & 1 \end{pmatrix} \tag{IV.22}$$

In the classical limit $\bar{R}(\lambda) = 1 - i\hbar r(\lambda) + O(\hbar^2)$. By "exponentiation" one gets [B8]

$$\bar{R}(\lambda-\mu)T(x_1,x_2;\lambda)\otimes T(x_1,x_2;\mu) = (\mathbb{1}\otimes T(x_1,x_2;\mu))\cdot(T(x_1,x_2;\lambda)\otimes \mathbb{1})\bar{R}(\lambda-\mu) \quad . \tag{IV.23}$$

This relation contains the algebra of (IV.21).

Theorem: The operator B^+ generates the **unnormalized** eigenstates

$$|k_1,\ldots,k_N\rangle = B^+(k_1),\ldots,B^+(k_N)|0\rangle \quad . \tag{IV.24}$$

Remark: This has been shown by <u>tedious</u> calculations explicitly. It is the essential step. One furthermore realizes the

Proposition: ln $A(\lambda)$ generates a commuting family of operators, among them the particle number operator N, the momentum operator P and the Hamiltonian H.

Proof: From the algebra (IV.21) one gets

$$A(\lambda)|k_1,\ldots,k_N\rangle = \prod_{j=1}^{N}\left(1 + \frac{ic}{\lambda-k_j+i\varepsilon}\right)|k_1,\ldots,k_N\rangle . \tag{IV.25}$$

The appropriate equation for ln $A(\lambda)$ can be expanded and shows this assertion.

One gets in addition

$$[H,A(\lambda)] = 0 , \quad [H,B^+(\lambda)] = \lambda^2 B^+(\lambda) . \tag{IV.26}$$

The interpretation of these relations is simple but interesting: $B^+(\lambda)$ creates particles with energy λ^2 and momentum λ.

Remarks:

α) One should emphasize that only for this simple model it is possible to work fully in the Fock space representation of the free field.

β) Since on the one hand

$$|k_1,\ldots,k_N\rangle = \prod_{1\leq\ell<m\leq N}\left(1 + \frac{ic}{k_m-k_\ell+i\varepsilon}\right)|k_1,\ldots,k_N,\text{in}\rangle \tag{IV.27}$$

holds for $k_1<\ldots<k_N$, but, on the other hand, the factors in front of the in-states can be obtained by applying A-operators to that state, one concludes that the operator analog of the reflection coefficient

$$R^+(\lambda) = B^+(\lambda) A^{-1}(\lambda) \tag{IV.28}$$

<u>creates</u> ordered in-states.

γ) It is remarkable that the algebra of R, R^+ is extremely simple

$$R^+(\lambda)R^+(\mu) = S_{\lambda\mu} R^+(\mu)R^+(\lambda)$$
$$R(\lambda)R^+(\mu) = \delta(\lambda-\mu) + S_{\lambda\mu} R^+(\mu)R(\lambda) \quad . \tag{IV.29}$$

It is clear that R^+ is not yet an in-field creation operator, but if one defines

$$\varphi_{in}^+(\lambda) = R^+(\lambda) e^{\int d\mu \Theta(\lambda-\mu) \ln S_{\lambda\mu} R^+(\mu)R(\mu)} \tag{IV.30}$$

this operator indeed fulfills

$$[\varphi_{in}(\lambda), \varphi_{in}^+(\mu)] = \delta(\lambda-\mu) \quad . \tag{IV.31}$$

These algebraic relations have been worked out by the author [16] and independently by Faddeev and Sklyanin [13].

δ) Altogether the chain of mappings

$$\psi^+(x) \rightarrow R^+(\lambda) \rightarrow \varphi in^+ = \Omega^+ \psi^+ \Omega \tag{IV.32}$$

give the Möller operator explicitly. It is clear that the inverse mapping is even of greater importance. It gives the interacting field operator in terms of in-field operators and allows to construct Green functions explicitly. Since the mapping (IV.30) can easily be inverted, the question is how to obtain the inverse of the first step of (IV.32) in the form of an expansion

$$\psi^+(\lambda) = R^+(\lambda) + (\text{terms with } R^+R^+R) + \ldots \quad . \tag{IV.33}$$

This can be obtained by quantizing the inverse step, as it has been first done by the author [16]; for further work in that direction, see [17].

4. Quantize Inverse Step:

The classical inverse problem has been described extensively. One has to solve GLM equations, which are in our case two times two matrix equations. It is interesting to note that a procedure similar to the quantization of the direct step works also here. One takes the classical equations, normal orders them in the sense that R^+ stands to the left of all R's and interprets all quantities as operators; this gives

$$-iK_2(x,y) = \sqrt{c} \int_x^\infty dz\, K_1(z,y)\, R(z+x)$$
$$-iK_1(x,y) = \sqrt{c}\, R^+(x+y) + \sqrt{c} \int_x^\infty dz\, R^+(x+z) K_2(z,y) \quad . \tag{IV.34}$$

The "potential" is equal to the field operator and given by

$$\psi(x) = -\frac{i}{\sqrt{c}}\, K_1(x,x) \quad . \tag{IV.35}$$

Iterating (IV.34) gives "normal ordered" expressions for ψ_1 and ψ_2. That (IV.34) is a correct operator equation has been shown by applying both sides to eigenstates of H [16]; an approach using dispersion relations was afterwards given in [17, B10].

The iteration of (IV.34) gives the following expansion for ψ^+

$$\psi^+(x) = R^+(x) + \sum_{n=1}^\infty \left(\frac{c}{4}\right)^n \int \prod_{j=1}^{2n} (ds_j \Theta(s_j - x)) R^+(x+s_1) R^+(s_1+s_3) \cdots$$
$$\cdots R^+(s_{2n}+x) R(s_{2n}+s_{2n-1}) \cdots R(s_1+s_2) = F(R^+, R) \tag{IV.36}$$

which gives ψ^+ expressed in terms of R^+ and R operators. The time evolution of ψ^+ is therefore trivially obtained by

$$\psi^+(t,x) = F\left(e^{iHt} R^+ e^{-iHt},\, e^{iHt} R\, e^{-iHt}\right) \tag{IV.37}$$

but R^+ and R are simple expression of in-field operators (IV.30) and their time evolution is the free one. Altogether one gets an operator analog of the inverse Möller mapping and Green's functions can easily be constructed.

<u>Remark</u>:

For relativistic field theories one has to do an essential further step; one has to change the representation, which means one has to go over to a new vacuum state. A similar problem occurs if one treats the finite density case of the NLS-equation [12]. One takes periodic boundary conditions and puts the system into a box of length L. The wave function

$$\chi(x;k) = <0|\psi(x_1) \ldots \psi(x_N)|k_1,\ldots,k_N> \qquad (IV.38)$$

has to fulfill therefore

$$\chi(x_1 = -\tfrac{L}{2},\ldots;k) = \chi(x_1 = \tfrac{L}{2},\ldots;k) . \qquad (IV.39)$$

This restricts the k_i's to solutions of

$$e^{ik_i L} = \prod_{j \neq i} S_{k_i k_j} = \prod_{j \neq i} e^{i\delta(k_j,k_i)} . \qquad (IV.40)$$

This set of equations can also be obtained from the algebra (IV.21). A more useful form of (IV.40) is obtained by taking the logarithm

$$k_i L = \sum_{j \neq i} \delta(k_j,k_i) + 2\pi n_j , \quad n_j \in \mathbb{N} . \qquad (IV.41)$$

The choice of the branch of the logarithm (or the choice of the n_j) determines the properties of the chosen state. To obtain the ground state one has to choose $n_{i+1} = n_i + 1$. Subtracting Eqs. (IV.41) for adjacent k_i's gives

$$k_{i+1} - k_i = \tfrac{1}{L} \sum_j \left[\delta(k_j,k_{i+1}) - \delta(k_j,k_i)\right] + \tfrac{2\pi}{L} . \qquad (IV.42)$$

For $L \to \infty$ the k_i's become infinitesimally spaced. If one defines a density $\rho(k)$ by

$$\lim_{L \to \infty} \frac{1}{L(k_{i+1} - k_i)} = \rho(k) \qquad (IV.43)$$

and the Fermi momentum by demanding that

$$\int_{-K_F}^{K_F} dk\, \rho(k) = n\,, \quad \int_{-K_F}^{K_F} dk\, k^2 \rho(k) = \varepsilon \qquad (IV.44)$$

where n is the density and ε the energy one finds that $\rho(k)$ fulfills an integral equation

$$1 = 2\pi\rho(k) + \int_{-K_F}^{K_F} dk'\, K(k-k')\rho(k'),\ K(k) = \frac{d\rho(k)}{dk} = \frac{2c}{k^2+c^2} \qquad (IV.45)$$

where the kernel K is expressed in terms of the time delay. Since (IV.45) is solvable, the finite density problem for the NLS equation can be solved exactly [12].

REFERENCES

[1] Kac, M., Amer. Math. Monthly 73 (1966) 1

[1] Kac, M. and Van Moerbeke, P., Proc. Nat. Acad. Sci. 71 (1974) 2350

[3] Gardner, C.S., Greene, J.M., Kruskal, M.D. and Miura, R.M., Commun. Pure and Appl. Math. 27 (1974) 97

[4] Glaser, V., Grosse, H. and Martin, A., Commun. Math. Phys. 59 (1978) 197

[5] Grosse, H., Acta Phys. Austr. 52 (1980) 89

[6] Lieb, E.H. and Thirring, W., in "Studies in Mathematical Physics" (Lieb, E., Simon, B. and Wightman, A., eds.), Princeton University Press 1976, p. 269

[7] Glaser, V. Grosse, H., Martin, A. and Thirring, W., in "Studies in Mathematical Physics" (like Ref. [6]), p. 169

[8] Ablowitz, M., Kaup, D.J., Newell, A.C. and Segur, H., Phys. Rev. Lett. 31 (1973) 125

[9] Grosse, H. Solitons of the Modified Korteweg-de Vries equation, Lett. Math. Phys. 8 (1984) 313

[10] Arnold, V.I., Mathematical Methods of Classical Mechanics (Springer 1978)

[11] Thirring, W., A Course in Mathematical Physics, Vol. 1: Classical Dynamical Systems (Springer 1978)

[12] Lieb, E.H. and Liniger, W., Phys. Rev. 130 (1963) 1605

[13] Faddeev, L.D. and Sklyanin, E.K., Doklady A N SSSR, 243 (1978) 1430

[14] Honerkamp, J., Weber, P. and Wiesler, A., Nucl. Phys. B152 (1979) 266

[15] Thacker, H.B. and Wilkinson, D., Phys. Rev. D19 (1978) 3660

[16] Grosse, H., Phys. Lett. 86B (1978) 267

[17] Cremer, D.B., Thacker, H.B. and Wilkinson, D., Phys. Rev. D19 (1979) 3660

TUNNELING IN ONE DIMENSION: GENERAL THEORY,
INSTABILITIES, RULES OF CALCULATION, APPLICATIONS

G. JONA-LASINIO
Dipartimento di Fisica, Università "La Sapienza"
Piazzale A. Moro 2, I-00185 Roma / Italy

G. MARTINELLI
Dipartimento di Matematica, Università "La Sapienza"
Piazzale A. Moro 2, I-00185 Roma / Italy

E. SCOPPOLA
Dipartimento di Fisica, Università "La Sapienza"
Piazzale A. Moro 2, I-00185 Roma / Italy

CONTENTS:

0. Introduction

1. Outline of the Approach: Deterministic and Probabilistic Methods

2. Some General Properties of the Logarithmic Derivative of the Ground State Wave Function

3. Double Wells: Relationship between Instability of Tunneling and Symmetry Breaking

4. Multiwell Potentials: Some "Laws" of Multiple Tunneling

5. Hierarchical Models: An Example of a Multiscale Analysis of Tunneling

6. Effective Potentials that Preserve the Qualitative Structure of Tunneling

7. Applications and Open Problems

0. Introduction

Tunneling, i.e. penetration of a particle in a classically forbidden region, is one of the most striking manifestations of Quantum Mechanics. Tunneling problems, however, become extremely difficult to solve as soon as one goes beyond a few very simple situations. For example, if we consider an arbitrary multiwell potential like in Fig. 0 exhibiting several equal minima (i.e. the classical state of lowest energy is degenerate) one is practically unable to tell where the ground state wave function will be localized in the limit $\hbar \to 0$, or to compute in the same limit the splittings of the ground state due to tunneling.

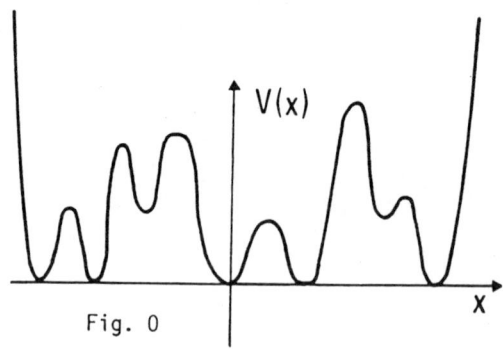

Fig. 0

The degeneration rate of these levels is expected to be exponentially small, $E_i - E_o \sim e^{-c_i/\hbar}$, but we do not have simple criteria to determine how many levels of this kind exist or to compute the constants c_i. Even assuming that we have been able to answer the above questions for a given potential, how does the answer vary upon small deformations of the profile of the potential?

Until a few years ago the only cases where tunneling was reasonably understood were the symmetric double well potential and the periodic potential. An important step towards the analysis of more complicated situations was made by the authors of the present report by giving a rather complete discussion [1] of the effects of small localized perturbations of the two cases just mentioned. The results were striking and unexpected and required the development of a new approach quite different from the usual one based on the WKB approximation to the Schrödinger equation. Our approach is based on a separation of the problem in two steps and is very close in spirit to stochastic mechanics. The first step, which is the most difficult, consists in a study of the logarithmic derivative of the ground state wave function. This quantity contains the essential information on tunneling and constitutes the input of the second step where the splittings of the tunneling levels and the corresponding wave functions are computed by probabilistic methods. Our results referred to one dimensional cases and have been recently rederived and extended to arbitrary dimension by Helffer and Sjöstrand [2] and by Simon [3] using functional analytic methods. Functional analytic methods were also used in [4] to rederive our results within conventional perturbation theory.

All the above treatments are limited to tunnelings among a finite number of wells. It is important to try to remove such a limitation. From the standpoint of physics this is particularly relevant because a detailed understanding of tunneling over large distances is a key to the analysis of disordered systems, e.g. the Anderson model [5,6], stochastic wave guides [7], etc., in the large disorder regime. Other examples where such an extension is required may be found in molecular physics. It is clear, however, that even for the more powerful techniques mentioned above the problem becomes intractable.

As a first step then it is interesting to examine intermediate situations, that is situations in which the complexity of the potential does not increase too much with the size of the region over which tunneling may take place. These examples, however, must be sufficiently "typical" to give a clue as to what happens in physically significant cases. A good example of what we mean by this are the hierarchical models [8] introduced by the authors and which have provided a basic idea in the proof of Anderson localization in dimension greater than one [6] (see later). The aim of this article is to provide an introduction to the basic techniques developed in [1] and to illustrate them by discussing a series of one-dimensional examples of increasing complexity that will culminate in the analysis of a one-dimensional hierarchical model. The examples treated go far beyond what can be found in the existing literature on tunneling and a number of general rules that emerge from our treatment constitute in our opinion veritable "laws" of tunneling phenomena.

We now describe in some detail the content of the following sections.

The first Section contains a general introduction to the method and describes the two steps mentioned earlier. The second Section is devoted to the description of a set of rules obeyed by the logarithmic derivative of the ground state wave function which are valid in any situation and were established in our previous work.

The third Section discusses double well problems and covers also material not included in our previously published work. The fourth Section may be considered the hard core of the paper. We discuss tunneling through sequences of two types of barriers considering situations of increasing complexity. In the analysis of our examples we develop inductive methods which allow to derive the behaviour in a long sequence from the analysis of short subsequences. We are not able to cover _every_ possible case but our results are sufficient to treat physically interesting problems. The material of this section is entirely new.

The fifth Section deals with a hierarchical model where tunneling can be analysed over all possible scales.

The final Section discusses in greater detail the relevance of this type of results in some physical application.

1. Outline of the Approach

The basic strategy of our approach is the following. Suppose that the potential in our Hamiltonian $H = -\frac{\hbar^2}{2}\Delta + V$ acting on $L^2(R^n)$

(we set the mass of the particle equal to 1) is such that the corresponding ground state has no nodes or nodal surfaces. In other words, $\psi_o \neq 0$ everywhere. Define the transformation

$$H \to U H U^{-1} = \psi_o^{-1} H \psi_o = -\hbar L + E_o \qquad (1.1)$$

where

$$L = \frac{\hbar}{2} \Delta + \underline{b} \, \underline{\nabla} \qquad (1.2)$$

acts on $L_2(R^n, \psi_o^2 \, dx)$ and

$$\underline{b} = \frac{\hbar}{2} \underline{\nabla} \ln \psi_o^2 \, . \qquad (1.3)$$

From (1.1) it follows, calling λ_k the eigenvalues of $-L$ and f_k the corresponding eigenfunctions, that

$$\begin{aligned} E_k - E_o &= \hbar \lambda_k \\ f_k &= \psi_k / \psi_o \end{aligned} \qquad (1.4)$$

The reader will recognize in L the generator of a diffusion process with diffusion constant $\sqrt{\hbar}$ and drift $\underline{b}(x)$. An equivalent way of describing the same diffusion process is by means of the stochastic differential equation

$$d\underline{x} = \underline{b}(x) dt + \sqrt{\hbar} \, d\underline{w}(t) \qquad (1.5)$$

where $\underline{w}(t)$ is the Wiener process.

From the Schrödinger equation and the definition of $\underline{b}$ it follows that $\underline{b}$ obeys

$$\hbar \, \text{div} \, \underline{b} + \underline{b}^2 = 2(V - E_o) \qquad (1.6)$$

with the additional condition that it must be a gradient, i.e. $\underline{b} = -\underline{\nabla} U$.

if we take naively the limit $\hbar \to 0$ in (1.6) we obtain the Hamilton-Jacobi equation

$$(\nabla U)^2 = 2(V(x) - E_0(\hbar = 0)) =$$
$$= 2(V(x) - \min_x V(x)). \tag{1.7}$$

The two steps of our approach mentioned in the Introduction are therefore

(1) Solution of Eq. (1.6) for $\hbar \to 0$

(2) Calculation of the spectrum of L in the same limit.

We can easily get rid of (2) which can be reduced to the application of a set of rules. The general theory behind these rules, which is probabilistic in nature, was developed by Ventzel and Freidlin [9] and has been adapted by the present authors to tunneling problems in Quantum Mechanics.

We now state these rules for the one-dimensional case. Let us denote $b^0(x) = \lim_{\hbar \to 0} b(x,\hbar)$ where $b(x,\hbar)$ is the solution of Eq. (1.6). Let x_i, $i = 1,2,\ldots,N$ be the absolute minima of the potential V. Let

$$V_{i,i+1} = \int_{x_i}^{x_{i+1}} |\min(b^0(x),0)|dx$$
$$V_{i+1,i} = \int_{x}^{x_{i+1}} \max(b^0(x),0)dx . \tag{1.8}$$

As it will be discussed in the next section, $|b^0(x)| = \sqrt{2V(x)}$ and $b^0(x)$ can change sign only once between x_i and x_{i+1} and goes from negative to positive values.

Let S be the set of N elements x_i of absolute minima of the potential and let W be a subset of S. By a W-graph we mean a graph consisting of arrows $x_i \to x_j$ with $j = i \pm 1$, $x_i \in S/W$, provided exactly one arrow issues from every point of S/W and the graph contains no cycle, that is between two neighbouring points of S there is at most one arrow.

Example

$$x_1 \longrightarrow \quad x_2 \longleftarrow \quad x_3 \quad x_4 \longleftarrow \quad x_5$$

$W = (x_2, x_4)$.

The collection of all W-graphs is denoted by $G(W)$. By $G^{(k)}$, $1 \le k \le N$ we mean the set of all W-graphs, where W runs through all possible k-element subsets of S. We shall use the letter g to denote the graphs. Define

$$V^{(k)} = \min_{g \in G^{(k)}} \sum_{(x_i \to x_j) \in g} V_{ij} . \qquad (1.9)$$

The rule for computing the lowest part of the spectrum of H is the following

$$E_1 - E_0 \approx \exp\left\{ -\frac{V^{(1)} - V^{(2)}}{2\hbar} \right\}$$

. . . .

$$E_n - E_0 \approx \exp\left\{ -\frac{V^{(n)} - V^{(n+1)}}{2\hbar} \right\}$$

. . . .

$$E_{N-1} - E_0 \approx \exp\left\{ -\frac{V^{(N-1)} - V^{(N)}}{2\hbar} \right\} \qquad (1.10)$$

where $\approx$ means

$$\lim_{\hbar \to 0} 2\hbar (E_n - E_0) = -(V^{(n)} - V^{(n+1)})$$

it can be seen that $V^{(N)} = 0$.

The above rule (1.10) solves completely the step (2). This type of rule is easily generalized to dimension greater than 1.

We now concentrate on step (1). The solution of (1.6) for $\hbar \to 0$ is a difficult problem of singular perturbation theory. The striking result which we obtained in our previous work is that the velocity field b solution of (1.6) may be extremely unstable under small deformation of the potential V. More precisely and contrary to naive intuition, the number M of stable critical points of $b^0(x)$, that is the points y_i, $i = 1,2,\ldots,M$ such that

$$b^0(y_i) = 0$$

$$b^0(y_i - \delta) > 0, \quad b^0(y_i + \delta) < 0$$

for an appropriate positive δ, is in general much smaller than the number of absolute minima of the potential. We can have $M = N$ only under very special symmetry or fine tuning conditions on the potential. However, these special situations are highly unstable. All this will be amply illustrated in the rest of the paper. In the next section, we describe some general properties of $b^0(x)$.

2. Some General Properties of $b^0(x)$

Eq. (1.6) in one dimension reduces to the Riccati equation

$$\hbar b' + b^2 = 2(V(x) - E_0) \,. \tag{2.1}$$

The properties listed below hold under the following assumptions on the potential V.

(1) $V(x) \geq 0$ and $V \in C^\infty$

(2) V has a finite number of zeros x_i, $i = 1,2,\ldots,N$ and

$$\lim_{x \to x_i} \frac{V(x)}{(x-x_i)^2} = \omega_i^2, \text{ that is V has quadratic minima}$$

(3) for $|x| > L$, $V(x)$ increases monotonically faster than x^2.

Under these hypotheses the following facts are a consequence of propositions (2.1), (2.2) and (2.3) of [1]. Denoting as before $b^o(x) = \lim_{\hbar \to 0} b(x,\hbar)$ we have

(i) for $x > x_N$ $b^o(x) = -\sqrt{2V(x)}$

         $x < x_1$ $b^o(x) = \sqrt{2V(x)}$

(ii) for $x \in [x_1, x_N]$ $|b^o(x)| = \sqrt{2V(x)}$

and $|b - b^o| < C\hbar$ uniformly in x and N

(iii) between two consecutive minima x_i, x_{i+1} there is at most one point y_i where $b^o(x)$ jumps from $-\sqrt{2V(x)}$ to $\sqrt{2V(x)}$, that is

$$b^o(x) = -\sqrt{2V(x)} \quad x \in [x_i, y_i)$$
$$b^o(x) = \sqrt{2V(x)} \quad x \in (y_i, x_{i+1}]\ .$$

If no such point exists, $b^o(x)$ has constant sign in the whole interval $[x_i, x_{i+1}]$. This excludes negative jumps. The above properties (i), (ii), (iii) determine $b^o(x)$ completely if we know the jump points y_i. This is the most difficult part of the problem and also the most substantial for the physical applications. In fact, the location of the jumps determines the localization properties of ground states and the order of magnitude of the low lying eigenvalues, according to the rules of the previous section. Simple rules can be given also for the localization properties of the low lying excited states.

A tool which allows to completely solve the problem in many concrete cases is given by the following formulae which are an immediate consequence of (2.1) [1]. Let x_o be an arbitrary point. Then for each y:

$$b(x) + b(2x_o - x) \equiv b(x) + \bar{b}(x) =$$

$$= (b(y) + \bar{b}(y))\, e^{-\int_y^x \frac{b(x')-\bar{b}(x')}{\hbar} dx'}$$

$$+ \frac{2}{\hbar}\int_y^x (V(x')-V(2x_0-x'))\, e^{\int_x^{x'} \frac{b(x'')-\bar{b}(x'')}{\hbar} dx''}\, dx' \qquad (2.2)$$

with some extra work we obtain from (2.2)

$$b(x)+\bar{b}(x) = -\frac{2}{\hbar}\int_x^\infty \Delta V(x')\, e^{\int_x^{x'} \frac{b(x'')-\bar{b}(x'')}{\hbar} dx''}\, dx' \qquad (2.3)$$

where

$$\Delta V(x) \equiv V(x) - V(2x_0-x).$$

3. Double Wells

The simplest situation is provided by the symmetric double well potential. In this case, by symmetry reasons, the drift $b^0(x)$ has a unique "jump" in the symmetry point of the potential $V(x)$. The precise form of the function $V(x)$ is not important, however, for definiteness, we shall refer to the function $V(x) = \rho(1-x^2)^2$. In this case by property (iii) we have:

$$b^0(x) = \sqrt{2\rho}\,(1-x^2) \qquad \forall x \in (0,1)$$

$$b^0(x) = -\sqrt{2\rho}\,(1-x^2) \qquad \forall x \in (-1,0). \qquad (3.1)$$

We then construct, following Section 1, the quantities $V_{-1,+1}$, $V_{+1,-1}$. Due to the symmetry of the problem:

$$V_{+1,-1} = V_{-1,+1} = 4\int_{-1}^{0} \sqrt{2V(x)}\, dx = \frac{8}{3}\sqrt{2\rho}\ .$$

By applying now the rules (1.10) we obtain for the splitting of the ground state energy:

$$\exp\left\{\left(-\frac{8}{3}\sqrt{2\rho}-\delta\right)/\hbar\right\} \le E_1 - E_0 \le \exp\left\{\left(-\frac{8}{3}\sqrt{2\rho}+\delta\right)/\hbar\right\} \quad (3.2)$$

where δ is a positive constant which goes to zero for $\hbar \to 0$.

A more interesting situation can be obtained if we consider an asymmetric double well potential obtained by a C^∞-deformation of the previous case such that the new potential still satisfies the Assumptions (1), (2), (3) of Section 2. We begin with the case in which

$$V(x) - V(-x) = \Delta V(x) > 0 \quad \forall x \in (a_1, a_2) \subset (0,1)$$

$$\Delta V(x) = 0 \quad \forall x \notin (a_1, a_2) \cup (-a_2, -a_1). \quad (3.3)$$

In order to specify the sign of $b^0(x)$ on the whole line we make use of the Formula (2.3) connecting the antisymmetric part of $b(x,\hbar)$ with the symmetric one with $x_0 = 0$. From this formula it follows immediately that $b^0(x) + \bar{b}^0(x) = 0$ $\forall x > a_2$. Besides by the positivity of $\Delta V(x)$, $x \in (a_1, a_2)$:

$$b^0(x) + \bar{b}^0(x) \le 0 \quad \forall x \in (1_1, a_2). \quad (3.4)$$

From:

$$b^0(x) = \sqrt{2V(x)} \quad \forall x \in (-\infty, -1)$$
$$b^0(x) = -\sqrt{2V(x)} \quad \forall x \in (1, +\infty)$$

and from (3.4) and (ii) of Section 2 applied to the interval (a_1, a_2) we get:

$$b^0(x) = -\sqrt{2V(x)} \quad \forall x \in (a_1, a_2).$$

By (iii) of Section 2, this holds for the whole interval $(-1, a_2)$. The situation is summarized in Fig. 1.

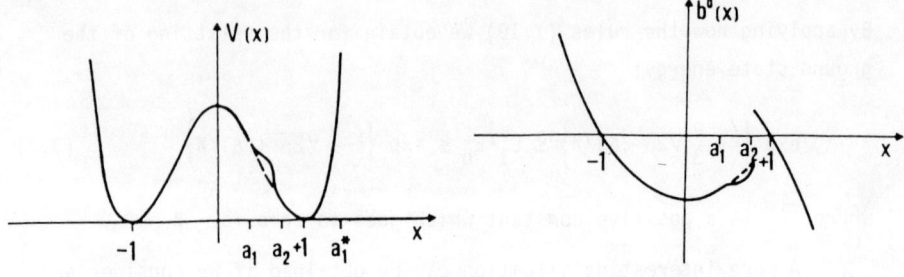

Fig.1

The case with $\Delta V(x) < 0$ $x \in (a_1, a_2)$ and $\Delta V(x) = 0$ $\forall x \notin (a_1, a_2)$ $(-a_2, -a_1)$ can be treated analogously and the result is that the jump of $b^o(x)$ occurs at the point $x = -a_2$ (see Fig. 2).

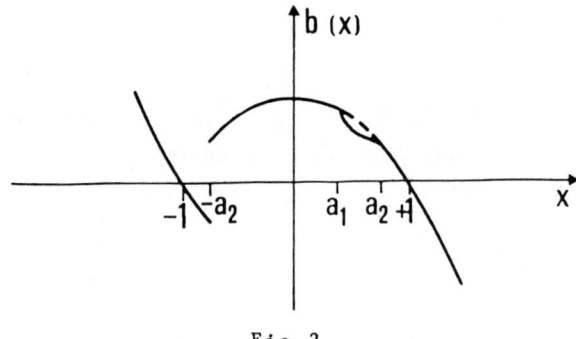

Fig.2

As for the splitting of the ground state is concerned in the first case $(\Delta V > 0)$ we obtain:

$$\exp\left\{-\frac{V_{+1,-1} - \delta}{\hbar}\right\} \leq E_1 - E_0 \leq \exp\left\{-\frac{V_{+1,-1} + \delta}{\hbar}\right\} \quad (3.5)$$

where

$$V_{+1,-1} = \min\ (V_{+1,-1},\ V_{-1,+1}) = 4 \int_{a_2}^{1} \sqrt{2V(x)}\ dx\ .$$

Using the identity

$$\frac{\psi_o(x)}{\psi_o(-1)} = \exp\left\{\int_{-1}^{x} \frac{b(x', \hbar)}{\hbar}\ dx'\right\}\ . \quad (3.6)$$

We get immediately that as $\hbar \to 0$ the ground state concentrates on the left hand side well. Analogous results can be obtained in the negative case $\Delta V < 0$. In this case the ground state wave function is concentrated on the right. The examples just discussed are called by Simon the "flea on the elephant". We now consider the more complicated situation in which we have local perturbations of both signs (see Fig. 3).

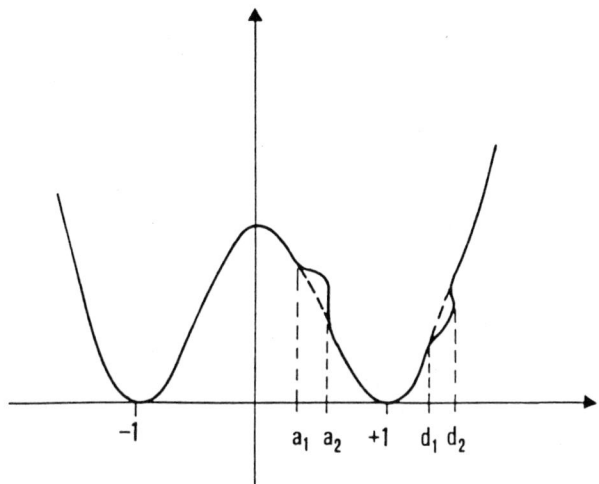

Fig.3

This is a particularly interesting case because, as we have seen, the perturbations tend to localize the wave functions in opposite ways. Therefore, one can envisage a situation such that the two effects compensate and tunneling takes place essentially as in the symmetric case. We now show that this compensation is in fact possible and we thus establish that symmetry of the potential is not a necessary condition for the delocalization of the states.

We apply Eq. (2.3) for $x \in [a_2, 1]$ and $x_0 = 0$.

$$(b+\bar{b})(x) = -\frac{2}{\hbar} e^{\int_x^{d_1} \frac{(b-\bar{b})(x')}{\hbar} dx'} \int_{d_1}^{d_2} \Delta V(x) e^{\int_{d_1}^{x'} \frac{(b-\bar{b})(x'')}{\hbar} dx''} dx' . \quad (3.7)$$

Since $\Delta V < 0$, $b+\bar{b} > 0$ and by (iii), Section 2, $b^o(x) > 0$ for $x \in [a_2, 1]$. From (3.7) it follows that $b^o + \bar{b}^o$ is exponentially small in $\hbar$ for all x such that

$$\int_x^1 \sqrt{2V(x')}\, dx' < \int_1^{d_1} \sqrt{2V(x')}\, dx' \ .$$

This implies that for such x $\bar{b}^o(x) = b^o(-x) < 0$; we now assume that:

$$\int_{d_2}^1 \sqrt{2V(x')}\, dx' > \int_1^{d_1} \sqrt{2V(x')}\, dx' \ .$$

This condition can also be formulated by saying that a_2 has a greater distance from 1 than d_1 in the Agmon metric [10].

We finally introduce our main assumption on ΔV:

$$\frac{2}{\hbar}\int_{d_1}^{d_2}\Delta V(x')\exp\left\{-\frac{2}{\hbar}\int_{d_1}^{x'}\sqrt{2V(x'')}\,dx''\right\}dx' \geq$$

$$\frac{C}{\hbar}\int_{d_1}^{d_1+\hbar}|\Delta V(x')|dx' \geq \hbar^\alpha \qquad 0 < \alpha < \infty$$ (3.8)

where C is a positive constant. Let now $\bar{d}_1$ be the point in $[a_2,1]$ such that its Agmon distance from $x=1$ is the same as that of d_1. Consider now $x \in [a_2, \bar{d}_1]$. We have:

$$\int_x^{d_1}(b^o-\bar{b}^o)(x')dx' = \int_x^{d_1}\sqrt{2V(x')} - \bar{b}^o(x')\,dx' \qquad (3.9)$$

in view of the previous discussion. From (3.7) and (3.8) it follows that $b+\bar{b}$ explodes for $\hbar \to 0$ unless $b^o(x) = +\sqrt{2V(x)}$ in $[a_2,\bar{d}_1]$. This implies that the jump of b^o is in $-\bar{d}_1$. A similar argument shows that if a_2 is closer to $x=1$ than d_1 in the Agmon metric, we fall back in the situation already discussed at the beginning of this section and b^o has a jump in a_2. By continuity we can conclude that for $\hbar$ small but fixed it is possible to find a $d_1(\hbar)$ such that the ground state is equally distributed between the two minima of the potential.

We conclude this section with some comments on the Condition (3.8).

This is satisfied for example if the k-th derivative of V in d_1 is different from zero and ΔV is independent of $\hbar$. However, it can be satisfied also when ΔV decreases with $\hbar$ as a power law. This is interesting because, as remarked in [4], it implies that tunneling is very sensitive to perturbations which are not only localized but also very weak.

4. Multiwell Potentials

In this section, we consider finite sequences of two types of barriers described by two smooth functions $V_1(x)$, $V_2(x)$ with support equal to [0,1] symmetric with respect to $x = 1/2$ with:

$$V_i(0) = V_i'(0) = 0 \qquad i = 1,2$$

$$V_i''(0) = \omega_i$$

$$V_1(x) \geq V_2(x) > 0 \qquad x \in (0,1) . \qquad (4.1)$$

A sequence of barriers will be specified by a sequence of numbers:

$$\{N_1 \; n_1 \; N_2 \; n_2 \; \ldots\ldots\ldots \; N_k\}$$

when N_i denotes a number of consecutive barriers of type V_1 and n_i denotes a number of consecutive barriers of type V_2

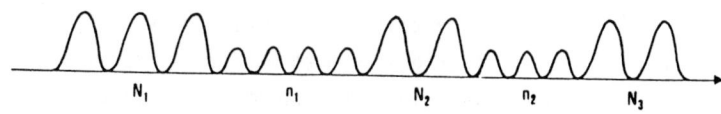

Fig. 4

At the end points of the sequence we put Dirichlet boundary conditions or we let the potential go to infinity very rapidly. We remark that our definition of V_i does not exclude that V_i exhibits several secondary maxima and minima. As in the case of the double well, our first problem consists in the approximate determination for $\hbar \to 0$ of the drift $b(x,\hbar)$ corresponding to the ground state. The general case of an arbitrary sequence is very complicated. However, there are situations where one can establish very simple rules for the calculations of b^o. Our rules are summarized in the following lemmas.

<u>Lemma 1</u>: Given a segment $\{N\ n\}$ such that there exists an interval in the support of the first barrier where $b^o(x) > 0$, then $b^o(x) \geq 0$ in the support of the remaining $(N-1)$ high barriers. Conversely, given a segment $\{n,N\}$ such that there exists an interval in the support of the last high barrier where $b_o < 0$, then $b^o(x) \leq 0$ in the support of the preceding $(N-1)$ barriers.

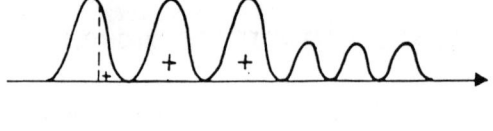

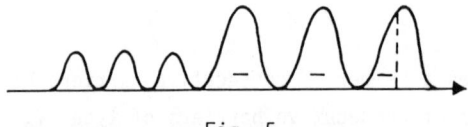

Fig. 5

The +,- signs in the pictures indicate the sign of b^o and we shall often use the expression "we have + or - in N_1 or N_2".

<u>Remark</u>: This lemma is a generalization of property (iii) of Section 2, i.e. a segment of high barriers behaves as far as the jumps of b_o are concerned like a single barrier. We also observe that the conclusion of Lemma 1 requires $\hbar$ small but is uniform in the length N.

<u>Lemma 2</u>: Consider a segment $\{N_1\ n\ N_2\}$ such that $b^o(x) > 0$ on an interval in the support of N_1 and $b^o < 0$ in an interval in the

support of N_2. Then $b^0(x)$ has jumps from $-\sqrt{2V(x)}$ to $\sqrt{2V(x)}$ at the symmetry points of the low barriers except the first and last one. We have in fact $b^0(x) = +\sqrt{2V(x)}$ in the support of the first and $b^0 = -\sqrt{2V(x)}$ in the support of the last one.

Remark: The meaning of this lemma is that the +,- conditions in b^0 at the end barriers of the segment are equivalent to a Dirichlet boundary condition for the Schrödinger equation as $\hbar \to 0$.

Lemma 3: Let $\{N_1 n_1 \ldots n_{i-1} N_i\}$ be a sequence such that:

(i) in the support of the first N_1 barriers there is an interval where $b^0 > 0$

(ii) in the support of the last N_i barriers there is an interval where $b^0 < 0$.

If there exists an n_m among the n_j such that $n_m > n_j \ \forall j \neq m$ then
in case (i) $b^0 > 0$ in the support of $N_j \ \ j = 1,2,\ldots,m$
in case (ii) $b^0 < 0$ in the support of $N_j \ \ j = m+1,\ldots,i$.

Corollary: Suppose that in a sequence $\{N_1 n_1 \ldots n_{i-1} N_i\}$ we have + boundary conditions in N_1 and - boundary conditions in N_i. If there exists a unique maximal segment n_m as in Lemma 3, then the ground state wave function of this sequence is exponentially localized on n_m.

The example treated in section 6 of Ref. [1] can be considered as an application of this corollary. A partial converse of this corollary is given by:

Lemma 4: Let $S = \{N_1 n_1 \ldots n_{i-1} N_i\}$ be a sequence with + boundary conditions in N_1 and - boundary conditions in N_i. Consider a segment $s = \{N_k n_k N_{k+1}\}$ contained in S. Then necessary and sufficient conditions for n_k to be maximal, that is $n_k \geq n_j \ \forall j \neq k$, is that there exists a + interval in N_k and a - interval in N_{k+1}. This lemma may be viewed as a stronger version of Lemma 3.

Lemma 5: Let $S = \{N_1 \ n_1 \ldots n_{i-1} N_i\}$ be a sequence with + boundary conditions in N_1 and - boundary conditions in N_i. Consider a segment $s = \{N_k \; n_k \; N_{k+1}\} \subset S$. Then b^o has jumps at the symmetry points of the low barriers in n_k except possibly the first or the last one, independently of the sign of b^o in N_k, N_{k+1}.

Remark: It is a consequence of the above analysis that the ground state wave function of an arbitrary sequence with (+,-) boundary conditions is exponentially localized in one or more segments of low barriers of maximal length. In other words, generalizing the remark after Lemma 1, we can say that each segment of the form $N_1 \; n \; N_2$ with n non-maximal, behaves, as far as the ground state is concerned, like a single potential barrier. This means that the restriction of the ground state wave function to the segment obtains its maximum at the boundary.

At this point we would like to be able to decide whether the ground state wave function is actually delocalized among the different maximal segments. This is in general a difficult problem. In fact, the localization of the g.s. depends on the detailed structure of the segments separating the maximal ones.

Proof of Lemma 1: We consider first the case $N = 2$.

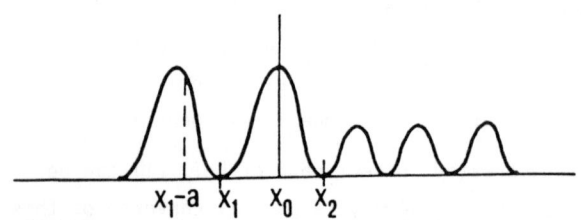

Fig. 6

We assume that $b^o(x) = +\sqrt{2V(x)}$ in $[x_1-a, x_1]$. We observe that by the property (iii) of Section 2 the interval over which $b^o > 0$ must be of the above form, i.e. its right end point must coincide with x_1. Using now the Formula (2.2) with $x_0 = (x_1+x_2)/2$ and $y \in [x_2, x_2+a]$

$$(b+\bar{b})(x) = -2 \int_x^y \frac{\Delta V(x')}{\hbar} \exp\left\{\int_x^{x'} \frac{(b-\bar{b})(x'')}{\hbar} dx''\right\} dx' +$$

(4.2)

$$+ (b+\bar{b})(y) \exp\left\{\int_x^y \frac{(b-\bar{b})(x'')}{\hbar} dx''\right\} .$$

We take $x \in [x_0, x_2]$. $(b+\bar{b})(y) > 0$ by construction and $\Delta V \leq 0$ in $[x,y]$. Therefore, $(b+\bar{b})(x) > 0$ and by (iii), Section 2, $b^0(x) = +\sqrt{2V(x)}$. $b^0(x)$ must also be greater than zero in order to avoid the exponential explosion of the right hand side of (4.2) when $\hbar \to 0$. The rest of the proof is by induction on N.

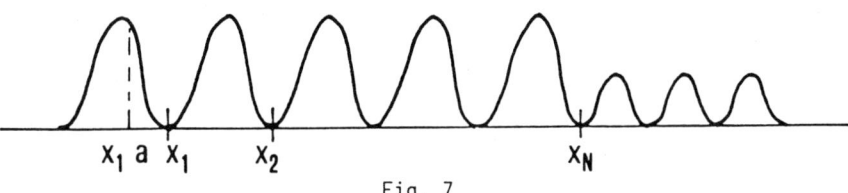

Fig. 7

Suppose the result valid for $N-1$ and take $x_0 = \frac{x_N + x_1}{2}$ and $y \in [x_N, x_N+a]$ in (4.2). Then $(b+\bar{b})(x) > 0$ in $[x_0, y]$ as in the previous case. We have to consider now two possibilities:

(1) $b^0(x) > 0$ for $x \in [x_1, x_1+\delta]$ $0 < \delta < 1$

(2) $b^0(x) < 0$ for $x \in [x_1, x_1+\delta]$ for any $1 \geq \delta > 0$.

In the first case the result follows from the induction. The second possibility must be excluded in order to avoid explosion in (4.2) since in this case $(b+\bar{b})(x) > 0$ in $[x_0, y]$ implies $b-\bar{b} > 0$ in $[x_N-\delta, x_N]$ for any $\delta > 0$. The converse part of the lemma is proved in a similar way.

Proof of Lemma 2:

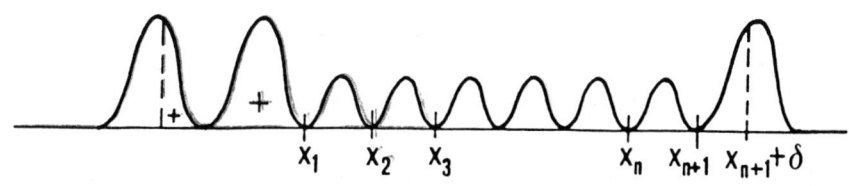

Fig. 8

We begin by showing that $b^o(x) = +\sqrt{2V(x)}$ for $x \in [x_1, x_2]$ and $b^o(x) = -\sqrt{2V(x)}$ for $x \in [x_n, x_{n+1}]$. We use Eq. (4.2) with $x_o = (x_1 + x_2)/2$ and $y \in [x_2, x_3]$. Therefore, by construction $(b+\bar{b})(x) > 0$ for $x \in [x_o, x_2]$ and thus, by (iii) of Sec. 2, $b^o(x) > 0$ in the same interval. If for $x \in [x_1, x_o]$ b^o were negative, then we would have explosion of $b+\bar{b}$ as $\hbar \to 0$. In conclusion $b^o(x) = +\sqrt{2V(x)}$ in $[x_1, x_2]$. A similar argument shows that $b^o(x) = -\sqrt{2V(x)}$ in $[x_n, x_{n+1}]$. Let now $\bar{x} = (x_{n+1} + x_1)/2$. We now take as reflection point x_o in (4.2) successively the symmetry points of the small barriers between x_2 and $\bar{x}$ and $\bar{x}$ and x_n, respectively. y is chosen in both cases in such a way that $(b+\bar{b})(y)$ has a definite sign by exploiting the boundary conditions on the high barriers. In this way one arrives easily at the following sign assignment for b^o:

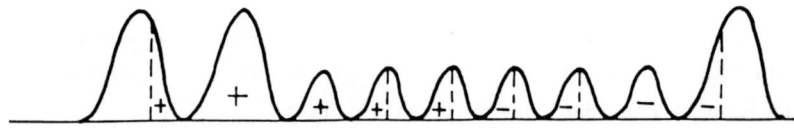

Fig. 9

Next we show that $b^o(x)$ is antisymmetric with respect to $\bar{x}$. We take $x_o = \bar{x}$ and $y \in [x_{n+1}, x_{n+1}+\delta]$. Then Formula (4.2) reduces:

$$(b+\bar{b})(x) = (b+\bar{b})(y) \exp\left\{\int_x^y \frac{(b-\bar{b})(x')}{\hbar} dx'\right\} \quad (4.3)$$

for $x \in [\bar{x}, x_{n-1}]$. Just by looking at Fig. 9 one realizes that:

$$\int_x^y (b-\bar{b})(x') dx' \leq -2 \int_{x_n}^{x_{n+1}} \sqrt{2V(x')} \, dx' \quad . \quad (4.4)$$

Therefore, $b+\bar{b}$ is exponentially small in this region and b^o is antisymmetric. We now show that there is a jump $(-,+)$ at the symmetry point of the first barrier to the right of $\bar{x}$. We choose in (4.2) $x_o = \bar{x}+1/2$ and y as before. If we assume now that the jump is to the right of $\bar{x}+1/2$ and we take $x = \bar{x}+1/2$ a discussion similar to the previous one and taking into account the antisymmetry with respect to $\bar{x}$, shows that

$$\int_x^y (b-\bar{b})(x')dx' < 0 \qquad (4.5)$$

and therefore $(b+\bar{b})$ $(\bar{x}+1/2)$ is exponentially small and thus the jump of b^o must coincide with $\bar{x}+1/2$. The proof of the lemma is completed by taking successively x_o coinciding with the symmetry point of the other barriers.

Proof of Lemma 3: We consider first a segment of the following form:

$$\{N_1 n_1 N_2 n_2 N_3\} \qquad \text{with} \qquad n_2 = n_1 + 1$$

and with $b^o > 0$ in the support of N_1.

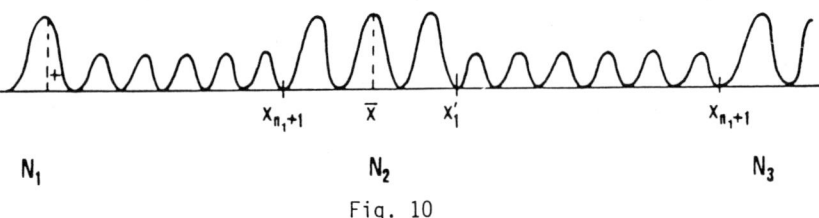

Fig. 10

We show that $b^o > 0$ in the support of N_2. Let $\bar{x}$ be the middle point of the segment N_2. Using (4.2) with $x_o = \bar{x}$ and y in the support of the last small barrier in n_2, we get $b^o(x) > 0$ and this implies, using Lemma 1, that $b^o > 0$ to the right of $\bar{x}$ in the support of N_2. In order to establish the positivity of b^o to the left of $\bar{x}$ we have to distinguish two cases:

(1) b^o is positive in the whole support of N_3.

(2) $\exists \delta > 0$ such that $b^o(x) < 0$ in $[x_{n_2+1}, x_{n_2+1}+\delta]$.

In case (1): We apply Eq. (4.2) with $x_o = \bar{x}+1/2$ and y in the support of the first barrier in N_3. Then $(b+\bar{b})(y) > 0$ and therefore, since $\Delta V \leq 0$, $(b+\bar{b})(x) > 0$ for x in the support of the first barrier in n_2. This implies $b^o > 0$ in the support of the first barrier of N_2. From Lemma 1 it follows that $b^o > 0$ in the support of N_2 (see Fig. 10).

<u>Case (2)</u>: Suppose that $b^o < 0$ in $[x_{n_1+1}, x_{n_1+1}+\delta]$. Then both segments of low barriers n_1 and n_2 are in the situation covered by Lemma 2 and we know the sign of b^o in them. We take now $\bar{x}$ as x_o in (4.2) and $y \in [x_{n_2}, x_{n_2+1}]$. With this choice the two terms in (4.2) have the same sign and:

$$\int_x^y (b-\bar{b})(x')dx' > 0 .$$

Therefore we have explosion for $\not{h} \to 0$. In conclusion, $b^o > 0$ in N_2 in every case. We now remark

(1) The proof becomes even simpler if $n_2 > n_1 + 1$.

(2) The same proof applies if between N_2 and n_2 we insert an arbitrary sequence of high and low barriers provided this contains only segments of low barriers of length less than n_2. The first statement of Lemma 3 now follows by repeated application of Remark (2) and Lemma 1 starting from the extreme left of the sequence. The proof of the second statement of Lemma 3 is identical starting from the extreme right.

<u>Proof of Lemma 4</u>: Necessity: We distinguish several cases:

(a) If $n_k \geq n_j + 1 \ \forall j \neq k$, that is it is the unique maximal segment in S, then the assertion follows from Lemma 3.

(b) There is more than one maximal segment. We have to exclude three possibilities:

    (i)   s has boundary conditions (+,+)
    (ii)   s has boundary conditions (-,-)
    (iii) s has boundary conditions (-,+).

(i) If between s and N_i there is no maximal segment of low barriers from Lemma 3 applied between N_k and N_i, we get a contradiction. In fact, the sign in N_{k+1} should be "-". Suppose next that to the right of s there is at least another maximal segment, possibly more than one. Then we claim that at least one of the maximal segments to the right of s must have (+,-) boundary conditions. This is proved by successive applications of Lemma 3. Let us denote by s',s",... the maximal segments of low barriers to the right of s.

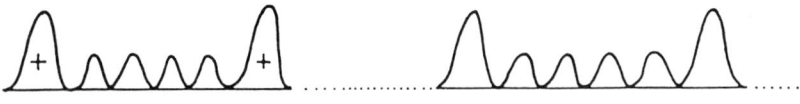

Fig. 11

By applying Lemma 3 to the subsequence which starts with the first high barrier in N_{k+1} and terminates with s' we conclude that s' has + boundary condition on its left.

If s' has "-" on its right, then we are finished.

If s' has "+" to the right, we repeat the argument between s' and s". This procedure must terminate, otherwise we get a contradiction at the end of the sequence where we have -boundary condition. We denote with $\bar{s}$ the segment with (+,-) boundary conditions. By Lemma 2, b_0 in $\bar{s}$ will have the signs indicated in Fig. 12.

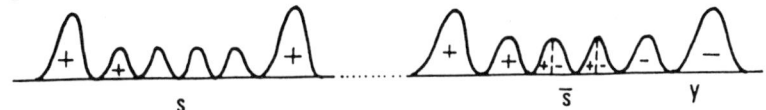

Fig. 12

We now show that this is in contradiction with the + sign to the right of s. We apply formula (4.2) with x_0 coinciding with the middle point between s and $\bar{s}$ and y as in Fig. 12. Then it is easy to see that $(b+\bar{b})(x)$ is exponentially small for x in $\bar{s}$. This implies that b_0 has jumps also at the symmetry points of the low barriers in s with the exception of the first, where we already know that it is positive, and the last one where it is negative. We now take x in the

support of the high barrier on the left of $\bar{s}$ and we observe that $b+\bar{b}$ is still exponentially small contradicting the (+,+) boundary condition on s.

(ii) To exclude this case, one repeats the above argument to the left of s.

(iii) We take x_0 and y as in the discussion of (i). With the present boundary conditions $(b+\bar{b})$ (y) < 0 and therefore (since $\Delta V = 0$) $(b+\bar{b})$ (x) < 0 for x in the support of the first high barrier to the left of $\bar{s}$. But this is in contradiction with the boundary conditions on s.

<u>Proof of Lemma 5</u>: The case (+,-) is covered by Lemma 2. The cases (+,+), (-,+), (-,-) are impossible if n_k is maximal. When n_k is not maximal, the proof for the cases (+,+) and (-,-) can be achieved with an argument very close to the one described in the proof of Lemma 4, case (i). It remains to discuss (-,+) and n_k not maximal. By arguments like those used in the proof of Lemma 4, there must be maximal segments s_L, s_R to the left and to the right of n_k with (+,-) boundary conditions. Consider then the situation to the right of n_k

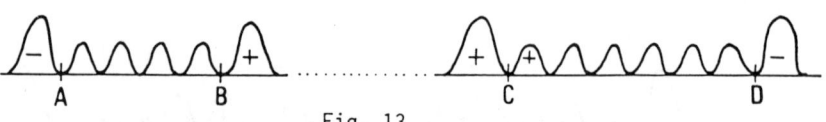

Fig. 13

Using Eq. (4.2) with $x_0 = (A+D)/2$, y in [D,D+1], $(b+\bar{b})$ (x) < 0 in $[D-n_k,D]$ which implies together with Lemma 2 $b_0 < 0$ for x in [A+1, A+3/2], [A+2,A+5/2], etc. By repeating the argument with the maximal segment to the left we conclude that $b_0 > 0$ for x in [A+3/2, A+2], [A+5/2, A+3], etc.

5. Hierarchical Models in One Dimension

In this section, we introduce a class of models which in terms of the notations of the previous section are characterized by the fact that $n_i = n_0$ $\forall i$; i.e. the segments of low barriers are all of the same length, while the lengths N_i of the regions of high barriers are arranged in a hierarchical way. In this way the difficulties mentioned in the remark after Lemma 5 in the previous section are not present and one can arrive, at least in cases with sufficient symmetries, at a rather complete description of b_0.

These models have been introduced for any dimension in [8], where several additional properties are analyzed including the temporal evolution.

We now describe a particular example characterized by special symmetries. The construction is inductive.

Let $S_0 = n_0$ be a basic segment of low barriers, we define

$$S_i = S_{i-1} \, N_i S_{i-1} \, N_i S_{i-1} \tag{5.1}$$

where N_i are segments of high barriers which increase very rapidly:

$$N_i = [d_i] \qquad d_i = n_0^{a^i} \qquad a > 1$$

where $[\cdot]$ denotes the integer part.

We want to determine the properties of the ground state of the sequence $N_{i+1} S_i N_{i+1}$ with $(+,-)$ boundary conditions, for large i.

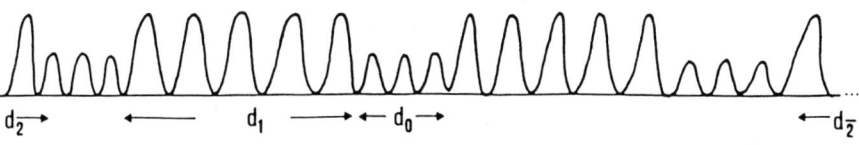

Fig. 14

We prove that for any sequence $N_{i+1}S_iN_{i+1}$ with $(+,-)$ boundary conditions, the ground state ψ_0 is delocalized in the following sense.

<u>Theorem</u>: Consider the normalization integrals $\int_{S_0} \psi_0^2\, dx$ and $\int_{S_0'} \psi_0^2\, dx$ where S_0 and S_0' are two different segments of n_0 low barriers contained in the sequence $N_{i+1}S_iN_{i+1}$. Then uniformly in i and independently of S_0 and S_0'

$$c_1 < \frac{\int_{S_0} \psi_0^2\, dx}{\int_{S_0'} \psi_0^2\, dx} < c_2 \tag{5.2}$$

the constants c_1 and c_2 for $\hbar$ small may be chosen independent of $\hbar$.

For convenience we center our sequence in 0

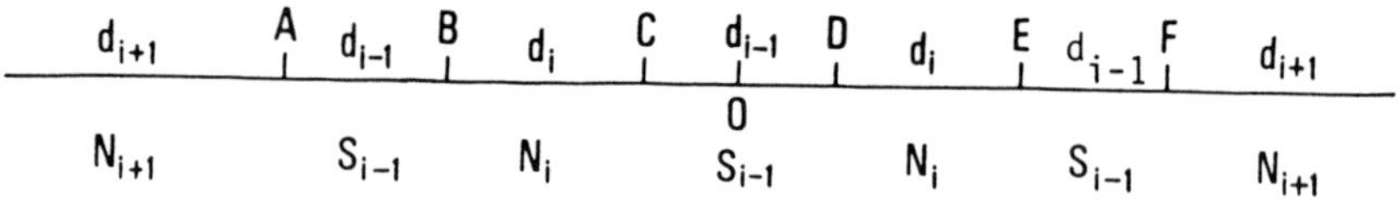

```
         A       B     C        D    E       F
 d_{i+1}   d_{i-1}  d_i   d_{i-1}  d_i   d_{i-1}   d_{i+1}
                             0
 N_{i+1}   S_{i-1}  N_i   S_{i-1}  N_i   S_{i-1}   N_{i+1}
```

Fig. 15

where A, B, etc. represent the end points of subsequences S_{i-1}. We first prove:

<u>Lemma 1</u>: If we have

$$|b - \sqrt{2V}| \leq C\hbar \quad \text{for} \quad x \in \left[A - \frac{[d_{i+1}]}{2}, A\right]$$

and $|b - \sqrt{2V}| \leq C\hbar$ for $x \in \left[F, F + \frac{[d_{i+1}]}{2}\right]$ then we have:

$$|b + \sqrt{2V}| \leq C\hbar \quad \text{for} \quad x \in \left[B, \frac{B+C}{2}\right] \cup \left[D, \frac{D+E}{2}\right]$$

and $|b - \sqrt{2V}| \leq C\hbar$ for $x \in \left[\frac{B+C}{2}, C\right] \cup \left[\frac{D+E}{2}, E\right]$

provided $\hbar < \varepsilon, \varepsilon$ and C are independent of the scale i.

Remark: In the present case we are interested in results that remain valid when we take the limit $i \to +\infty$. It is important therefore that the semiclassical requirement $\hbar$ small does not depend on the scale considered. This is indeed the case.

We shall see that in the proof of the previous lemma we use only Lemma 1 of the previous section whose conclusions, as we remarked, are independent of the length of the sequence.

Proof of the Lemma: First we observe that due to the symmetry of the problem with respect to the origin, $b(x)$ is antisymmetric.

We use formula (2.2) with $x_0 = \frac{B+C}{2}$ and $y = F + [d_{i+1}]/2$:

$$(b+\bar{b})(x) = (b+\bar{b})(y) \exp\left\{ \int_x^y dx' \frac{(b-\bar{b})}{\hbar}(x') \right\}$$

$$- \frac{2}{\hbar} \int_x^y \Delta V(x') \exp\left\{ \int_x^{x'} dx'' \frac{(b-\bar{b})}{\hbar}(x'') \right\} dx' \tag{5.3}$$

$\Delta V \leq 0$ for $x \in [E,F]$ and zero otherwise.

The first term is exponentially small of order $\exp\left\{-C\frac{d_{i+1}}{\hbar}\right\}$ because in the integral $\int_x^y (b-\bar{b}) dx'$ the leading contribution, which is negative, comes from the scale d_{i+1}. The second term can be estimated from below as follows:

$$- \frac{2}{\hbar} \int_x^y dx' \, \Delta V(x') \exp\left\{ \int_x^{x'} \frac{(b-\bar{b})}{\hbar} dx'' \right\} \geq$$

$$\geq \frac{c}{\hbar} \exp\left\{-\frac{2}{\hbar} \int_x^F \sqrt{2V(x')} \, dx'\right\} \tag{5.4}$$

and this is clearly larger than the first terms since $|F-x_0| \simeq O(d_i)$. Therefore $(b+\bar{b})(x) \geq 0$ for $x \in \left[\frac{B+C}{2}, C\right]$. Using Lemma 1 and the antisymmetry the conclusion of the lemma follows in the intervals $\left[\frac{B+C}{2}, C\right]$ and $\left[D, \frac{D+E}{2}\right]$. One now proves that b is almost antisymmetric with respect to x_0.

By this we mean that $b+\bar{b}$ is exponentially small in $\hbar$. From this will follow the conclusion of the lemma in the other two intervals $\left[B, \frac{B+C}{2}\right]$ and $\left[\frac{D+E}{2}, E\right]$. We apply Formula (5.3) with x_0 as before and $y = \frac{D+E}{2}$. In this case $\Delta V = 0$ in $[x_0, y]$; therefore for $x \in [x_0, C]$:

$$(b+\bar{b})(x) = (b+\bar{b})(y) \exp\left\{\int_x^y dx' \frac{(b-\bar{b})}{\hbar}(x')\right\}. \tag{5.5}$$

We want to show that the integral in the exponent is negative at least for $x \geq x_0 + \delta$, with $\frac{1}{2} > \delta > 0$ and independent of $\hbar$. We have:

$$\int_x^y (b-\bar{b}) \, dx' = \int_x^{x^*} (b-\bar{b}) \, dx' + \int_{x^*}^y (b-\bar{b}) \, dx'$$

where x^* is the point symmetric of x with respect to the origin. Then $\int_x^{x^*} b(x') dx' = 0$ by antisymmetry. Therefore

$$\int_x^y (b-\bar{b}) dx' = -\int_{\bar{x}^*}^{\bar{x}} b \, dx' + \int_{x^*}^y (b-\bar{b}) dx'$$

where $\bar{x}$ and $\bar{x}^*$ are symmetric of x and x^* with respect to x_0. We now show that the first contribution is negative. For this we need to apply again (5.3) with symmetry point in $\frac{A+B}{2}$ and taking y for example in $(\frac{B+C}{2} + C)/2$. Then $(b+\bar{b})_{x_0 = \frac{A+B}{2}}(x)$ with an obvious meaning of the notation, is positive for $x \in \left[\frac{A+B}{2}, y\right]$ we now remark that

$$\int_{\bar{x}^*}^{\bar{x}} b \, dx' = \int_{\frac{A+B}{2}}^{\bar{x}} (b+\bar{b}) \Big|_{x_0 = \frac{A+B}{2}} dx' \geq 0$$

from which the assertion follows. Then

$$\int_x^y (b-\bar{b}) dx' \leq \int_{x^*}^y (b-\bar{b}) dx' \leq - \int_{x^*}^y 2\sqrt{2V} \, dx' + O(\hbar). \qquad (5.6)$$

The proof is finished.

In the lemma just proved we used at several points the antisymmetry of b with respect to the origin. It is clear that this condition cannot be true for a subsequence $S_{i'}$, $i' = i-1$ contained in S_i which is not in the origin. However, we can convince ourselves easily that b is "almost" antisymmetric with respect to the center of $S_{i'}$, in the sense that $b+\bar{b}$, taking as x_0 the center of $S_{i'}$ is $O(\exp\{-c \frac{d_i+1}{\hbar}\})$. This follows as usual from the comparison equation (5.3) which gives

$$(b+\bar{b})(x) = (b+\bar{b})(y) \exp\left\{\int_x^y \frac{b-\bar{b}}{\hbar}\right\}$$

where we take as y the middle point of the barrier $N_{i'+1} = N_i$ to the right of $S_{i'}$. From Lemma 1 it follows in fact that the sign of b is "+" for a segment of length $\frac{N_{i'+1}}{2}$ to the left of $S_{i'}$ and "-" for a segment of length $\frac{N_{i'+1}}{2}$ to the right of $S_{i'}$. The lemma can then be reapplied to $S_{i'}$ on the scale $i' = i-1$ using approximate antisymmetry instead of exact antisymmetry. This procedure can be iterated until we reach the smallest scale n_0. Thus we have proved:

<u>Corollary</u>: For each segment of high barriers contained in S_i we have $b_0 = -\sqrt{2V}$ to the left of its middle point and $b_0 = +\sqrt{2V}$ to the right. Furthermore, the $\lim_{\hbar \to 0} b = b_0$ is uniform in i.

We have now all the tools to prove the theorem. We first show that if on the scale i we take x contained in the middle subsequence S_{i-1} and x^* is the symmetric of x with respect to the middle point of the left barrier N_i, then

$$\exp(C) \geq \frac{\psi_o(x)}{\psi_o(x^*)} = \exp\left\{\frac{i}{\hbar}\int_{x^*}^{x} b \, dx'\right\} > 1 \tag{5.7}$$

with C independent of $\hbar$. Using the same argument that led to Eq. (5.6), we have for a suitable constant C

$$0 < \int_{x^*}^{x} b \, dx' < \int_{x_o}^{x} (b+\bar{b})_{x_o} \, dx' \leq \int_{0}^{\infty} e^{-Cx/\hbar} dx$$

where the inversion point x_o as usual is the middle point of the left barrier N_i. By induction over lower scales, which is possible because the above argument does not depend on the scale i, we can obtain (5.7). The whole argument can be compared with the general discussion in arbitrary dimension given in sec. 4 of Ref. [8].

6. Effective Potential

In Sec. 4 we remarked that subsequences of barriers may behave, as far as the localization properties of the ground state are concerned, like a single barrier. This suggests quite naturally that an originally complicated potential can be replaced by one of simpler structure while keeping fixed the qualitative properties of the ground state wave function. In view of the general philosophy of our approach, discussed in Sec. 1, this implies that also the low lying tunneling states in a complicated potential can be studied with the help of a simpler "effective" potential. We want to discuss in some detail this problem in connection with the hierarchical models of the previous section. The simplest choice of an effective potential is the one made in Ref. [8]. This consists in substituting the segments of length n_o of low barriers with segments of the same length of zero potential, while the regions N_i are replaced by square barriers of equal height λ **such** that

$$\sqrt{2\lambda} = \int_0^1 \sqrt{2V_1}\, dx \tag{6.1}$$

where V_1 is the function describing the high barriers. It is easy to see that the structure of b_0 is the same as in the original model (see also [8]) and therefore the qualitative structure of the spectrum for energies E such that

$$E - E_0 \le O\!\left(\exp\left\{-\frac{1}{2\hbar}\sqrt{2\lambda}\, d_1\right\}\right) \tag{6.2}$$

is maintained. Of course we lose the structure of the ground state over distances less than d_1 and correspondingly the spectrum of tunnelings over the smaller scales. For this effective potential the structure of the spectrum has been determined in [8] and is as follows. Suppose we have determined the spectrum for a sequence $N_{i+1}\, S_i N_{i+1}$. Then when we go to the next scale and we consider a sequence $N_{i+2}\, S_{i+1}\, N_{i+2}$ each energy level in the region (6.2) is split into three new levels and the order of the splitting is $\exp\{-\frac{1}{2\hbar}\sqrt{2\lambda}\, d_{i+1}\}$ (see Fig. 16).

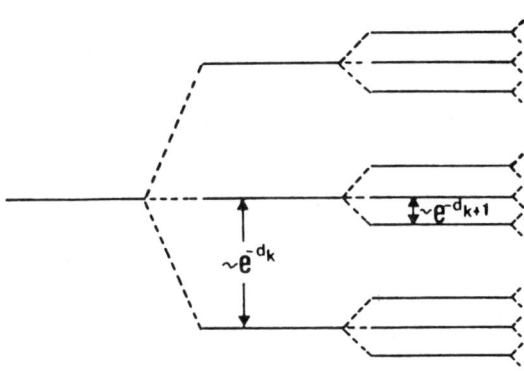

Fig.16

7. Applications

In this final section we would like to comment briefly on the physical relevance of the previous results.

The symmetric double well is a model that appears in several situations. In particular it has been used in molecular physics to describe in the Born-Oppenheimer approximation pyramidal molecules like NH_3, AsH_3, etc. and their substitution compounds. In these molecules the heavy atom, N, As, etc., according to the symmetric double well model should be delocalized over the two minima corresponding to opposite positions with respect to the H_3 plane. However, there is evidence that this is not always true and in particular NH_3 and AsH_3 seem to behave differently, AsH_3 being localized in only one configuration.

A possible explanation of this fact can be based on the instability of tunneling illustrated for the double well in Section 3.

In fact, the observed molecules are never isolated and the interaction perturbing the symmetric double wells may favor through a cooperative effect states in which the single molecule is localized.

This possibility has been examined in [11] where it was found that the order of magnitude of perturbations in the gaseous states of NH_3 and AsH_3 is such that a different behaviour with respect to localization should be expected. AsH_3 is much more semiclassical than NH_3, so that our instability effect should be relevant.

We now turn our attention to disordered systems. There are cases where the diffusion of an electron in a disordered crystal is essentially determined by the possibility of tunneling through "effective potential barriers" of arbitrary length.

As this work illustrates, the underlying mechanism behind localization at low energy is the extreme instability of tunneling which makes travelling over large distances very unlikely. This was already observed in [12]. By this we do not mean that Anderson localization follows immediately from results of the kind proved in Section 4. These results deal with cases which are already rather complicated but still of finite "complexity" if compared with typical stochastic configurations of the potential appearing in the Anderson model. To fill this gap it is necessary to introduce a coarse grained description of tunneling through a multiscale analysis of the problem. The first step in this direction was taken in [5] where it was proved that the diffusion constant for the Anderson model is zero at low energy or large disorder. The next step was the proof in [6] that all states are exponentially localized. This step required the introduction of new ideas. The success of the coarse graining procedure depends on the possibility of showing that localization properties of the eigenfunctions on a given scale are approximately independent of other scales. This type of property is automatically guaranteed in hierarchical models for geometrical reasons (see Section 5 and especially [8]). In the Anderson model localization at high disorder or low energy was proved in [6] by showing that the structure of tunneling for typical configurations of the potential is the same as in the stochastic hierarchical models of [8]. Hierarchical models and the Anderson model are the only two cases where the coarse graining procedure has been successfully implemented.

The results of Section 4 and the discussion of Section 6 give an indication on how to construct a coarse graining in other particular non-stochastic situations.

Clearly, if one is given an arbitrary sequence of high and low barriers, one cannot hope to analyze tunneling in terms of simple criteria. However, the problem can become more tractable if some measure of the complexity of the sequence is known e.g. the Kolmogorov complexity of the binary sequence which can be associated to it.

References

[1] G. Jona-Lasinio, F. Martinelli, E. Scoppola, "New Approach to the Semiclassical Limit of Quantum Mechanics. I. Multiple Tunneling in One Dimension", Comm. Math. Phys. $\underline{80}$, 223 (1981)

[2] B. Helffer, J. Sjöstrand, "Multiple Wells in the Semiclassical Limit I", Comm. P.D.E. $\underline{9}$, 337 (1984)

B. Helffer, J. Sjöstrand, "Puits Multiples en Limite Semi-classique II", Ann. Inst. Henri Poincaré $\underline{42}$, 127 (1985)

[3] B. Simon, "Semi-classical analysis of low-lying eigenvalues IV. The flea on the elephant, J. Funct. Anal. $\underline{63}$, 123 (1985)

[4] S. Graffi, V. Grecchi, G. Jona-Lasinio, "Tunneling Instability via Perturbation Theory", J. Phys. $\underline{A17}$, 2935 (1984)

[5] J. Fröhlich, T. Spencer, "Absence of diffusion in the Anderson tight binding model for large disorder or low energy", Comm. Math. Phys. $\underline{88}$, 151 (1983)

[6] J. Fröhlich, F. Martinelli, E. Scoppola, T. Spencer, "Constructive proof of localization in the Anderson tight binding model", to appear in Comm. Math. Phys.

[7] F. Martinelli, "On the wave equation in random domains: localization of the normal modes in the small frequency region", to appear in Ann. Inst. H. Poincaré

[8] G. Jona-Lasinio, F. Martinelli, E. Scoppola, "Multiple tunnelings in d-dimension: a quantum particle in a hierarchical potential", Ann. Inst. H. Poincaré $\underline{42}$, 73 (1985)

[9] A.D. Ventzel, M.I. Freidlin, "Fluctuations in Dynamical Systems under the Action of Small Random Perturbations", Moskva 1979 An English translation has appeared in the Springer Series: Grundlehren der Mathematischen Wissenschaften, Vol. $\underline{260}$ (1984)

[10] S. Agmon, "Lectures on exponential decay of solutions of second order elliptic equations", Princeton Math. Notes $\underline{29}$, Princeton University Press (1983)

[11] P. Claverie, G. Jona-Lasinio, "Instability of Tunneling and the concept of Molecular Structure in Quantum Mechanics", Preprint LPTHE 84/42

[12] G. Jona-Lasinio, "Qualitative Theory of Stochastic Differential Equations and Quantum Mechanics of Disordered Systems", Helvetica Physica Acta $\underline{56}$, 61 (1983)

GEOMETRIC ASPECTS OF BRS AND ANOMALIES

Daniel KASTLER
Dept. of Physics and CPT - C.N.R.S.
Luminy - Case 907
F-13288 Marseille, France

Anomalies and (classical BRS transformations can be detached from their gauge quantum field context to be described as purely differential-geometric objects - in fact, items in the theory of smooth principal bundles (whose mathematical exegesis as such should, in our opinion, be further pursued). We here sketch the description of these aspects, as a preversion of a work in collaboration with R. Stora, in which details and proofs will be given [3].

Chiral anomalies of gauge fields have been discovered fifteen years ago. They are one of the important items in perturbative gauge field theories, through which the analysis of consistency of perturbative renormalization of gauge fields have come to a satisfying state.

Let $S(\varphi)$ be a classical action of the renormalizable type, φ a "matter field" transforming linearly under a compact Lie group G with Lie algebra L. And let $\Gamma(\varphi)$ be the quantum vertex functional

$$\Gamma(\varphi) = S(\varphi) + \sum_{n=1}^{\infty} \hbar^n \Gamma^{(n)}(\varphi) . \qquad (1)$$

If G leaves S invariant, i.e. if

$$W_{cl}(u)S(\varphi) = 0 , \quad u \in L , \qquad (2)$$

(with W_{cl} the representation of L induced by that of G), then one can arrange the renormalized perturbative series for $\Gamma(\varphi)$ in such a way that

$$W(u)\Gamma(\varphi) = 0, \quad u \in L, \tag{3}$$

with W a representation of L:

$$[W(u), W(u')] = W([u,u']), \quad u,u' \in L. \tag{4}$$

Now, using "minimal coupling" one replaces the action $S(\varphi)$ by an action $S(\varphi,a)$ involving a classical gauge field (connection form) a, invariant under the corresponding gauge group G with Lie algebra L, i.e. such that

$$W_{cl}(\Omega) \, S(\varphi,a) = 0, \quad \Omega \in L, \tag{5}$$

where W_{cl} is a representation of L:

$$[W_{cl}(\Omega), W_{cl}(\Omega')] = W_{cl}([\Omega,\Omega']), \quad \Omega,\Omega' \in L. \tag{6}$$

The problem is now whether the corresponding vertex function $\Gamma(\varphi,a)$ can be chosen in such a way that

$$\begin{cases} W(\Omega) \, \Gamma(\varphi,a) = 0 \\ [W(\Omega), W(\Omega')] = W([\Omega,\Omega']) \end{cases}, \quad \Omega,\Omega' \in L. \tag{7}$$

This is in fact not the case in the presence of chiral spinors φ. Instead of the first Equation (7), one has now

$$W(\Omega) \, \Gamma(\varphi,a) = \int_M A(\Omega,a) \tag{8}$$

with the <u>anomaly</u> $A(\Omega,a)$ a differential 4-form on space-time M, linear in Ω, polynomial in a and its derivatives.

The turning point which initiated the understanding of the algebraic nature of anomalies was the <u>Wess-Zumino consistency condition</u>, obtained as follows: the fact that W is a Lie-algebra representation implies that

$$W(\Omega) \int A(\Omega',a) - W(\Omega') \int A(\Omega,a)$$
$$- \int A([\omega,\omega']) = 0 \qquad (9)$$

from which it follows[o)]

$$W(\Omega) \, A(\Omega',a) - W(\Omega') \, A(\Omega,a)$$
$$- A([\Omega,\Omega']) = 0 \qquad (10)$$

where one recognizes in the r.h.s. the expression of a Lie-algebra Chevalley coboundary operator applied to the local functional A. This reveals the nature of anomalies as elements of the first cohomology group $H^1(L,\Gamma^{loc})$ of L with values in local functionals of potentials. We now turn to the mathematical description of BRS relations and anomalies.

$\mathbb{P} = (P \to M, G)$ denotes a smooth principal bundle with total space P, base M and structural (Lie) group G (acting on the right on P). $\mathcal{G}$ denotes the corresponding gauge group (a Frêchet Lie group) and we denote L = Lie G, $\mathcal{L}$ = Lie $\mathcal{G}$. $\mathcal{G}$ is defined as the group of automorphisms of $\mathbb{P}$ inducing the identity on the base, thus acting (on the left) on P as follows

$$gz = zg(z), \quad z \in P, \quad g \in \mathcal{G} \qquad (11)$$

where $z \to g(z)$ is a smooth map: $P \to G$ ad-equivariant in the sense

$$g(zs) = s^{-1}g(z)s, \quad z \in P, \quad s \in G. \qquad (12)$$

This allows to consider $\mathcal{G}$ as the set of these maps with the product in $\mathcal{G}$ the pointwise product of these maps. Correlatively, $\mathcal{L}$ consists of the smooth maps $\Omega: P \to L$ Ad-equivariant in the sense

$$\Omega(zs) = s^{-1}\Omega(z)s, \quad z \in \mathbb{P}, s \in G, \qquad (13)$$

---

o) We denote W by ρ in the rest of the text.

with the bracket of L the pointwise bracket in L and exponentiation given by $e\Omega(z) = e^{\Omega(z)}$, $z \in P$.

We denote by $\Lambda^*(P,\mathbb{C}) = \oplus \Lambda^p(P,\mathbb{C})$ the de Rham complex of P, a graded commutative differential algebra (GCDA) under the wedge product $\wedge$ of differential forms and the exterior differential d of P. There are canonical, mutually commuting left actions r, resp. ρ, of the groups G, resp. $\mathcal{G}$, on $\Lambda^*(P,\mathbb{C})$ obtained by pull back (straight pull back for G acting to the right on P, pull back of the inverse for $\mathcal{G}$ acting on the left on P). Both actions are by GCDA-automorphisms.

A central object for us is the L-valued de Rham complex of P, $\Lambda^*(P,L)$, alias

$$\Lambda^*(P,L) = L \otimes \Lambda^*(P,\mathbb{C}) \tag{14}$$

to which the above operators are extended as follows

$$d = id_L \otimes d \tag{15}$$

$$r(s) = Ad s \otimes r(s), \quad s \in G \tag{16}$$

$$\rho(g) = id_L \otimes g . \tag{17}$$

With the <u>Schouten product</u> [$\wedge$], defined as

$$[u \otimes \alpha \wedge v \otimes \beta] = [u,v] \otimes (\alpha \wedge \beta), \quad u \otimes \alpha, v \otimes \beta \in \Lambda^*(P,L) , \tag{18}$$

$\Lambda^*(P,L)$ then becomes a differential graded Lie algebra (DGL)[1], acted upon by G and $\mathcal{G}$ through DGL-automorphisms. Defining as usual

$$\rho(\Omega) = \frac{d}{dt}\bigg|_{t=0} \rho(e^{t\Omega}) \tag{19}$$

---

1) i.e. one has the commutation rule $[\mu \wedge \lambda] = (-1)^{pq}[\lambda \wedge \mu]$ and the graded Jacobi identity $(-1)^{pr}[\lambda \wedge [\mu \wedge \nu]] + \text{circ} = 0$ for $\lambda \in \Lambda^p(P,L)$, $\mu \in \Lambda^q(P,L), \nu \in \Lambda^r(P,L)$.

we then obtain a Lie algebra representation ρ of L on $\Lambda^*(P,L)$ - by zero-grade graded derivations of this DGL. Note that the set $A^*(P,L) = \oplus A^p(P,L)$ of fixpoints of $\Lambda^*(P,L)$ under $r(s)$ as defined in (6) (= Ad equivariant L-valued differential forms on P) is a sub-DGL of $\Lambda^*(P,L)$ stable under the representation ρ of G (resp. L).

We now consider the Chevalley cohomology of L with values in the representation space $A^*(P,L)$. The latter acts in the space $A^{**} = \oplus A^{p,\alpha}$, where

$$A^{p\alpha} = \Lambda^p(L, A^\alpha(P,L)) = A^p(P,L) \otimes \Lambda^\alpha L^* , \qquad (20)$$

$\Lambda^\alpha L^*$ the αth exterior power of the $C^\infty(M)$-module L^* dual of the $C^\infty(M)$-module L - note that $\Lambda^* L^* = \Lambda^\alpha L^*$ carries the wedge product of these exterior powers. We now equip A^{**} with the following structure: we define the exterior derivation as

$$(dU)(\Omega_1,\ldots,\Omega_p) = d\{U(\Omega_1,\ldots,\Omega_p)\} \qquad (21)$$

further the <u>Lie algebra coboundary</u> as

$$(\delta U)(\Omega_0,\ldots,\Omega_\alpha) = \sum_{i=0}^{\alpha} (-1)^i \rho(\Omega_i) U(\Omega_0,\ldots,\hat{\Omega}_i,\ldots,\Omega_\alpha)$$
$$+ \sum_{0 \leq i \leq j} U([\Omega_i,\Omega_j],\ldots,\hat{\Omega}_i,\ldots,\hat{\Omega}_j,\ldots,\Omega_\alpha) \qquad (22)$$

where $U \in \Lambda^{p\alpha}$ is considered as a $A(P,L)$-valued alternate α-linear form on L - the caret $\hat{}$ meaning omission of the corresponding argument; finally the product $[\wedge]$ as

$$[\lambda \otimes \phi \wedge \mu \otimes \psi] = (-1)^{\alpha q}[\lambda \wedge \mu] \otimes (\phi \wedge \psi) \qquad (23)$$

for $\lambda \times \phi \in A^{p\alpha} = A^p(P,L) \otimes \Lambda^\alpha L^*$, and $\mu \otimes \psi \in A^{q\beta} = A^q(P,L) \otimes \Lambda^\beta L^*$.

---

2) We set as usual $\Lambda^0 L^* = \mathbb{C}$. The tensor product in (20) is w.r.t $C^\infty(M)$.

We then have the following structure:

(i) (A^{**}, d, δ) is a double complex with horizontal differential d and vertical differential δ, i.e. one has

$$d^2 = \delta^2 = d\delta - \delta d = 0 \tag{24}$$

d, resp. δ, increasing by one the <u>order of the form</u> p, resp., the <u>ghost number</u> α. The corresponding <u>total complex</u> $(^*A, \Delta)$ is then traditionally defined by setting

$$^nA = \bigoplus_{p+\alpha=n} A^{p\alpha} \;, \tag{25}$$

and

$$\Delta = d + s \;, \tag{26}$$

where

$$\begin{cases} s = (-1)^p \delta \\ \text{i.e.} \quad sU = (-1)^p \delta U, \quad U \in \Lambda^{p\alpha} \end{cases} \tag{27}$$

with the effect that

$$s^2 = sd + ds = 0 \;, \tag{28}$$

hence

$$\Delta^2 = 0 \;. \tag{29}$$

(note that Δ increases by one the <u>total grading</u> $n = p+\alpha$).

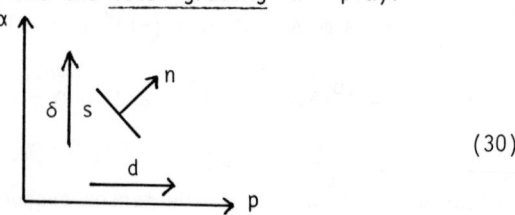

(30)

(ii) $^*A$ (i.e. A^{**} with the total grading n) equipped with the differential Δ and the product (bracket) [ ] becomes a DGL.

We are now able to display the BRS relations

$$\begin{cases} sa = -d\omega - [a \wedge \omega] \\ s\omega = -\frac{1}{2} [\omega \wedge \omega] \end{cases} \quad (31)$$

involving an arbitrary connection L-form a and the Maurer-Cartan form ω of G. This goes as follows: let A be the set of one-forms of principal connections of $\mathbb{P}$: A consists of the $a \in A^1(P,L)$ with vertical values

$$a_z(L_{z*_e} u) = u, \quad z \in P, \quad u \in L \quad (32)$$

where L_{z*_e} is the tangent map at the unit $e \in G$ of L_z defined by $L_z s = z^s$, $z \in$, $s \in G$. Identifying A with $A \otimes 1$, $1 \in \wedge^0 L^* = C^\infty(M)$ we thus have that

$$a \in A^{10} . \quad (33)$$

On the other hand, the Maurer-Cartan form ω of G is here considered under its tautological form, namely[3]

$$\omega = id_L$$

and we perform the following further piece of interpretation: since, according to (3), we have the identification $L = A^0(P,L)$ $\omega: L \to A^0(P,L)$ is taken as an element of A^{01}:

$$\omega \in A^{01} . \quad (34)$$

---

[3] This corresponds to restricting a (right or left) invariant one form on L with values in L to its value at the unit of G .

Now, with (33) and (34), the r.h.s. of the relations (31) actually makes sense - and in fact these relations hold true.

We conclude with a short comment on anomalies and the algorithm for computing them. Gauge field anomalies have been revealed by the Wess-Zumino compatibility condition [1] to be element of the one-cohomology $H^1(L,\Gamma^{loc})$ of L with values in local functionals of the potentials (connection one-forms). This cohomology is tightly related to the Chevalley cohomology of L with values in $A^*(P,L)$ described above (with boundary operator δ described in (22)) - the latter in fact serving to obtain anomalies in a purely algebraic way, which we now sketch. We denote by A the following element of ^1A:

$$A = a + \omega, \qquad (35)$$

and set, for $B \in {}^1A$

$$F^B = \Delta B + \frac{1}{2}[B \wedge B] \qquad (36)$$

and

$$\mathcal{D}^B \lambda = \Delta \lambda + [B \wedge \lambda], \quad \lambda \in {}^*A \qquad (37)$$

with the ensuing "Bianchi identity"[4]

$$\mathcal{D}^B F^B = 0. \qquad (38)$$

One has in particular

$$\mathcal{D}^{tA} F^{tA} = 0 \qquad (39)$$

and easily computes that

$$\mathcal{D}^{tA} A = \frac{d}{dt} F^{tA} \qquad (40)$$

and

---

4) (36), (37) and (38) pertain in fact to the general theory of GDL's, and thus apply to $(*A, \Delta, [\wedge])$.

$$F^A = P^a = da + \frac{1}{2}[a \wedge a] \tag{41}$$

(the last relation following from the BRS relations). With P an Ad-equivariant polynomial of order k on L it then follows that

$$k\Delta \int_0^1 P(A, F^{tA}, \ldots, F^{tA}) dt = P(F^a, \ldots, F^a) \tag{42}$$

(with an abbreviated notation implying antisymmetrization under the polynomial). This writes

$$\Delta Q^{2k+1} = P(F^a, \ldots, F^a) \tag{43}$$

denoting by ^{2k+1}Q the 2k+1-form

$$^{2k+1}Q = \int_0^1 P(A, F^{tA}, \ldots, F^{tA}) dt . \tag{44}$$

Decomposing ^{2k+1}Q in the double complex (30):

$$^{2k+1}Q = Q^{2k+1,0} + Q^{2k,1} + \ldots + Q^{0,2k+1} \tag{45}$$

then yields

$$\begin{cases} sQ^{2k+1,0} = F^a \\ sQ^{2k+1,0} + dQ^{2k,1} = 0 \\ sQ^{2k,1} + dQ^{2k-1,2} = 0 \end{cases} \tag{46}$$

Assuming $\mathbb{P}$ trivial we have, choosing 2k=dim M, after pulling back by some section and integration over the base M assumed retractable (using Stockes' Theorem)

$$s \int d^4x \, Q^{2k,1} = 0 \tag{47}$$

identical with the Wess-Zumino compatibility: this yields a purely algebraic computation of anomalies.

REFERENCES

[1]   J. Wess and B. Zumino, Phys. Lett. 37B (1971) 95

[2]   R. Stora, 1976 Cargèse Lectures, Eds. M. Levy, P. Miller, Nato ASI Series, Plenum 1977

[3]   D. Kastler, R. Stora, A differential-geometric setting for BRS Transformations and Anomalies I and II, to appear

COMPUTER QUANTUM FIELD THEORY[+]

C.B. Lang
Institut für Theoretische Physik
Universität Graz
A-8010 Graz, Austria

CONTENTS

1. INTRODUCTION

1.1 Preliminaries
1.2 Lattice Quantum Field Theory
1.3 Analytic Methods

2. COMPUTER METHODS

2.1 Generalities
2.2 Configuration Sampling
    2.2.1 "Ten Minutes in Monte Carlo"
    2.2.2 Microcanonical Approach
    2.2.3 Stochastic Differential Equations

2.3 Problems and Limitations
    2.3.1 Finite Size
    2.3.2 Local Updating
    2.3.3 Finite Amount of $'s

3. PURE GAUGE THEORY

3.1 Action
3.2 Observables
3.3 Critical Structure and Continuum Limit
3.4 Some Results for Pure Gauge Theory

4. GAUGE THEORY WITH MATTER FIELD

4.1 Action
4.2 Gauge-Higgs System Without Radial d.f.
    4.2.1 Higgs Field in the Fundamental Representation of SU(2)
    4.2.2 Higgs Field in the Adjoint Representation of SU(2)
    4.2.3 Results for Other Gauge Groups

4.3 Systems with Dynamical Radial Mode
4.4 QCD and Scalar QCD

[+] Lectures given at ZiF, Univ. Bielefeld within "Project No. 2: Mathematics + Physics" in March 1984; updated version completed in January 1985.

1. INTRODUCTION

1.1 Preliminaries

Infinitely many degress of freedom may be necessary to describe few physical quantities. Many different set-ups of a system often describe identical physics. This appears to be true especially for renormalizable Quantum Field Theories, which will be the topic of my lecture. In this case, as in many others, the reduction of the effective number of parameters from infinity to a few is due to the fixed-point structure on the critical surface of the theory. Prototypes of such models are the spin models of statistical physics and we rely heavily on this analogy. Most of the methods to deal with relativistic QFT's on discretized space-time have been developed in that context.

In the first Part we avoid to go to the subtleties of the lattice formulation of gauge theories; after the Introduction we discuss in Chapter 2 those methods of structural analysis where computers have been necessary (maybe not always sufficient) tools. In the subsequent chapters we concentrate first on pure gauge theory on the lattice in Chapter 3. In the 4th Part we present results for the engagement of matter fields (mainly bosonic) with gauge fields.

In view of the variety of reviews on this subject (Drouffe 1978a, Drouffe 1983a, Creutz 1983a, Creutz 1983b, Hasenfratz 1983, Kogut 1979, Kogut 1983, Rebbi 1983, Rebbi 1984) I choose to be quite restrictive in the topics to discuss. The emphasis lies on results obtained by computer methods and especially on pure gauge and gauge-Higgs systems.

1.2 Lattice Quantum Field Theory

The object of our desire is the Wick-rotated, Euclidean QFT in four dimensions (3-space, 1-time). The theory is regularized being formulated on a lattice, in most discussions this is a regular hypercubic one: $\Lambda \in \mathbb{Z}^4$. There are investigations of the more exotic simplicial lattice (Drouffe 1983b) and a formulation on a random-set of points

(Christ 1982). Up to now conclusions drawn from them do not disagree with what we have learned from the simply hypercubic system.

The Lagrangian density $L(\Phi)$ gives rise to a total action $S(C)$ for any given configuration C of field variables $\Phi(x)$,

$$S(C) = \sum_{x \in \Lambda} L(\Phi(x) \in C), \quad C = \{\Phi(x), x \in \Lambda\} \ . \tag{1.1}$$

Quantization via functional integration is analog to the sum over configurations in the macrocanonical formulation of statistical physics where the partition function is given through

$$Z = \sum_{\substack{\text{all configurations} \\ C = \{s(x), x \in \Lambda\}}} \exp[-(C)] \tag{1.2}$$

for the total energy, say

$$H(C) = \frac{1}{kT} \sum_{\substack{<xy> \\ \text{nearest neighbours}}} s(x)s(y), \tag{1.3}$$

for a system of spins $s(x)$. Observables are weighted averages over configuration space

$$<O> = \frac{1}{Z} \sum_C O(C) \exp[-H(C)], \tag{1.4}$$

with, e.g.

$$O(C) = \sum_{x \in \Lambda} s(x) s(x+n) \tag{1.5}$$

for a correlation function.

Obviously the analogy is as follows.

<u>Spin system</u>

spins $s(x)$

<u>Lattice QFT</u>

field variables $\Phi(x)$, $\psi(x)$, $U_{x,\mu}$ etc.

configuration $C = \{s(x), x \in \Lambda\}$	configuration $C = \{\Phi(x),\ldots, x \in \Lambda\}$
energy $H(C)$	action $S(C)$
inverse temperature $1/kT$	inverse coupling, e.g. for gauge theories $1/g^2$
sum over configuration $\sum_C$	sum over configuration $\sum_C$ corresponding to the lattice regularization of the functional integral $\int [d\Phi][\ldots]$, regularized $[d\Phi] = \prod_{x \in \Lambda} d\Phi_x$, with Haar measure $d\Phi_x$
correlation $<s(0)\, s(n)>$	propagator $<\Phi^+(0)\Phi(n)> \sim \exp(-m_\Phi \|n\|)$ for large distance $\|n\|$
order parameters, e.g. $<s>$	order or disorder parameters, e.g. $<\Phi>$ for Φ^4-theory, string tension for gauge theory, etc.
near phase transition: critical indices, universality	near higher order phase transition: $\xi/a \to \infty$ (lattice spacing a, correlation length ξ in physical units) $\to$ continuum limit.

The last point is essential: the lattice formulation of a QFT is a regularization and physics should eventually become independent of the regularization. This is the case at a phase transition of higher order where the correlation length in lattice units tends to infinity although in physical units it should remain constant. Physics should be independent of the scale - it ought to be the same as in the continuum. As will be discussed in more detail later, this explains that in lattice

QFT our prime interest lies in the behaviour at the critical point. In spin models there lies still physics in the behaviour away from the critical point, the lattice spacing may remain unequal zero.

Although we need some time before we turn to the definitions of gauge theory, I would consider it unfair to go ahead without introducing the variables at least.

(a) <u>Matter fields</u>

Bosons $\Phi(x)$, $\Phi^+(x)$ or fermions $\Psi(x)$, $\bar{\Psi}(x)$, live on the sites of the lattice (Fig. 1), elements of a finite dimensional vector space carrying a unitary or orthogonal representation of e.g. the gauge group. Fermions have the additional spinor character (i.e. another index) and are anticommuting objects generating a Grassmann algebra (Berezin 1966).

For a theory of free bosons the kinetic part of the continuum Lagrangian translates to the lattice like

$$(\partial_\mu \Phi)^+ (\partial_\mu \Phi) \to \frac{1}{a^2} (\Phi_{x+\mu} - \Phi_x)^+ (\Phi_{x+\mu} - \Phi_x)$$
$$= \frac{1}{a^2} \left[\Phi_x^+ \Phi_x + \Phi_{x+\mu}^+ \Phi_{x+\mu} - (\Phi_{x+\mu}^+ \Phi_x + \Phi_x^+ \Phi_{x+\mu}) \right] \quad (1.6)$$

from where we recognize the origin of the nearest neighbour coupling.

(b) <u>Gauge fields (vector fields)</u>

Gauge theories (cf. Abers 1973) are characterized by the invariance of the action under the gauge transformation

$$\Phi_x \xrightarrow[\text{g.tr.}]{} \Phi'_x = g_x \Phi_x , \quad (1.7)$$

g_x is element of the gauge group G (e.g. $N \times N$ unitary matrix for $SU(N)$) $= \exp(-i\varepsilon^\alpha(x)\tau^\alpha)$.

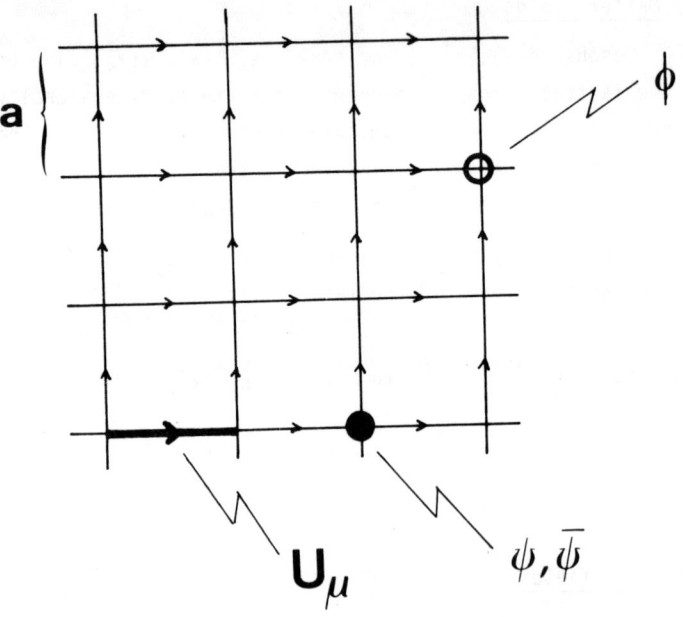

Fig. 1: Bosons (Φ) and fermions (Ψ) live on the sites and gauge U_μ on the links of a lattice.

This requires the introduction of gauge fields $A_\mu(x)$ and the replacements

$$\partial_\mu \phi_x \sim (\partial_\mu + ig A_\mu(x))\phi_x \qquad (1.8)$$

$$\phi_x^+ \phi_y \sim \phi_x^+ \underbrace{[P \exp ig \int_x^y A_\mu(x)dx^\mu]}_{U_{x \to y} \in G} \phi_y \qquad (1.9)$$

in order to obtain a gauge invariant action and operator. For nearest neighbour coupling this involves

$$U_{x \to x+a\mu} \equiv U_{x,\mu} \in G \ .$$

Wilson (1974) (and in the context of spin models already Wegner 1971) suggested a formulation of lattice gauge theory with this field operator, which (unlike $A_\mu(x)$) assumes its values on the compact group manifold. Introducing positive orientations of all such link operators we have

$$U_{x,\mu}^+ = U_{x+\mu,-\mu} \ . \qquad (1.10)$$

The gauge transformation formulated on the lattice is

$$U_{x,\mu} \xrightarrow[g.tr.]{} U'_{x,\mu} = g_x U_{x,\mu} g_{x+a\mu}^+ \qquad (1.11)$$

leaving the nearest-neighbour coupling term invariant.

Later we shall discuss the construction of the action and other gauge-invariant operators.

As you see, lattice fields are either site variables like spins (e.g. complex 3-component vectors for fields in the fundamental representation of $SU(3)$) or link variables (e.g. 3×3 unitary matrices for $SU(3)$).

1.3 Analytic Methods

We shall be content with the Lagrangian formulation, where the overwhelming majority of computer simulations has been done, and shall not discuss the Hamiltonian approach for continuous time. The Lagrangian should obey the Osterwalder-Schrader (on the lattice: Osterwalder-Seiler, cf. Osterwalder 1975, Osterwalder 1978, Seiler 1982) positivity properties that guarantee the possibility to undo the Wick-rotation (i.e. the existence of a transfer matrix formalism and a quantum mechanical state space with positive definite scalar product).

The problem is to find for some operators of physical interest the expectation value

$$<O> = \frac{\sum_C O(C) \exp[-S(C)]}{\sum_C \exp[-S(C)]} \qquad (1.12)$$

with sufficient accuracy for sufficiently large lattices.

As long as we do not have to give the configurations explicitly, we may keep the lattice volume arbitrarily large. Let me briefly mention some of the analytic methods, all inherited from statistical physics. This is the domain of rigorous results and I advise interested readers to the expert review of Seiler (1982).

(a) Weak coupling expansion

This is the well-known perturbation series in g for the non-compact fields $A_{x,\mu}$, however with technical complications due to the lattice geometry. In the asymptotically free gauge theories we know that terms like $\exp(-c/g^2)$ are relevant but are not picked up by the (at most asymptotic) weak coupling series. There is a point, however, where careful expansion is necessary: whenever one wants to establish the numerical connection to the regularized continuum theory one has to determine the ratios of the regularization dependent scale parameters of different schemes (e.g. $\Lambda^{mom}/\Lambda^{lattice}$, see Chapter 3.3).

(b) Strong coupling expansion

This is the expansion in $1/g^2$ ("high temperature expansion"); there are some rigorous results on the convergence domains that lead to general statements like: a lattice gauge theory confines static charges that represent the center of the group non-trivially. In practice there exist expansions to quite high orders (e.g. $(1/g^2)^{16}$ for the free energy, cf. Drouffe 1980) for some quantities of interest. Technically, the coefficients of the series amount to sums over e.g. closed surfaces in 4-dimensional space. Unfortunately, the unknown analytical structure away from the real $1/g^2$-axis prevents convergence towards smaller g^2-values. The quality of resummation techniques, however, depends strongly upon the information on the singularity structure.

Strong coupling expansion has been useful to check the computer results but otherwise not been particularly predictive.

(c) Saddle point methods and mean field approximations

Their reliability depends crucially on the dimensionality of the system and its order parameters. Whereas $d=4$ is the boundary for spin systems above which the approximation becomes exact, for gauge theories this approach provides usually only rough information on the possible position of critical points. The method can hardly decide the order of a phase transition or even whether it may be just a crossover phenomenon. More about this and related topics you find in the review of Drouffe and Zuber (Drouffe 1983a).

For finite size lattice and discrete gauge groups one may even think about **explicit summation**. This is out of reach. Take a group of order m and a lattice volume of N^4 points: there are m^{4N^4} configurations, an enormous number even for $m=2$ (group $Z(2)$) and $N=5$. However, do we really need this? Most of the configurations will have an action much larger than the rest, therefore contribute little to the partition function. This leads us to the ideas discussed in the next chapter.

2. COMPUTER METHODS

2.1 Generalities

A configuration of fields on a finite size lattice may be represented by a finite set of numbers. We want to determine the expectation value of some operator averaged over all possible configurations with Boltzmann weight.

$$<O> = \frac{1}{Z} \int [d\Phi]\, O(\{\Phi\})\, \exp[-S(\{\Phi\})]$$

$$= \frac{1}{Z} \sum_{\text{all } C} O(C)\, \exp[-S(C)]\,. \qquad (2.1)$$

Naive Monte-Carlo integration amounts to find an estimator for this sum by selecting randomly a finite number of configurations

$$<O> \approx \sum_{\substack{M \text{ randomly chosen} \\ C\text{'s with measure } [d\Phi]}} O(C) e^{-S(C)} \bigg/ \sum_{M\ldots} e^{-S(C)} \qquad (2.2)$$

sampled according to the measure $[d\Phi]$.

Note, however, that usually most of the configurations contribute little due to the suppression from the Boltzmann factor. To accelerate the procedure one samples the configurations according to their importance, i.e. one takes $\exp[-S(C)]\,[d\Phi]/Z$ as probability distribution in the Monte Carlo approach:

$$<O> \approx \frac{1}{M} \sum_{\substack{M \text{ configurations } C \\ \text{chosen according} \\ \exp[-S(C)][d\Phi]/Z}} O(C) \qquad (2.3)$$

This is, of course, only possible if the Boltzmann factors are real and non-negative. If this is not the case, there exist Langevin-equation type techniques to generate samples of configurations.

The computer approach therefore has two parts:

(A) generate large enough samples of configurations in thermal equilibrium
(B) perform measurements, block-spin transformations etc. on these configurations.

2.2 Configuration sampling

2.2.1 "Ten minutes in Monte Carlo"

We want to construct a Markov chain of configurations

$$C \xrightarrow{P(C \to C')} C' \quad ,$$

such that the desired Boltzmann-distribution $\exp[-S(C)]$ is an eigenvector of the transition probability, i.e.

$$\sum_C e^{-S(C)} P(C \to C') = e^{-S(C')} \quad , \qquad (2.4)$$

an equilibrium distribution transforms into itself. Most approaches use detailed balance

$$\frac{P(C \to C')}{P(C' \to C)} = \frac{e^{-S(C')}}{e^{-S(C)}} \qquad (2.5)$$

which saturates the eigenvector condition and is therefore a sufficient condition (c.f. Binder 1979). Out of the possible implementations of this framework two are best-known, and I want to discuss them now briefly.

(a) <u>Metropolis-method (Metropolis 1953)</u>

One proceeds according to the scheme:

(i) given some configuration

$$C = \{\Phi_1, \Phi_2, \ldots, \Phi_i, \ldots\}$$

choose

$$C' = \{\Phi_1, \Phi_2, \ldots, \Phi'_i, \ldots\}$$

with an a priori probability

$$P_o(C \to C') = P_o(C' \to C) .$$

This means for a Φ assuming its values in a group that the probability that Φ' lies a distance away in some direction in group space should be equal to that in the opposite direction.

Usually these changes concern only a single variable or even component of this variable - they are "local".

(ii) Determine the change in the action $(S'-S)$ due to the trial change $\Phi \to \Phi'$; now accept the new configuration with the probability

$$P(C \to C' \text{ with } S' \leq S) = 1$$

$$P(C \to C' \text{ with } S' > S) = \exp(-(S'-S)).$$

Lowering of the action is always accepted, an increase only with exponential probability. In practice, the distribution $e^{-(S'-S)}$ is obtained by choosing some random number $r \in (0,1)$

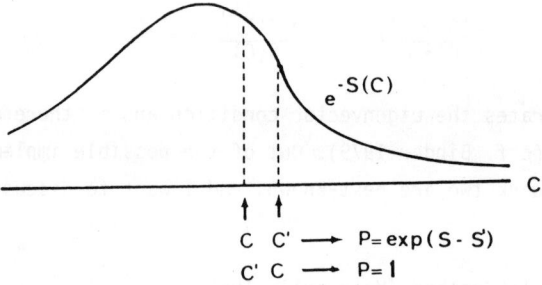

$C \;\; C' \longrightarrow P = \exp(S - S')$
$C' \;\; C \longrightarrow P = 1$

uniform and accepting C' if $r \leq \exp(-(S'-S))$.

(iii) go back to (i) and repeat the algorithm for all variables many times.

The updating of the variables may be performed in sequential or random order, or even simultaneous if they are independent. One iteration of all variables is called one "sweep" through the lattice. Sufficiently many sweeps have to be done such that the system forgets its initial values. As an example for the implementation of this algorithm on a computer see Bhanot (1982).

One may save some time, if one performs a couple of updates ("hits") for a variable before proceeding to the next one. For a large number of hits this is equivalent to another method, the

(b) <u>heat bath-method</u>

From a given configuration

$$C = \{\Phi_1, \Phi_2, \ldots, \Phi_i, \ldots\}$$

one chooses a new configuration

$$C' = \{\Phi_1, \Phi_2, \ldots, \Phi_i', \ldots\}$$

according to the probability $\exp[-S(\Phi_i, \text{all other fixed})]$; this is repeated for all variables in one "sweep". The applicability of this method depends crucially on the possibility to generate Φ_i's according to this distribution, i.e. on the complexity of the action. Most fast programs used today prefer a "multi-hit" Metropolis alogrithm.

2.2.2 Microcanonical Approach (Callaway 1982a,b)

The usual expectation value of some operator may be written equivalently

$$<O>_{can} = \frac{\int [d\Phi] O \exp(-S)}{\int [d\Phi] \exp(-S)} \tag{2.6}$$

$$= \frac{\int [d\phi][dp] O \exp(-S-\Sigma p^2/2)}{\int [d\phi][dp] \exp(-S-\Sigma p^2/2)}$$

where we have introduced an additional set of variables p, as many as there are ϕ's. Identifying

$$S + \Sigma p^2/2 \equiv \beta H$$

we recognize that $<O>$ is the expectation value of the operator of a canonical ensemble in a heat bath of temperature $1/\beta$. To describe a microcanonical ensemble one replaces $\exp(-\beta H)$ by $\delta(H-E)$; the system lives in phase space $\{\phi,p\}$ on a surface of constant $H = E$. The microcanonical expectation value is given by

$$<O>_{micro} = \frac{\int [d\phi][dp] O \, \delta(H-E)}{\int [d\phi][dp] \, \delta(H-E)} \qquad (2.7)$$

and may be obtained equivalently by following the development of a system described by the Hamilton equations

$$\dot{\phi}_i = p_i \;, \quad \dot{p}_i = -\frac{\partial S/\beta}{\partial \phi_i} \;. \qquad (2.8)$$

Thus the configurations are obtained by the iterative solution of a large set of coupled differential equations. For large systems $\exp(-\beta H)$ becomes increasingly peaked so that the microcanonical expectation value approaches the canonical one. One can show

$$<O>_{micro} = <O>_{can} (1 + O(\tfrac{1}{N})) \qquad (2.9)$$

where N denotes the number of degress of freedom.

In this approach one needs no random numbers, it is essentially the complexity of the system that accounts for the erratic behaviour. Parallel processing of the variables is possible which is especially helpful for non-local actions.

Another method baptized "microcanonical demon" by Creutz (1983c) is actually more like a heat bath. A "demon" travels from variable to variable carrying a bag of energy with him. Random changes of the variable are only accepted if the demon gains energy from the total S or if it has enough energy in its bag to afford a loss (cf. also Bhanot 1983b).

2.2.3 Stochastic Differential Equations

The idea is to construct a Markov process by which some given distribution density in configuration space $\rho(\Phi,\tau)$ approaches for $\tau \to \infty$ the distribution $\rho_\infty(\Phi) = \exp(-S(\Phi))/Z$ in equilibrium. In practice τ is the computer time.

The solution of a Fokker-Planck equation

$$\frac{\partial \rho(\Phi,\tau)}{\partial \tau} = \frac{\partial}{\partial \Phi_i}\left(\frac{\nu}{2}\frac{\partial}{\partial \Phi_i}\rho(\Phi,\tau) - \rho(\Phi,\tau)u_i(\Phi,\tau)\right) \qquad (2.10)$$

for the drift velocity

$$u_i = \frac{\nu}{2}\frac{\partial \ln \rho_\infty(\Phi)}{\partial \Phi_i} = -\frac{\nu}{2}\frac{\partial S}{\partial \Phi_i} \qquad (2.11)$$

has this property. The static equation is

$$u_i(\Phi)\rho(\Phi) = \frac{\nu}{2}\frac{\partial \rho(\Phi)}{\partial \Phi_i} \qquad (2.12)$$

giving the desired equilibrium distribution. One may construct a sequence of configurations with the help of the corresponding Langevin-equation (e.g. Klauder 1983, Parisi 1981, Fucito 1981, Risken 1984)

$$\dot{\Phi}_i = u_i(\Phi) + \xi_i(\tau) \; , \qquad (2.13)$$

where ξ_i is a random variable producing Gaussian white noise

$$\langle \xi_i(\tau)\xi_j(\sigma)\rangle = \nu \delta_{ij}\, \delta(\tau-\sigma) ,$$
$$\langle \xi(\tau)\rangle = 0 . \qquad (2.14)$$

Then the subsequent configuration is obtained via

$$\phi_i^{(n+1)} = \phi_i^{(n)} + u_i(\phi^{(n)})\Delta\tau + \xi_i^{(n)}\Delta\tau \qquad (2.15)$$

and expectation values are evaluated through

$$\langle O\rangle = \lim_{T\to\infty} \frac{1}{T}\int_0^T O(\Phi(\tau))d\tau = \lim_{\tau\to\infty}\int O(\Phi(\tau))\rho(\Phi,\tau)[d\Phi] \quad (2.16)$$

(cf. Fig. 2).

Once again parallel processing is possible and only changes of S enter in the updating procedure. This means that unlike the Monte-Carlo methods e^{-S} needs not to be positive since it is not considered a probability distribution and S may even be complex (cf. Klauder 1983).

In all the discussed approaches one starts with an arbitrary (e.g. completely random ≡ hot, or completely ordered ≡ cold) configuration and discards sufficiently many configurations. As soon as one is reasonably sure to obtain configurations in equilibrium distribution one starts measuring. To decide when, is one of the most subtle points in practice.

It is common to all approaches that one usually introduces no explicit gauge fixing. Since the distribution is automatically normalized one does not bother about the superfluous additional variables, they even are helpful for randomization.

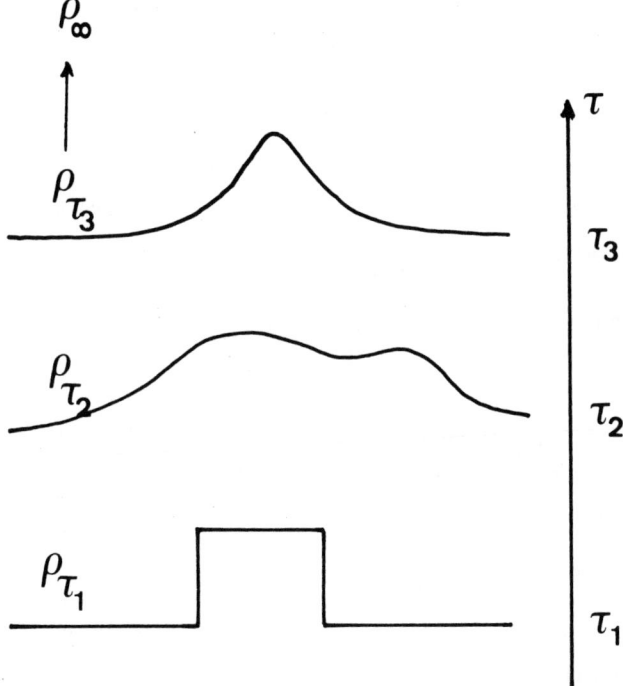

Fig. 2: Development of the distribution density with the (computer) time τ.

2.3 Problems and Limitations

2.3.1 Finite Size

As will be discussed in Chapter 3.3, the most interesting point of our theory usually is where a phase transition with diverging correlation length occurs: $\xi/a \to \infty$. It is there where one hopes to obtain continuum-like results. Obviously this situation cannot be reproduced on a finite lattice. Measurements will be reliable in a region

$$\text{lattice constant } a \ll \text{measurement region} \lesssim \xi \ll \text{linear lattice size } N\,a. \quad (2.17)$$

In actual calculations performed up to now the corresponding numbers are typically

$$0.1 - 0.2 \text{ fm} \lesssim 0.2 - 1.5 \text{ fm} \lesssim \xi \lesssim 1.6 - 3.2 \text{ fm}.$$

One is in a situation like between Scylla and Charybdis (Homer -750): if the lattice spacing is too big, the results are not close enough to the continuum situation; if it is small enough, the whole lattice becomes too small compared to the physical size of the object measured. I shall come back to these problems at the end of Chapter 4.

Meanwhile let us mention some positive aspects of finite size effects. Since the maximal value the correlation length may assume is proportional to the lattice extension, one may obtain information on the critical exponents from this dependence (<u>finite size scaling</u>, cf. Barber 1983). Another approach is to study the dependence of the system on the type of boundary conditions. Observables like the Wilson-loop are bounded

$$\langle O \rangle_{\text{free b.c.}} \leq \langle O \rangle_{\text{infinite lattice}} \leq \langle O \rangle_{\text{fixed b.c.}} \quad (2.18)$$

In the framework of LGT this behaviour has been studied by Mütter (1982).

In most calculations one uses periodic b.c. (for bosonic variables, antiperiodic for fermionic variables). We know from the experience with spin models (Binder 1979) that periodic b.c., if compared to others, usually give results closest to the thermodynamic limit.

There is a situation where one is interested explicitly in a finite size, i.e.

$$aN_\tau \text{ finite} \ll a N_s \to \infty .$$

Such a system with large space volume, finite time size and periodic (for bosons) b.c. in τ-direction may be interpreted as a system at a non-zero physical temperature $1/aN_\tau$ (Gross 1981). Phenomenology of QCD at finite temperature has been studied by many groups (Kuti 1981, McLerran 1981), extensively by the Bielefeld group (Celik 1983, Satz 1984a,b, cf. also Ch. 3).

2.3.2 Local Updating

The algorithms to produce configurations as presented above are intrinsically local. A new configuration is obtained from the old one by many small changes. Imagine now some configuration or a set of configurations getting trapped in a sort of local maximum of the Boltzmann factor. Eventually it will be able to escape due to the Markov property; however, the actual measurements consider only a finite number of configurations and may come out biased in such a situation. It is possible that different local maxima are related by global transformations leaving the action invariant.

Such an example is the Ising model without external field at large $1/kT$. An overall spin-flip leaves the total energy invariant. On a finite lattice the expectation value of the magnetization vanishes always; investigation of the time evolution shows long periods of seemingly stable positive or negative values of the configuration magnetization values. Another such example is the thermal loop (Polyakov loop, cf. Ch. 3) in gauge theories. In these cases one profits from the local updating; the local maxima are disconnected in the thermodynamic limit. The results in the long (CP-time) periods of positive or negative magnetization are close to those for infinite volume. Allowing for global flips would destroy this feature.

A situation where such effects are not welcome is the following. Imagine a specific configuration in a gauge-Higgs theory (cf. Ch. 4) where all link variables in a direction are equal to some group element unequal unity and all other links equal unity. This configuration has unit plaquette-variables. However, due to the periodicity it cannot be gauge-transformed into one with only unit link variables! Higgs fields living on such a configuration feel the difference since the observable $TR[\mathbb{1} - \phi_x^+ U_{xy} \phi_y]$ cannot assume its true minimum 0 (cf. Lang 1981b).

Further examples are systems where ferromagnetic couplings compete with anti-ferromagnetic ones (Bhanot 1983a, Stamatescu 1983). Such locking-in phenomena might result in uncomplete consideration of the contributions from different topological sectors in configuration space (Fucito 1984).

2.3.3. Finite Amount of $'s

In stochastic methods the accuracy improves with $1/(CP\text{-time})^{1/2}$ ~ $1/(\text{money})^{1/2}$. At the same time one has to fight finite size problems and the required CP-time is proportional to the lattice volume.

Let me mention some numbers. One update of a SU(3) gauge field configuration on a 16^4 lattice ($\approx 3.10^6$ variables) with a 10-hit Metropolis method takes roughly 60 sec. on a CDC 7600 or 10 sec. on a CYBER 205. In order to obtain reliable estimates of some observables, one needs at least several thousand configurations, let alone the problem with fermions (Ch. 4).

Although this might be sufficient reason for pessimism the actual development in the last years shows almost exponential growth in the lattice size (Fig. 3). This is related to the improved accessability to supercomputers.

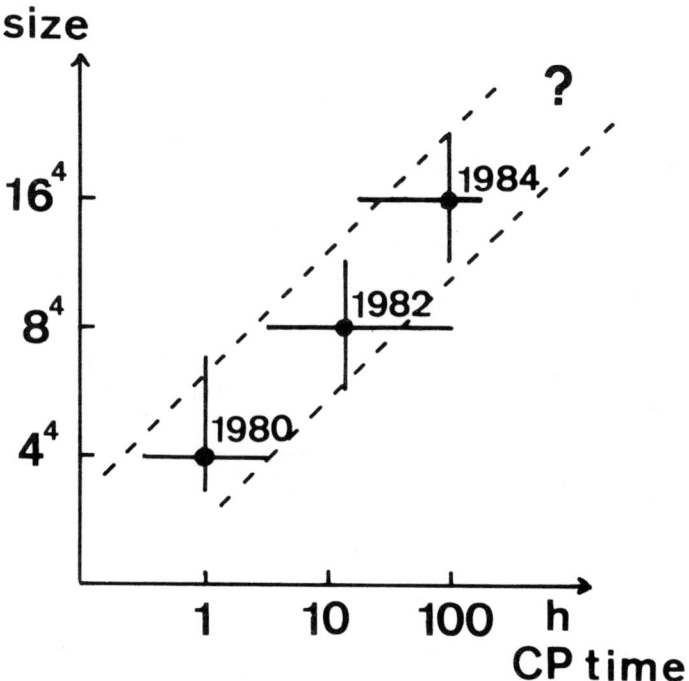

Fig. 3: Sketch of the amount of CP-time on a fast vector-processor (like CRAY I or CYBER 205) used for typical projects in lattice gauge theory; the typical values for lattice size used in the last years is indicated, too.

3. PURE GAUGE FIELD THEORY

3.1 Action

The gauge fields $U_{x,\mu}$ live on links. A gauge transformation amounts to choose group elements for a set of sites: g_{x_1}, g_{x_2}... and transform the link variables according to (1.11). Then, under such a gauge transformation a product of link variables along some polygon $P_{i \to m}$ transforms according to

$$\prod_{P_{i \to m}} U \equiv U_{ij} U_{jk} \cdots U_{nm} \longrightarrow$$

$$g_i U_{ij} g_j^+ g_j U_{jk} g_k^+ g_k \cdots g_n^+ g_n U_{nm} g_m^+ \qquad (3.1)$$

$$= g_i \left(\prod_{P_{i \to m}} U \right) g_m^+ \;.$$

Considering closed polygons (i=m) we find that the class function

$$\chi \left(\prod_{P_{i \to i}} U \right) \to \chi(g_i \prod_{P_{i \to i}} U \, g_i^+) = \chi \left(\prod_{P_{i \to i}} U \right) \qquad (3.2)$$

is invariant under a gauge transformation. This function χ may be any combination of characters of the group.

The action should be gauge invariant. We therefore write it as a linear combination

$$S = \sum_{\substack{\Lambda \text{ different} \\ \text{loop-shapes} \\ P_\ell}} \sum_{\substack{\text{all irred.} \\ \text{representations} \\ i}} c_i^{(\ell)} \chi_i (\prod_{P_\ell} U) \qquad (3.3)$$

$$= \sum_\Lambda c_f^\square \chi_f(U_\square) + \cdots$$

(f denotes the fundamental representation).

The simplest non-trivial closed curve is the plaquette

$$U_\Box = U_{x,\mu} U_{x+\mu,\nu} U^+_{x+\nu,\mu} U^+_{x,\nu} \qquad (3.4)$$

The action originally proposed by Wilson (1974) reads

$$S = \sum_{\Box \in \Lambda} S_\Box , \quad S^{Wilson}_\Box = -\frac{1}{g^2} Tr(U_\Box + U^+_\Box). \qquad (3.5)$$

As an example consider compact QED: gauge group $U(1)$.

$$U_{x,\mu} \equiv \exp(i\theta_{x,\mu}),$$
$$U_\Box = \exp(i\theta_{x,\mu\nu}), \quad \theta_{x,\mu\nu} \equiv \theta_{x,\mu} + \theta_{x+\mu,\nu} - \theta_{x+\nu,\mu} - \theta_{x,\nu}, \qquad (3.6)$$
$$S^{Wilson}_\Box = -\frac{2}{g^2} \cos\theta_{x,\mu\nu}.$$

There are infinitely many other choices, some of them of merits due to their simple form. Manton (1980) suggested

$$S^{Manton}_\Box = \frac{1}{g^2} d(U_\Box, \mathbb{1})^2 \qquad (3.7)$$

where $d(U,\mathbb{1})$ denotes the distance between U and the unit element on the group manifold. For $U(1)$ e.g.

$$S^{Manton}_\Box = \frac{1}{g^2} \theta^2_{x,\mu\nu} \qquad (3.8)$$

i.e. the first non-trivial term in a θ-expansion of S^{Wilson}.

Another choice is (Drouffe 1978, Menotti 1981) $S^{Heat\ Kernel}$:

$$\exp(-S^{HK}_\Box (U_\Box, N/g^2)) = K(U, N/g^2)/K(\mathbb{1}, N/g^2) \qquad (3.9)$$

(N for SU(N)) where K is the fundamental solution of the diffusion equation of the group manifold

$$\frac{\partial K}{\partial t} = \Delta K \qquad (3.10)$$

(Δ is the Laplace-Beltrami operator on the group). For U(1) one has

$$K(U, 1/g^2) = \sum_{n=-\infty}^{\infty} \exp(-\frac{1}{g^2}(\theta + 2n\pi)^2). \qquad (3.11)$$

S^{Manton} and S^{HK} have been studied in the Monte Carlo approach for SU(2) first by Lang (1981a, 1982b).

Further choices are

$$S_\square^{Mixed} = -\beta_f \chi_f(U_\square) - \beta_a \chi_a(U_\square), \qquad (3.12)$$

(f and a denote fundamental and adjoint representation), cf. Bhanot (1981), and

$$S_\square^{TI} = -\beta_f(\frac{5}{3}\chi_f(U_\square) - \frac{1}{12}\chi_f(U_{\rectangle})) \qquad (3.13)$$

where $\rectangle$ denotes the rectangular double-plaquettes of size 1×2. The latter action is chosen such as to improve asymptotic scaling by removing the leading log corrections (Symanzik 1982, 1983a,b, Weisz 1983, Curci 1983, Belforte 1983).

Naively ("formally") we might take a continuum limit by letting $a \to 0$ in the action; correctly we have to do this in the full, integrated theory (cf. the discussion in Ch. 3.3). In the naive limit

$$U_{x,\mu} \to \exp(i\,a\,g\,A_{x,\mu})$$

$$U_{x,\mu\nu} \to \exp(i\,a^2 g\,(F_{\mu\nu} + O(a)))$$

$$\text{where } F_{\mu\nu} = \partial_\mu A_\nu - \partial_\nu A_\mu - ig[A_\mu, A_\nu]$$

$$S \to \int d^4x \, \text{Tr}(F_{\mu\nu}^2) + \text{terms of higher order in } a \, . \tag{3.14}$$

In lowest order all actions have the same formal continuum limit and all of them contain the fundamental representation, i.e. the center group (Z(N)) for SU(N)) is represented faithfully.

Universality would require that all choices give the same physics. Up to a rescaling one should obtain the same results in the continuum limit (cf. Ch. 3.3).

3.2 Observables

Gauge-invariant observables may be constructed out of class functions of the ordered product of link variables along a closed curve $\chi(\prod_P U)$.

(a) $<\text{Tr}U_\square>$

This is the simplest observable corresponding to the <u>internal energy</u>

$$E = \partial \ln Z / \partial \beta \tag{3.15}$$

of our system ($\beta = 2N/g^2$ for SU(N)). Correlations of such plaquette variables give the specific heat

$$C = -\beta^2 \frac{\partial E}{\partial \beta} = \beta^2 N_\square (<(\text{Tr}U_\square)^2> - <\text{Tr}U_\square>^2) \tag{3.16}$$

where $N_\square$ denotes the number of plaquettes on the corresponding lattice.

(b) Wilson loop $W(R,T) = <\text{Tr} \sqsupset_R^T>$

This quantity is proportional to the expectation value to find a static charge along the rectangular path ($R \times T = n_R a \times n_T a$) (Wilson 1974).

For large T one expects

$$W(R,T) \sim \exp(-E_{q\bar{q}}(R)T) \tag{3.17}$$

It can be shown rigorously (Wilson 1974, Osterwalder 1978, Seiler 1982) that for gauge groups with non-trivial center (like $U(1)$ or $SU(N)$) there is a convergent strong coupling expansion (around $1/g^2 = 0$) with

$$W(R,T) \underset{\text{large } n_R, n_T}{\lesssim} \exp(\underbrace{-\sigma RT}_{\sigma a^a n_R n_T}) \,. \tag{3.18}$$

This is the so-called "area law", implying $E_{q\bar{q}}(R) \sim R$, a linearly rising confining potential.

In MC-calculations one extracts from the expectation values W estimates for σa^2 at different values of $1/g^2$. We identify this "string tension" as an order parameter

$$\sigma \begin{cases} \neq 0 & \text{confinement phase} \\ & \text{("area law")} \\ \\ = 0 & \text{unconfined phase} \\ & \text{("perimenter law")} \end{cases} \tag{3.19}$$

(c) Thermal loop $L = \langle \text{Tr} U_\uparrow \rangle$

Here one takes the product of link variables running through a point in 3-space $\vec{x}$ along a line in τ-direction through the whole period, closed only by virtue of the periodicity in τ. One expects

$$L \sim \exp(-E_q a N_\tau) \tag{3.20}$$

for a lattice of time-extension N_τ. E_q is the gluonic energy of a static charge at $\vec{x}$. This expectation value is an order parameter sensitive to the physical temperature (not to be mixed up with g^2) of the system (cf. Ch. 3.2). One has

$$L \begin{cases} = 0 & \text{confinement} \quad T < T_c \\ \\ \neq 0 & \text{no confinement} \quad T > T_c \\ & \text{(Debye screening} \\ & \text{of static charges)} \end{cases} \tag{3.21}$$

This thermal phase transition occurs also for theories with confinement permanent with respect to $1/g^2$, since it is due to the finite size in τ-direction (cf. Celik 1983, Satz 1984a,b).

From correlations between thermal loops at different spatial points

$$\langle L_{\vec{x}} L^*_{\vec{y}} \rangle \sim \exp(-E_{q_{\vec{x}} \bar{q}_{\vec{y}}} \, a \, N_\tau) \, , \tag{3.23}$$

we obtain the energy of a system of a static charge and an anticharge separated by the spatial distance $|\vec{x}-\vec{y}|$; this may be interpreted as static potential.

(d) Loop-loop correlations

Products of Wilson-loops separated by some distance should behave like

$$\langle \, \square \underset{n_t a}{\longleftrightarrow} \square \, \rangle \sim \exp(-m_g \, a \, n_t) = \exp(-a \, n_t/\xi_{phys}) \tag{3.24}$$

where ξ_{phys} denotes the correlation length and m_g the mass gap. For non-abelian gauge theories one expects a purely gluonic state called "glueball" with non-vanishing mass m_g.

3.3 Critical Structure and Continuum Limit

The continuum limit should be taken at values of the coupling constant where the physics (the physically meaningful observables) becomes independent of the lattice constant scale. There one expects the same behaviour for each scale size, thus $\xi_{phys}/a \to \infty$. That behaviour is characteristic for a 2nd order phase transition. For this reason such a critical point is essentially the only point of interest in our whole approach. In spin systems the actual lattice spacing is never really zero, the behaviour of the system away from the critical point is therefore also of interest. In QFT it is only the critical behaviour that leads us to the desired continuum quantities.

Let us try to understand universality, independence of physical results from the specific form of the action. In the parameter space of all possible couplings we expect a domain of criticality (critical surface) where $\xi/a = \infty$. Let us assume that we apply a scale changing operation to the system (e.g. a block-spin transformation by integrating out local interactions). Then if we started on the critical surface a change of scale will transport us to another point on it. Repetition will bring us to some fixed point, its position depending on the type of transformation. However, everywhere in the domain of attraction of this f.p. we find the same physics, the same long distance behaviour, the same critical exponents. Different theories corresponding to different starting points thus have universal properties, they belong to one universality class. Directions that keep us within the critical surface are called irrelevant, corresponding to irrelevant contributions to the actions.

Starting somewhat away from the surface the transformation would transport us away towards the trivial f.p. at $\xi/a = 0$, into a "relevant" direction. The speed of that move away is controlled by the corresponding critical exponent. In general we expect only few (often only one) relevant operators.

Non-abelian gauge theories are asymptotically free, their critical surface is given by $g = 0$. The leading relevant operator is $F_{\mu\nu}^2$, however there are many possible terms in the lattice action contributing to it. Their relative weight at the f.p. will depend on the actual BST applied and is not universal.

For a continuum limit (removal of the regularization parameter a) we require independence of physical quantities P on a.

$$\frac{\partial P(g,a)}{\partial \log a} = \left(\frac{\partial}{\partial \log a} + \frac{\partial g}{\partial \log a} \frac{\partial}{\partial g} \right) P(g,a) = 0 , \qquad (3.24)$$

where

$$-\frac{\partial g}{\partial \log a} = \beta(g) = -\beta_0 g^3 - \beta_1 g^5 + O(g^7) , \qquad (3.25)$$

may be determined from perturbation theory in g. (The RHS of Eq. (3.24) is actually more precise $O(\frac{a^2}{\xi^2} \log \frac{a}{\xi})$ on the lattice.) For the gauge group SU(N) and n_f flavours of massless fermions (fundamental representation) one obtains

$$\beta_0 = \frac{1}{(4\pi)^2} \left(\frac{11}{3} N - \frac{2}{3} n_f \right)$$

$$\beta_1 = \frac{1}{(4\pi)^4} \left(\frac{34}{3} N - \frac{10}{3} N n_f - \frac{N^2-1}{N} n_f \right).$$

(3.26)

Thus we see that the change of g with the scale away from the f.p. g = 0 is non-linear in g. A linear approximation would simulate marginality (one stays on the critical surface), higher order corrections move one away into the relevant direction (Gross 1973, 1976; Politzer 1973, Kogut 1974, Hasenfratz 1984).

Integration of the above differential equation gives

$$a = \frac{1}{\Lambda} f(g) \tag{3.27}$$

$$f(g) = (\beta_0 g^2)^{-\beta_1/2\beta_0^2} \exp(-\frac{1}{2\beta_0 g^2}) (1 + O(g^2)). \tag{3.28}$$

Here the integration constant fixes the scale; its value can be determined only non-perturbatively. It depends on the details of the regularization scheme, i.e. for lattice theories also on the specific form of the action. The ratio of $\Lambda_{lattice}$ for a given action to $\Lambda_{continuum}$ for a given continuum regularization scheme may be determined by perturbative expansion at $g \to 0$ (up to $O(g^2)$). For $n_f = 0$ Hasenfratz (1980) obtained

$$\frac{\Lambda_{MOM}}{\Lambda_{lattice, Wilson}} = \begin{cases} 57.5 & \text{for SU(2)}, \\ 83.3 & \text{for SU(3)}, \end{cases} \tag{3.29}$$

values verified and extended to $n_f \neq 0$ in subsequent work (Dashen 1981, Weisz 1981, Kawai 1981, Celmaster 1982). Ratios of Λ parameters for different lattice actions have been determined and checked in Monte Carlo calculations by various authors (e.g. Lang 1981a, 1982b, Gonzales-Arroyo 1982, cf. Creutz 1983a, Berg 1984).

Therefore, we expect $a \to 0$ for $g \to 0$ if the lattice approximation gives the right continuum theory. Actually, this behaviour has been confirmed in MC-calculations (see Ch. 3.4).

Different actions differ in their results for $g > 0$, i.e. $a > 0$. Such a difference has been observed e.g. in Fig. 4 where we sketch the phase structure in a subspace of 2 couplings (fundamental and adjoint character of $U_\square$) for S^{Wilson}, S^{Manton} and S^{HK} of Chapter 3.1. For $SU(N)$, $N \geq 4$ we find even a first order phase transition for S^{Wilson}: the line of phase transitions from the upper left corner crosses the Wilson-line. However, the order parameter (string tension) does not vanish, the phase transition may be circumvented by choosing another action. Only the limit $g \to 0$ should be universal.

Physical quantities like masses should asymptotically scale

$$m_L = m \, a(g) = m \frac{1}{\Lambda_L} f(g) \tag{3.30}$$

where m_L denotes the dimensionless mass as determined by lattice calculations. Ratios between different physical quantities should become independent of g^2 in the scaling regime

$$\frac{m_L}{\tilde{m}_L} = \frac{m \, a(g)}{\tilde{m} \, a(g)} = \frac{m}{\tilde{m}} \quad (\text{up to } O(\tfrac{a^2}{\xi^2} \log \tfrac{a}{\xi}) O(g^2)). \tag{3.31}$$

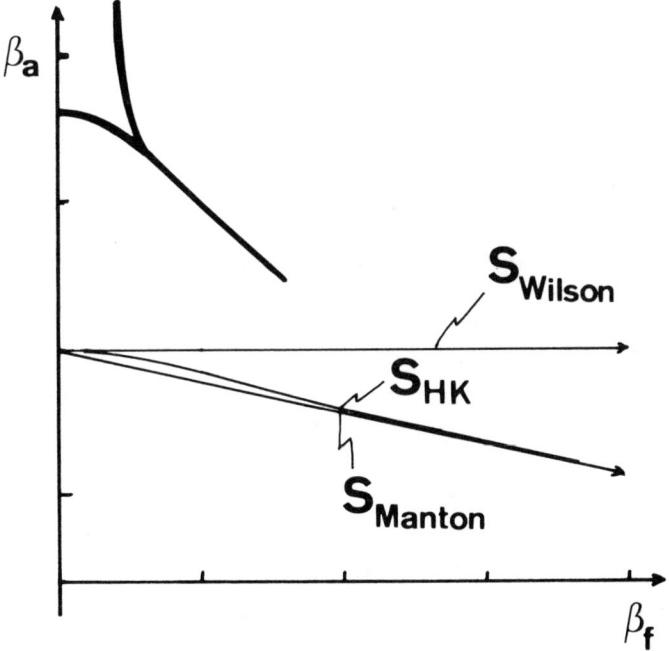

Fig. 4: Projections of some action forms discussed in Ch. 3.1 into the (β_f, β_a)-plane together with phase transition lines in that plane found by Bhanot (1981).

Let us assume that we work at a value of g sufficiently small (to be checked afterwards by consistency of the results). How can we determine $a(g)$? In Fig. 5 you can see how it is done. Results for, say, a propagator at different values of the coupling constants g are rescaled such that they coincide. This gives the functional dependence $a(g)$ up to a constant, the scale parameter Λ. It may be fixed by trading its value for e.g. an experimentally determined physical mass value.

In pure gauge theory the most popular quantity is the expectation value of Wilson-loops

$$W(R,T) \sim \exp(-\sigma_{phys}\, a^2(g)) , \qquad (3.32)$$

and indeed it turned out that there is a regime (cf. Fig. 6) where

$$\sigma_L = \sigma_{phys}\, a^2(g) \sim \frac{\sigma_{phys}}{\Lambda_L^2}\, f^2(g) ,$$

$$\sigma_L / f^2(g) \xrightarrow[g \to 0]{} \sigma_{phys} / \Lambda_{L,Wilson}^2 \approx 1.1 \times 10^4 \pm 10\% , \qquad (3.33)$$

(first determined by Creutz 1979a, recently on a lattice of size $16^3 \times 32$ by Barkai 1984). The string model relates σ to the Regge slope (Goddard 1973) and gives

$$\sigma_{phys} \approx 1/2\Pi\alpha' \approx 0.159\, \mathrm{GeV}^2$$

$$\sim \Lambda_{L,Wilson}^{SU(3)} \approx 3.8\, \mathrm{MeV} . \qquad (3.34)$$

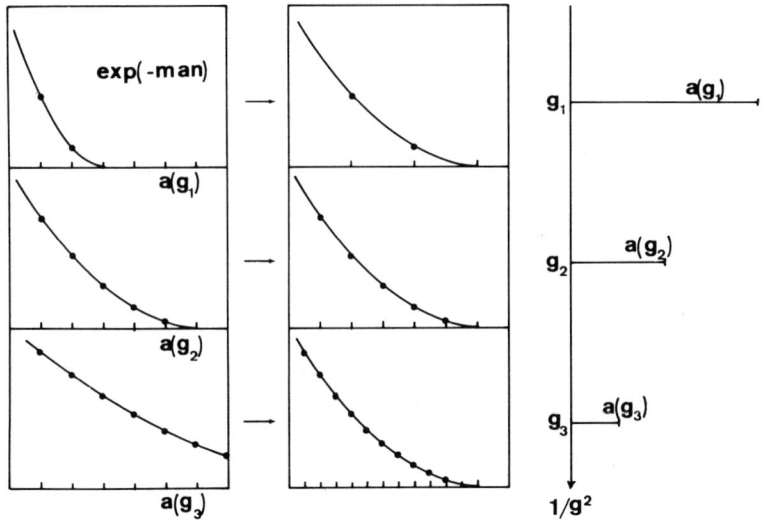

Fig. 5: Sketch of the way to determine a(g). The left column gives results that might have been obtained for some correlation function decaying like exp(-amn). In the middle column the scale a(g) has been modified such that the three curves obtained for different g-values agree. The resulting values a(g) are plotted in the left-most part.

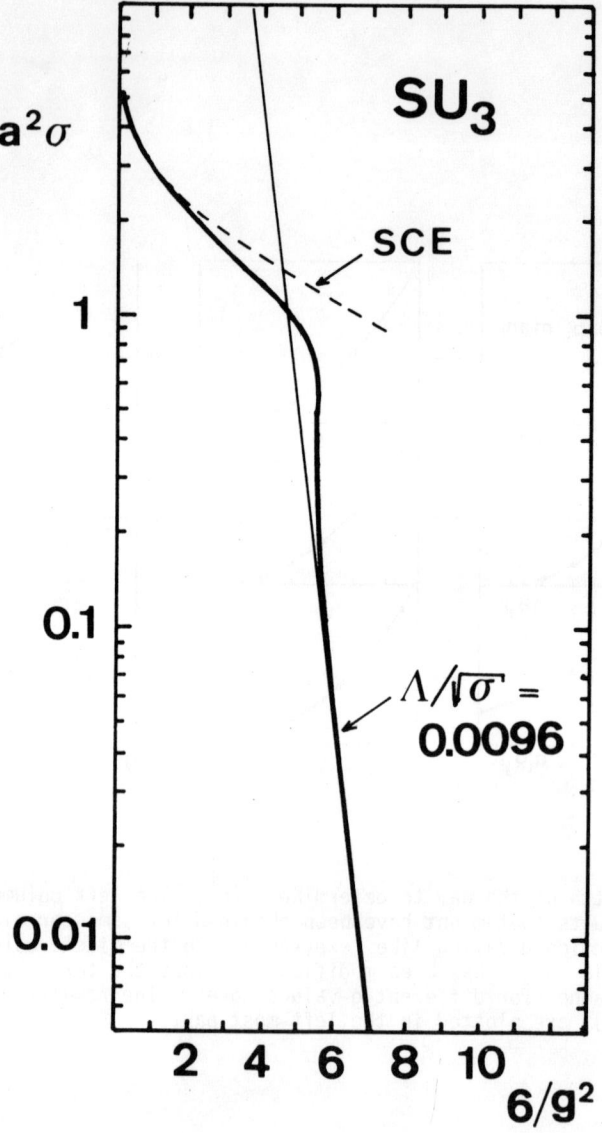

Fig. 6: The string tension for SU(3) pure gauge theory as obtained by MC-calculations.

3.4 Some Results for Pure Gauge Theory

Let me give you some examples which are typical for the type of calculations performed. Obviously, I cannot be complete and refer you to the original work and other reviews for more details. For the notation cf. Ch. 3.1.

ad (a)

The internal energy variable has been evaluated in Monte Carlo calculations usually in "thermal cycles" continuously varying from low to high values of the coupling and back. In Fig. 7 (Creutz 1979b) one finds clear hysteresis signals for $Z(N)$ gauge theories indicating phase transitions. For $N \to \infty$ one approaches the gauge group $U(1)$ and investigations of the specific heat exhibited a peak near $1/g^2 = 1$ that increases for increasing lattice size (Fig. 8). This is a signal for a 2nd order phase transition (Lautrup 1980). Subsequent calculations lead to a somewhat more refined picture: there appears to be a line of phase transitions in the fundamental-adjoint plane for $U(1)$, but for the Wilson action the order is still weak 1st. There may be tricritical structure in the space of possible coupling constants (Jersak 1983, Evertz 1985).

For $SU(2)$ and $SU(3)$ one also finds slope maxima of the energy but no finite size scaling.

ad (b)

In Fig. 9 one sees that for $U(1)$ the string tension σa^2 vanishes at some coupling value $1/g^2 \approx 1$ thus for compact QED we find a confinement phase and a free Coulomb phase as expected from general arguments (Guth 1980, Fröhlich 1982a, cf. Seiler 1982).

Results for the string-tension for non-abelian gauge groups (Fig. 6 for $SU(3)$) show different behaviour. There it seems that σa^2 vanishes no-where and behaves as expected from the renormalization group arguments in Ch. 3.3. Recently, there has been some discussion on the onset of scaling for $SU(3)$ (cf. Hasenfratz 1984). Most authors would agree with the value

$$\Lambda_{L,Wilson}^{SU(3)} \approx 0.095 \sqrt{\sigma_{phys}} \pm 10\% \approx 3.8 \text{ MeV} . \qquad (3.35)$$

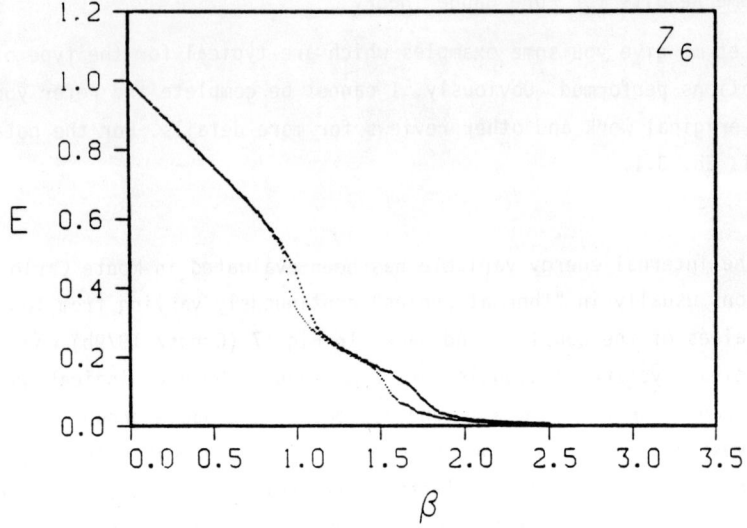

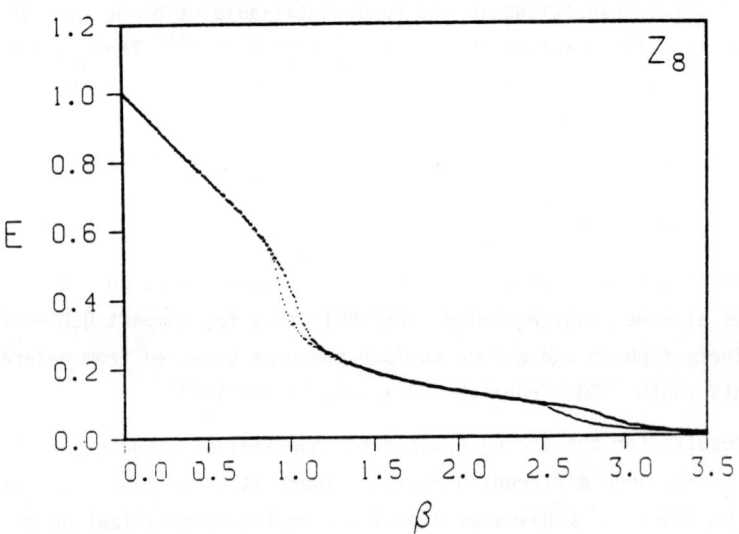

Fig. 7: Results by Creutz (1979b) for $Z(N)$ gauge theories; the phase transition at $\beta \approx 1$ remains at this value for $N \to \infty$, i.e. for $U(1)$, whereas the crystallizing transition moves to infinity.

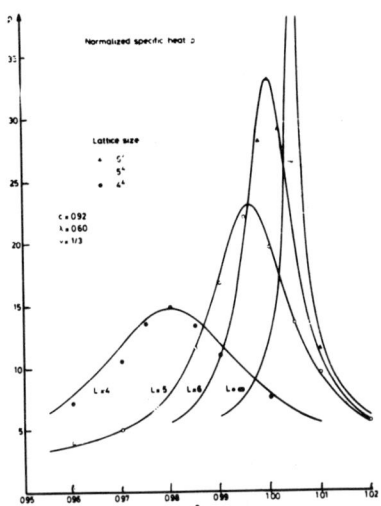

Fig. 8: The specific heat for U(1) gauge theory and lattice size 4,5 and 6 (Lautrup 1980) - the scaling peak seems to indicate a 2nd order phase transition, see, however, Fig. 9.

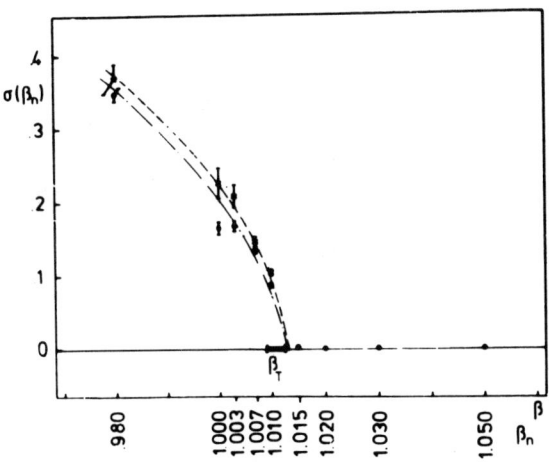

Fig. 9: The string tension for U(1) gauge theory has a small discontinuity at the transition to the Coulomb-phase indicating that it is actually weak 1st order for the Wilson-action (Jersak 1983).

We conclude that MC-results indicate confinement for all values of g as well as asymptotic freedom.

ad (c)

The above results are meant to approximate the situation of infinite lattice volume and physical temperature $T=0$. For $N_\tau \ll N_s$ one obtains information on systems with finite $T \neq 0$. In Fig. 10 (Gavai 1983) we find that the corresponding order parameter L deviates from zero above some temperature. The actual phase transition where thermal deconfinement takes place is somewhat obscured due to the finite space volume. For SU(3) one finds a deconfinement temperature

$$T_c \approx 0.65 \sqrt{\sigma_{phys}} \approx 260 \text{ MeV} . \qquad (3.36)$$

As mentioned, the correlations of the thermal loops at different positions in 3-space lead to the energy of a pair of static charges. This may be interpreted as static potential and for N_τ sufficiently large (T small) in order to remain in the confinement phase one finds a shape consistent with

$$aV(n_r) \sim \sigma a^2 n_r + \frac{c}{n_r} , \qquad (3.37)$$

i.e. a Coulomb term plus a linear rising confining term (cf. Fig. 11 by Stack 1984) in agreement with the qualitative expectations.

We may evaluate the information obtained for $V(|\vec{x}-\vec{y}|)$ to find

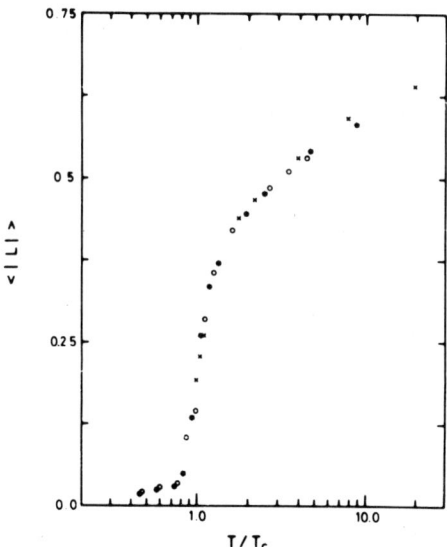

Fig. 10: The order parameter for thermal deconfinement is plotted versus the temperature; these results have been obtained for different types of actions for SU(2) gauge theory by Gavai (1983).

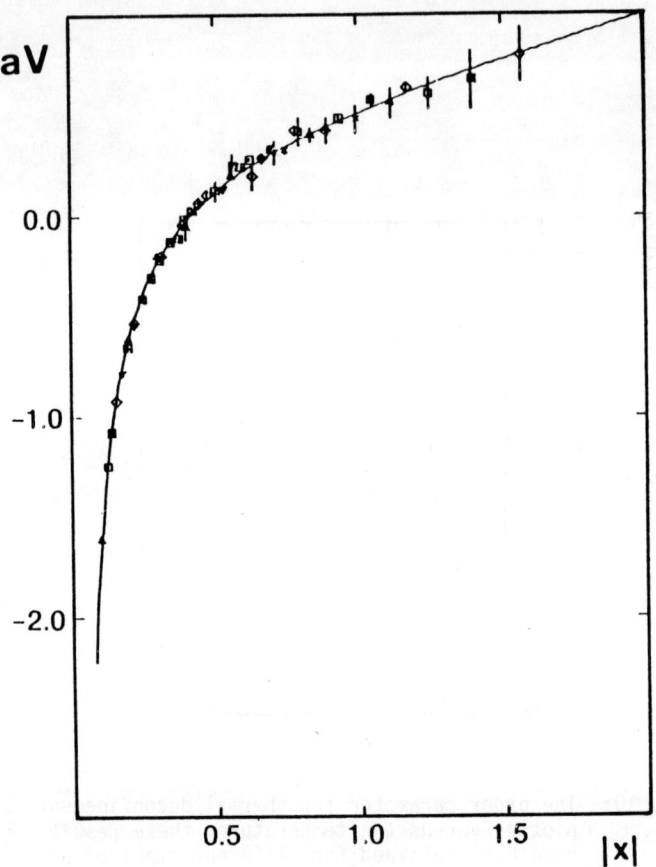

Fig. 11: The potential between two static SU(3) sources for pure gluon gauge theory (Stack 1984).

out about possible restoration of rotational invariance. For strong coupling ($1/g^2 \approx 0$) we know that equipotential curves (say, in the xy-plane) are squares (Fig. 12a). One finds in MC-calculations that for decreasing coupling g ($1/g^2 > 1/g^2_{crossover}$) rotational invariance improves and eventually becomes restored (Fig. 12b, c obtained for SU(2) by Lang 1982c).

One may also investigate the behaviour of the energy of more static charges depending on their spatial distribution (Lang 1983, Markum 1985). One finds that the colour forces are saturated quickly and the residual gluonic forces between e.g. "static mesons" are extremely small.

ad (d)

The signal for the correlations between small Wilson loops is comparatively weak and one is plagued with all sorts of statistical problems. It seems to be hard to obtain results for glueball propagation over distances exceeding two lattice spacings. Most authors would agree with the following mass for the lowest lying glueball, 0^{++} state:

$$m_0 \approx 2.66 \sqrt{\sigma_{phys}} \pm 20\% \approx 1060 \text{ MeV} \pm 20\% . \qquad (3.38)$$

For a detailed account see Berg (1983, 1984).

Of course there has been done much more than I could mention here. One interesting recent development is the application of MC-block spin techniques to lattice gauge theories (cf. Hasenfratz 1984) which seems to give promising results for the β-function. It is not clear at the moment whether this approach will be helpful in removing the restrictions from the finite lattice size.

Fig. 13 gives the dependence a(g) as obtained in the scaling

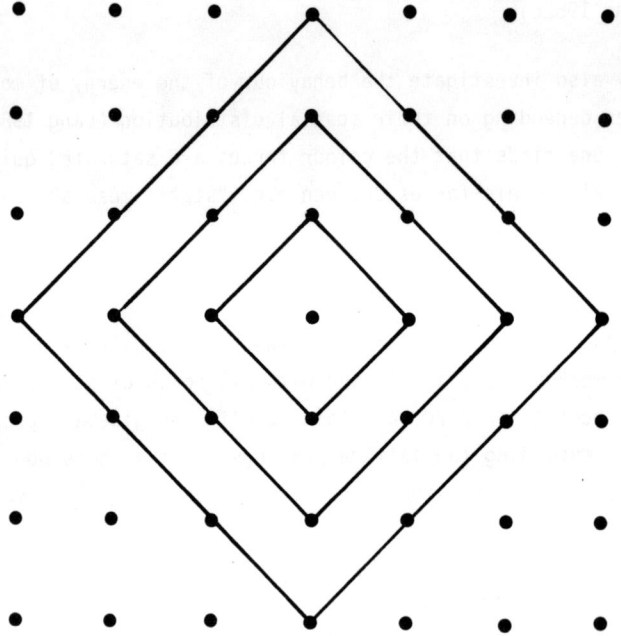

Fig. 12: Restoration of rotational invariance; the equipotential lines obtained for SU(2) gauge theory are drawn for $\beta = 0$(a), $\beta = 2$(b), and $\beta = 2.25$(c) (Lang 1982c).

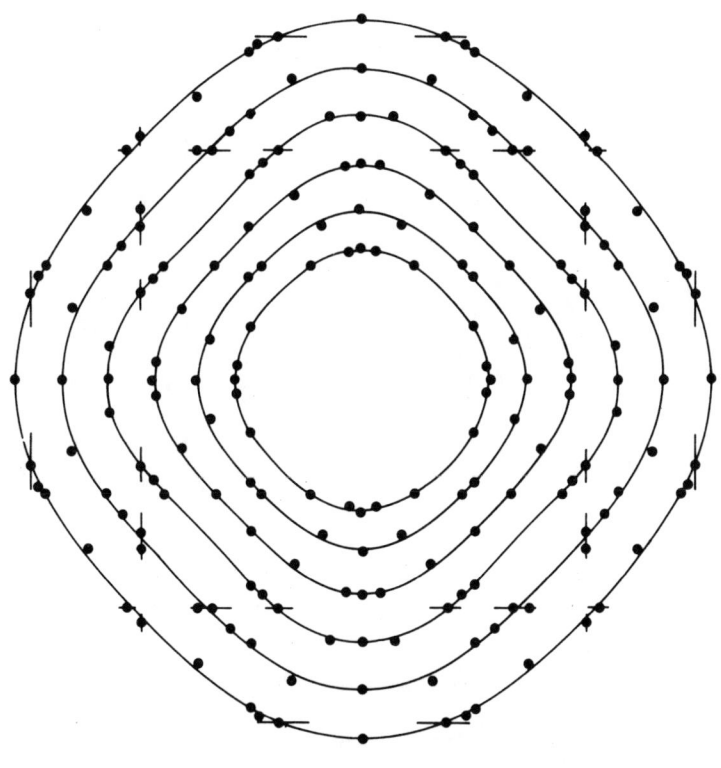

12b

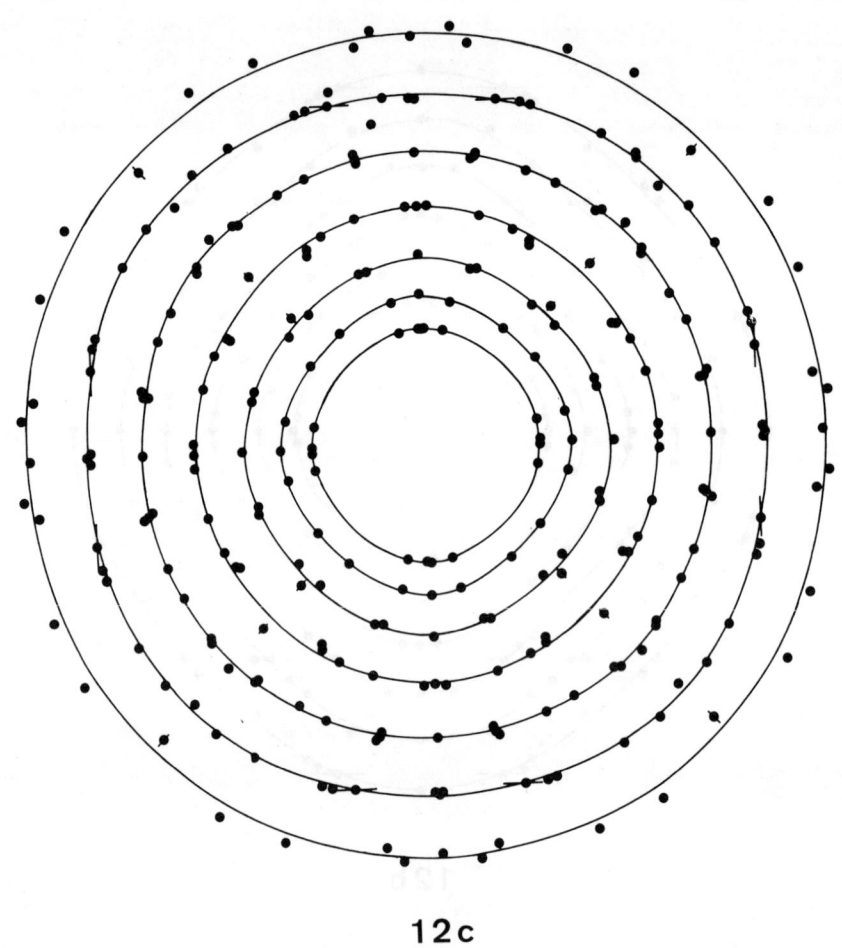

12c

regime and Fig. 14 summarizes the information obtained for SU(3) **pure gauge theory**.

4. GAUGE THEORY WITH MATTER FIELDS

4.1 Action

The fermions $\psi(x)$, $\bar{\psi}(x)$ generate a Grassmann algebra; they carry indices of a unitary representation of the gauge group as well as spinor indices. Furthermore, quarks seem to occur in families; their gauge field interaction is independent of the family structure.

We may also work with bosons $\phi(x)$, $\phi^+(x)$, with similar properties (except for the Grassmann nature), also spanning a vector space carrying a unitary representation of the gauge group, e.g. a 2-component vector for the fundamental representation of SU(2).

The action may be written

$$S = S^{(gauge)} + S^{(fermion)} + S^{(boson)} \qquad (4.1)$$

where $S^{(gauge)}$ has been discussed in Ch. 3.1.

The action for fermions in the fundamental representation is bilinear in the fields

$$S^{(f)} = \sum_{x,y} \bar{\psi}_x Q^{(f)}_{xy} \psi_y , \qquad (4.2)$$

whereas the boson-action may also contain a potential term

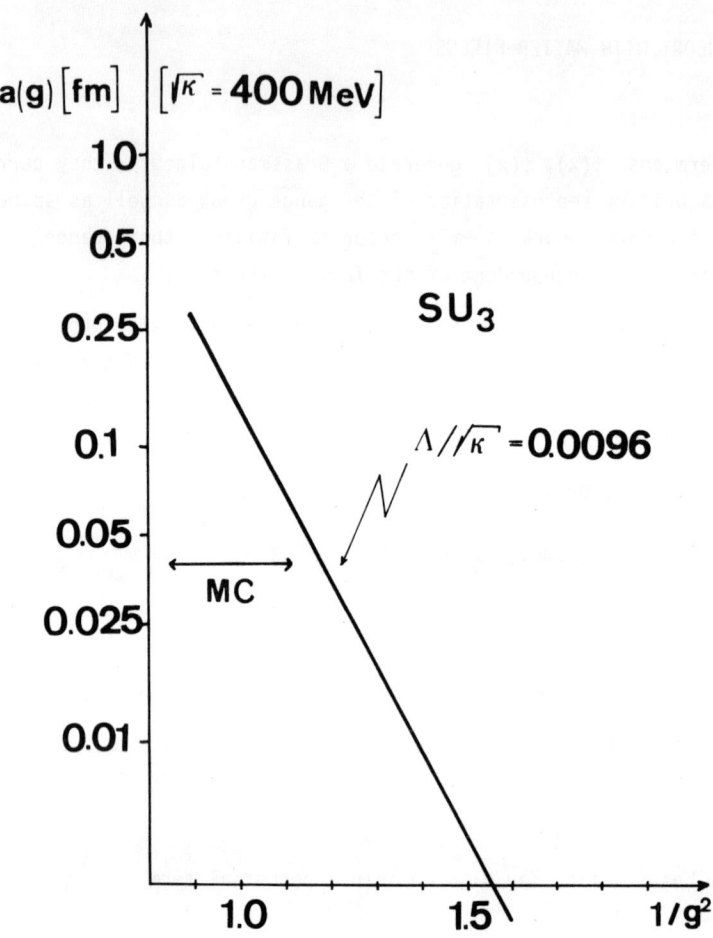

Fig. 13: The lattice spacing a(g) in physical units; in this figure the physical string tension is denoted by κ and one assumes the value $\sqrt{\kappa} \approx 400$ MeV (cf. Eq. 3.34) from the string model. The results in the scaling regime of Fig. 6 then allow this determination of a(g).

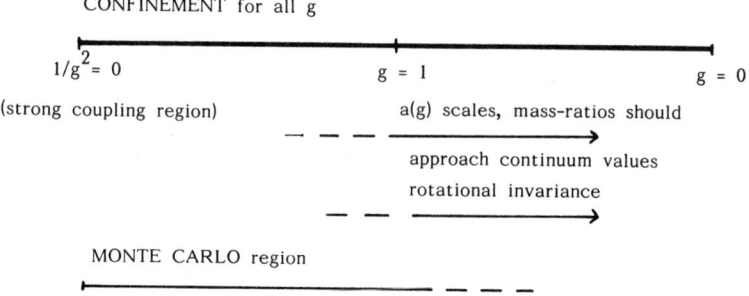

Fig. 14: Summary of the results for pure gauge theory (non-abelian gauge group).

$$S^{(b)} = \sum_{x,y} \phi_x^+ Q_{xy}^{(b)} \phi_y + \sum_x V(\phi_x^+ \phi_x) \ . \tag{4.3}$$

As an explicit example let me write down the matrix Q for the fermion action introduced by Wilson (1974, 1977)

$$Q_{xy}^{(f)} = \delta_{xy} - K[(r+\gamma_\mu)U_{x,\mu}\delta_{x,y-\mu} + (r-\gamma_\mu)U_{x-\mu,\mu}^+ \delta_{x,y+\mu}] \ ,$$

$$Q^{(f)} \equiv \mathbb{1} - KM^{(f)} \ . \tag{4.4}$$

The fermions are here in the fundamental representation of the gauge group; antiperiodic b.c. guarantee OS-positivity (Osterwalder 1975, 1978).

Wilson proposed to choose $r=1$; in this form (generally for $r \neq 0$) the action explicitly breaks chiral symmetry. On the other hand, the naive choice $r=0$ leads to a 16-fold degeneracy of the fermion states in the system. This dilemma is due to well-known theorems stating the incompatibility of chiral symmetry with regularization schemes (Adler 1969, Bell 1969, Nielsen 1981a,b, cf. also Creutz 1983a,b, Hasenfratz 1983). There are other forms of the action, notably that of Kogut and Susskind (Kogut 1975, Banks 1977, Susskind 1979, Sharatchandra 1981) in which the fermionic d.f. are distributed on different lattice sites.

For bosons in the fundamental representation one may write

$$Q_{xy}^{(b)} = \delta_{xy} - K[U_{x,\mu}\delta_{x,y-\mu} + U_{x-\mu,\mu}^+ \delta_{x,y+\mu}] \ ,$$

$$Q^{(b)} \equiv \mathbb{1} - KM^{(b)} \ , \tag{4.5}$$

and the b.c. have to be periodic. More general, the nearest neighbour coupling for bosons in the vector representation $R(\phi)$ is

$$R(\phi_x^+) \ R(g(U_{x,\mu} \ ; \ \phi_{x+\mu})) + \text{h.c.} \tag{4.6}$$

where $g(U_{x,\mu}; \phi_{x+\mu})$ denotes the group element $\phi_{x+\mu}$ gauge transformed with $U_{x,\mu}$.

Here we shall concentrate on the results obtained for bosons -- the gauge-Higgs systems -- and we shall mention only briefly the problems with fermions.

4.2 Gauge-Higgs Systems Without Radial d.f.

Think of a potential like $\lambda(\phi^+\phi - 1)^2$ for $\lambda \to \infty$, freezing the radial d.f. on the bosonic field. As example let me discuss the situation for the non-abelian gauge group SU(2) in the next two subchapters.

4.2.1 Higgs Fields in the Fundamental Representation of SU(2)

One may write

$$\phi = \begin{pmatrix} \phi_1 \\ \phi_2 \end{pmatrix} \quad \phi_1, \phi_2 \in \mathbb{C}, \quad |\phi_1|^2 + |\phi_2|^2 = 1, \tag{4.7}$$

and introduce the SU(2) matrix

$$\sigma = \begin{pmatrix} \phi_1 & -\phi_2^* \\ \phi_2 & \phi_1^* \end{pmatrix}, \tag{4.8}$$

to find

$$S^{(b)} = -K_f \sum_{x,\mu} \mathrm{Tr}(\sigma_x^+ U_{x,\mu} \sigma_{x+\mu}) + \text{const.} \tag{4.9}$$

From that representation it is simple to discuss the following cases.

(i) $\underline{K_f = 0}$: one is left with pure gauge theory, the Higgs fields are decoupled.

(ii) $\underline{1/g^2 \to \infty}$: the gauge field configuration has to have $U_\square = \mathbb{1}$, thus it is equivalent to $U_{x,\mu} = \mathbb{1}$ up to a gauge transformation. One is left with a pure SU(2) 4-dim. spin system and expects a 2nd order phase transition for some critical value of K. For

$K < K_{crit}$ one is in the disordered (high temperature) phase, for $K > K_{crit}$ spontaneous symmetry breaking (magnetization $\neq 0$) occurs due to the strong nearest neighbour interaction.

(iii) $1/g^2 = 0$: the pure gauge field part of S does not contribute. For each ϕ-configuration one may perform a variable transformation

$$\sigma_x^+ U_{x,\mu} \sigma_{x+\mu} \rightarrow U'_{x,\mu} \qquad (4.10)$$

leaving invariant the measure $dU_{x,\mu} = dU'_{x,\mu}$. Thus

$$S^{(b)} = - K_f \sum_{x,\mu} \text{Tr } U_{x,\mu} \qquad (4.11)$$

and the system factorizes and may be integrated explicitly (analytically). It is continuous for all K_f without a phase transition.

(iv) $\underline{K_f \rightarrow \infty}$: apply the transformation discussed above ("U-gauge"). One finds a freezing

$$\text{Tr } U_{x,\mu} \rightarrow \text{Tr } \mathbb{1}, \qquad (4.12)$$

the stability group is trivial, the gauge group is completely broken.

It can be shown that the clustering properties give the following analyticity domains (Osterwalder 1978, Fradkin 1979, Seiler 1982...)

$$\begin{array}{ll} \{K_f, g^2 \mid K_f \text{ and } 1/g^2 \text{ small}\} & \text{(area law)} \\ \{K_f, g^2 \mid K_f g^2 \text{ large enough }\} & \text{(area law)} \end{array} \qquad (4.13)$$

This leads to the expected behaviour shown in Fig. 15a.

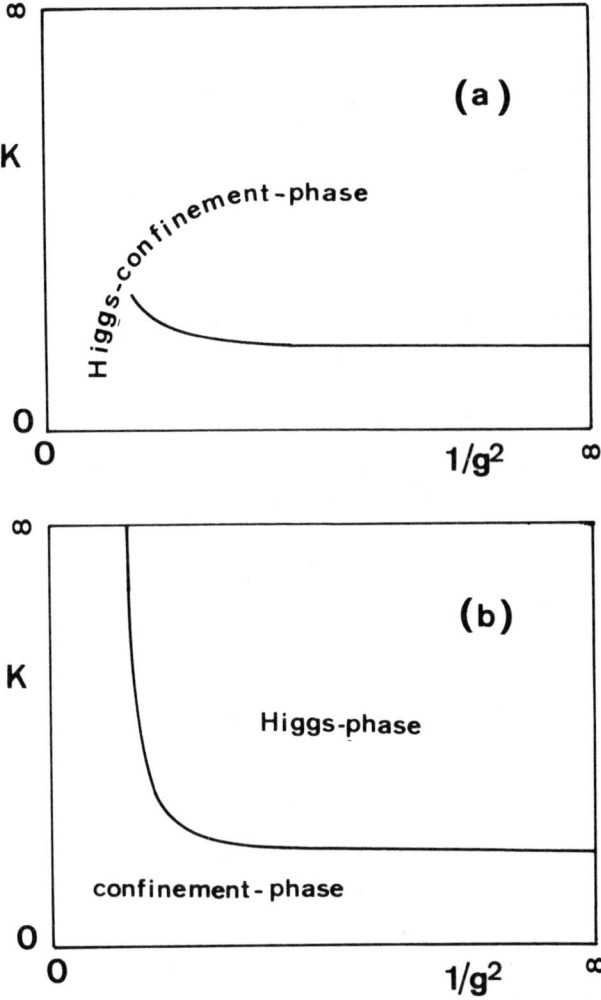

Fig. 15: The phase structure of the SU(2) gauge-Higgs system in the $(K, 1/g^2)$-phase. The Higgs-fields have fixed length and are in the (a) fundamental or (b) adjoint representation of the gauge group. The phase-transition along the line is 2nd order.

The MC-analysis has confirmed these expectations and given a more detailed picture. As an example, see Fig. 16 from Lang (1981b) where the hysteresis cycles for the plaquette and the link observables are exhibited. That analysis was actually performed for the 120-element group $\tilde{Y}$, the discrete icosahedral subgroup of SU(2) for lattice size 4^4. The phase transition appears to be 2nd order everywhere. Meanwhile the quality of MC-investigation has improved and in Chapter 4.3 you will find more detailed results.

4.2.2 Higgs Fields in the Adjoint Representation of SU(2)

The Higgs field may be written

$$\vec{\phi} = (\phi_1, \phi_2, \phi_3), \quad \phi_i \in \mathbb{R}, \quad \vec{\phi} \cdot \vec{\phi} = 1, \qquad (4.14)$$

i.e. the $\vec{\phi}$ live on S^2, the surface of a 3-d sphere. It is possible to represent them again by SU(2) matrices since

$$V_x = \cos\vartheta + i\,\vec{\sigma} \cdot \vec{\phi}_x \sin\vartheta \qquad (4.15)$$

for fixed ϑ utilizing the isomorphism $S^2 \sim S^3/S^1$. In this form

$$S^{(b)} = -\frac{K_a}{2\sin^2\vartheta} \sum_{x,\mu} \mathrm{Tr}\,(V_x^+ U_{x,\mu} V_{x+\mu} U_{x,\mu}^+) \qquad (4.16)$$

due to the property of gauge transformation in the adjoint representation. Again the boundaries of the phase diagram $(1/g^2, K_a)$ are easy to obtain.

(i) $K_a = 0$: leaves us with pure gauge theory.

(ii) $1/g^2 \to \infty$: again $U_{x,\mu} = 1$; one has an $O(3)$ spin model with a 2nd order phase transition.

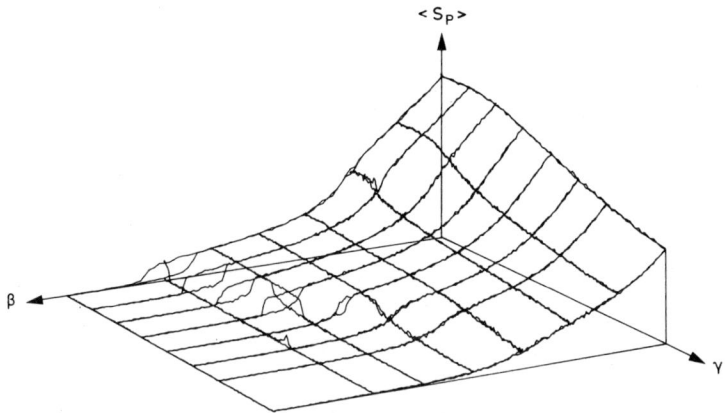

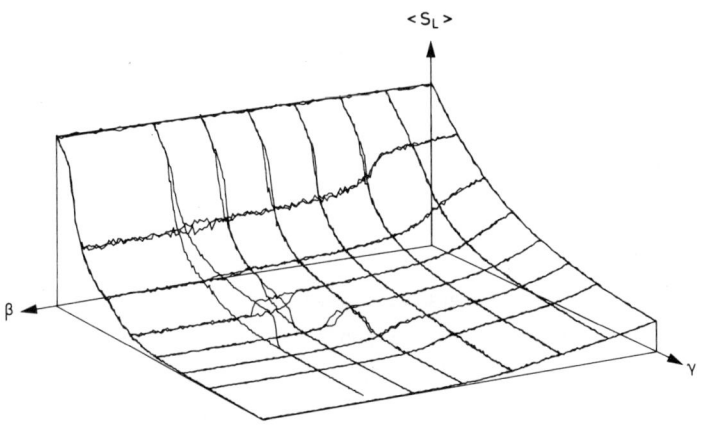

Fig. 16: Hysteresis-curves obtained for the plaquette (S_P) and link (S_L) variables for a SU(2) gauge-Higgs system with Higgs-fields in the fundamental representation (Lang 1981b).

(iii) $1/g^2 = 0$: now one may transform away the Higgs fields only up to some group element V_0,

$$S^{(b)} \to -\frac{K_a}{2\sin^2\vartheta} \sum_{x,\mu} \text{Tr}(V_0^+ U_{x,\mu} V_0 U_{x,\mu}^+), \qquad (4.17)$$

again the functional integral decouples and may be solved explicitly - no phase transition.

(iv) $K_a \to \infty$: the system tries to minimize $\text{Tr}(V_0^+ U_{x,\mu} V_0 U_{x,\mu}^+)$. This is achieved for $V_0^+ U V_0 \sim U$, i.e. for elements U invariant (up to a constant) under a rotation around the direction given by V_0. This stability group is $U(1)$ for one such $SO(3)$ Higgs-field (the adjoint representation of $SU(2)$); one is left with a $U(1)$ gauge theory. For more Higgs fields one needs simultaneous invariance under $V_0^+ U V_0$ and $V_1^+ U V_1$, therefore the stability group reduces to $Z(2)$.

Rigorous results give the analyticity domains

$$\{K_a, g^2 \mid K_a \text{ small}, 1/g^2 \text{ small}\} \quad \text{(area law)}$$
$$\{K_a, g^2 \mid K_a \text{ large}, g^2 \text{ small}\} \quad \text{(perimeter law)}. \qquad (4.18)$$

The resulting phase diagram as obtained for $SU(2)$ by Lang (1981b) and Brower (1982) is shown in Fig. 15b. The phase transition line appears to be 2nd order. Recently, however, it has been shown that the $U(1)$ phase transition for the Wilson action is actually first order (Jersak 1983, Evertz 1985). It is unclear at present how this is felt in the gauge-Higgs phase diagram and whether there is a tri-critical point. Furthermore, the dependence of the phase diagram on the specific form of the action (the content of higher representations) is unknown. This system is the only gauge-Higgs system where the thermal deconfinement has been studied (Karsch 1983, Drouffe 1984).

4.2.3 Results for Other Gauge Groups

For the U(1) gauge group Higgs fields of charge n_q have the representation

$$\phi_x = \exp(i\, n_q\, \vartheta_x)\,, \tag{4.19}$$

$$S^{(b)} = -K \sum_{x,\mu} \cos\left[n_q(\vartheta_{x+\mu} - \vartheta_x + \vartheta(U_{x,\mu}))\right]. \tag{4.20}$$

Now the pure gauge theory (K = 0) has a phase transition and for $n_q = 1$ one obtains a phase diagram as given in Fig. 19 for $\lambda \to \infty$. For $n_q \geq 2$ there is a stability group $Z(n_q)$ for $K \to \infty$; therefore the phase transition line runs through up to $K = \infty$ (cf. Creutz 1980b). The SU(3) gauge-Higgs system has been studied by Kigugawa (1982).

4.3 Systems with Dynamical Radial Mode

Again we take SU(2) as our example. Here most results have been obtained for fields in the fundamental representation

$$\phi = \begin{pmatrix} \phi_1 \\ \phi_2 \end{pmatrix}, \quad \phi_1, \phi_2 \in \mathbb{C}\,, \tag{4.21}$$

and

$$V(\phi_x^+ \phi_x) = \lambda(\phi_x^+ \phi_x - 1)^2\,. \tag{4.22}$$

This specific form is chosen to expose a simple limit for $\lambda \to \infty$. Other choices closer to the continuum Lagrangian are as good and related via trivial parameter transformations (Kühnelt 1984, Munehisa 1984b).

Choosing the notation

$$\phi = \rho \begin{pmatrix} \sigma_1 \\ \sigma_2 \end{pmatrix},\quad \sigma_1, \sigma_2 \in \mathbb{C},\ \rho \in \mathbb{R}^+,\ |\sigma_1|^2 + |\sigma_2|^2 = 1, \tag{4.23}$$

and

$$\sigma = \begin{pmatrix} \sigma_1 & -\sigma_2^* \\ \sigma_2 & \sigma_1^* \end{pmatrix} \in SU(2) \;, \qquad (4.24)$$

the action becomes especially simple (Kühnelt 1984)

$$S^{(b)} = \lambda \sum_x (\rho_x^2 - 1)^2 - K \sum_{x,\mu} \rho_x \rho_{x+\mu} \, Tr(\sigma_x^+ U_{x,\mu} \sigma_{x+\mu}) \qquad (4.25)$$

and the measure

$$d\phi d\phi^+ \to \rho^3 d\rho \, d\sigma \qquad (4.26)$$

where $d\sigma$ is the SU(2) Haar measure.

Note that in addition to the local SU(2) gauge symmetry

$$\phi_x \to g_x \phi_x \;, \quad U_{x,\mu} \to g_x U_{x,\mu} g_{x+\mu}^+ \;, \qquad (4.27)$$

there is also a global SU(2) symmetry

$$\phi_x \to \phi_x g \;. \qquad (4.28)$$

Bound states will be multiplets under this symmetry, e.g.

$$\text{singlet } 0^{++}: \; \rho^2 \quad \text{and} \quad \sum_{\mu=1}^{3} \rho_x \rho_{x+\mu} \, Tr(\sigma_x^+ U_{x,\mu} \sigma_{x+\mu}) \qquad (4.29a)$$

$$\text{triplet } 1^{--}: \; \rho_x \rho_{x+\mu} \, Tr(\vec{\sigma} \, \sigma_x^+ U_{x,\mu} \sigma_{x+\mu}) \;, \qquad (4.29b)$$

where $\vec{\sigma}$ denotes the Pauli-matrices.

Further symmetries of the phase diagram are $K \leftrightarrow -K$ and for $K = 0$: $\beta \leftrightarrow -\beta$.

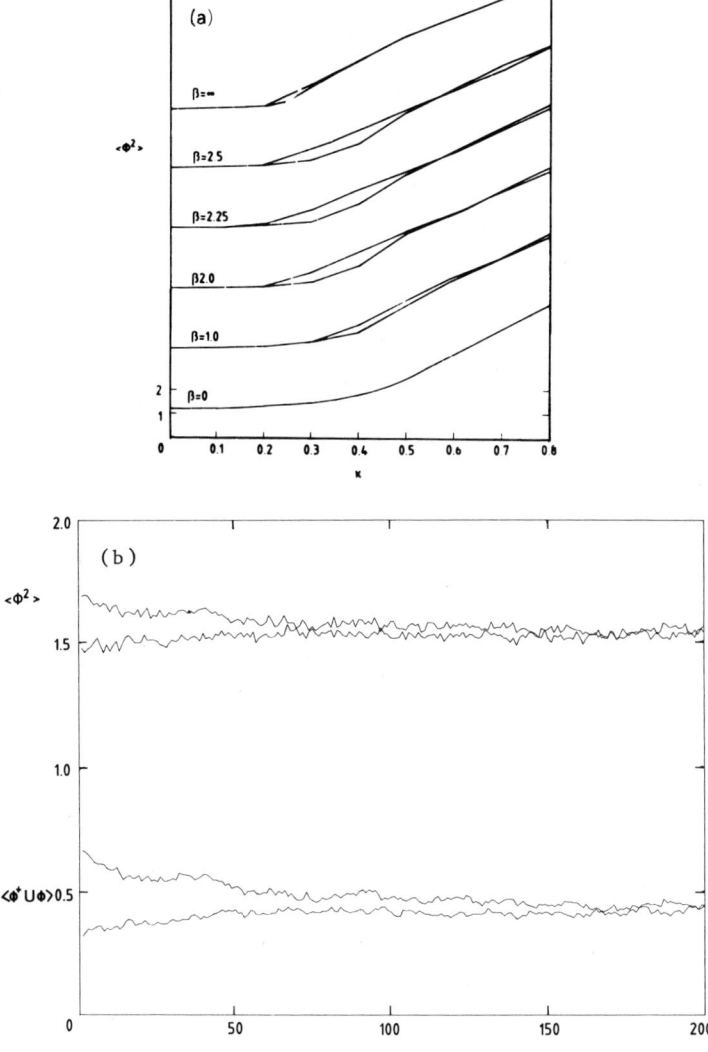

Fig. 17 (a) Hysteresis curves for $<\phi^+\phi>$ (obtained by Kühnelt 1984) for different values of β; the gauge system contains SU(2) Higgs-fields in the fundamental representation with radial d.f.
(b) Relaxation runs by the same authors to demonstrate that the phase transition is 2nd order.

Fig. 17a shows some hysteresis runs obtained for specific β-values and λ = 0.5, and Fig. 17b exhibits the relaxation behaviour at a critical point indicating the second order of the phase transition.

In Fig. 18 the phase diagram for different values of λ (from Kühnelt 1984) is drawn. Recent results (Munehisa 1984b, Gerdt 1984, Jansen 1985b) indicate that the order of this phase transition changes with λ; for $\lambda \gtrsim 0.5$ it is 2nd order (although a very weak 1st order transition cannot be excluded from the MC-data), below 0.5 it becomes 1st order. This behaviour would reconcile with the perturbative result by Coleman (1974) indicating a 1st order phase transition for large β < ∞ and small λ.

A possible continuum theory has to be obtained at a 2nd order phase transition. Here the candidate is the point at $K_{crit}(\lambda)$ at $1/g^2 = \infty$. Block-spin methods (Kogut 1974, Swendsen 1984, Callaway 1984a, Lang 1985, Hasenfratz 1984) may allow to follow the renormalization flow and identify the fixed point. Recent MC-results (Lang 1985) demonstrate that the infrared fixed point of ϕ^4-theory in d = 4 is Gaussian, independent of the bare λ coupling. This confirms earlier expectations from perturbative arguments (Wilson 1974), rigorous proofs in d = 4 + ε (Aizenman 1981, 1982, Fröhlich 1982a, Sokal 1982) or for small coupling (Gawędzki 1985), and MC-results (Freedman 1982, Callaway 1984b). This means that the only continuum theory that may be obtained as a limit of the regularized ϕ^4-theory is a non-interacting free theory! It may be possible that introduction of gauge fields changes the situation and allows to construct a non-trivial continuum theory. This is one of the most fascinating aspects of scalar QCD or QED.

Leaving aside for the moment the block spin methods one could follow another approach pursued e.g. in lattice QCD. One determines dimensionless masses m_1, m_2, m_3 ... corresponding to asymptotic (bound)

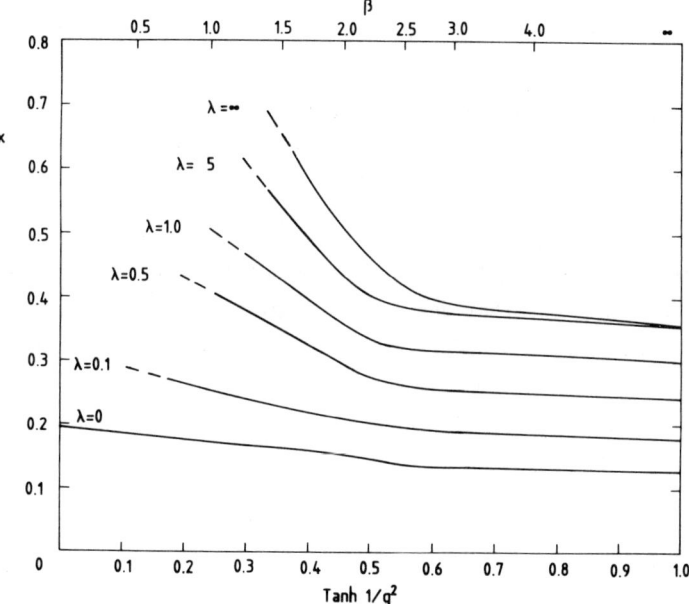

Fig. 18: The phase diagram of the SU(2) gauge-Higgs system (Higgs-fields in the fundamental representation with dynamical radial d.f.) for various values of the ϕ^4-coupling λ (Kühnelt 1984).

states of the theory and determines trajectories in the phase diagram where m_1/m_2 and m_1/m_3 are constant. This leads to a line $K(g^2)$, $\lambda(g^2)$ in coupling constant space (K,λ,g^2). If there is a sensible continuum limit all other physical quantities (or better: ratios of them) should become constant along this line, approaching their continuum values for $g^2 \to 0$. Investigations of these aspects are under progress (Montvay 1985a,b, Jansen 1985b).

Fig. 19 gives results obtained for various λ values for $U(1)$ gauge-Higgs theory ($n_q = 1$) by Jansen (1985a) (cf. also Gerdt 1983, Munehisa 1983a,b, 1984a,b). There has been also work on the $SU(3)$ adjoint Higgs model by Gupta (1984).

4.4 QCD and Scalar QCD

For $\lambda = 0$ the phase diagram obtained for the gauge-Higgs systems in Ch. 4.3 shows that one looses completely the phase above the critical line. The reason for this feature is obvious. The nearest neighbour interaction competes with the mass-term $\phi^+\phi$ and wins this competition at $K = K_{crit}(g^2)$. For $\lambda > 0$ there is still the ϕ^4-potential to prevent unboundedness of the action from below, for $\lambda = 0$ the system becomes unphysical for $K > K_{crit}$.

For bosons Monte Carlo techniques as discussed may be applied. For fermions this is not possible, there is no possibility to treat Grassmann variables on the computer like bosonic ones. However, both for bosons and for fermions the bilinearity of the action (in ϕ^+,ϕ or $\bar{\psi},\psi$) allows another approach. One may explicitly integrate over the variables η^+,η (Berezin 1966, Matthews 1954)

$$\int [d\eta\, d\eta^+]\, e^{-\eta^+ Q \eta} = \det Q^{-c_\eta}, \qquad (4.30)$$

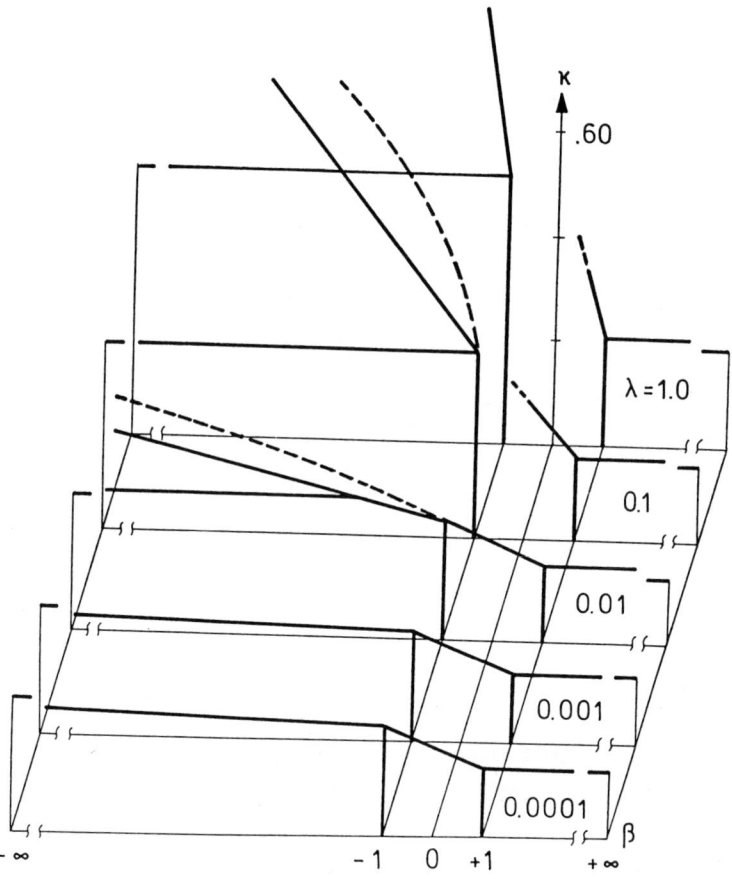

Fig. 19: The phase diagram of the U(1) gauge-Higgs system (Higgs-fields of charge one with dynamical radial d.f.) for various values of the ϕ^4-coupling λ (Jansen 1985a).

$$\int [d\eta\, d\eta^+]\, \eta_i^+ \eta_j\, e^{-\eta^+ Q \eta} = (Q^{-1})_{ji}\, \det Q^{-c_\eta},$$

$$\int [d\eta\, d\eta^+]\, \eta_i^+ \eta_j\, \eta_k^+ \eta_\ell\, e^{-\eta^+ Q \eta} = \left[(Q^{-1})_{ji}(Q^{-1})_{\ell k} + c_\eta (Q^{-1})_{jk}(Q^{-1})_{\ell i}\right] \det Q^{-c_\eta},$$

with

$$c_\eta = \begin{cases} +1 & \text{for } \eta^+,\eta \equiv \phi^+,\phi \quad \text{(bosons)} \\ -1 & \text{for } \eta^+,\eta = \bar\psi,\psi \quad \text{(fermions)} \end{cases}$$

Let us expand these expressions around the pure gauge theory, i.e. in the variable K

$$\begin{aligned} \det Q^{-c_\eta} &= \exp(-c_\eta\, \text{Tr}\,\log(1 - KM)) \\ &= \exp(c_\eta \sum_{n=1}^{\infty} \frac{K^n}{n}\, \text{Tr}\, M^n) \\ &= \exp(c_\eta \sum_\ell K^{2\ell} \sum_{\text{all } P_{2\ell}} \text{Tr}(\prod_P U)\, \text{Tr}(\prod_P (r \pm \gamma))) \\ &= \exp(-S_{\text{eff}}). \end{aligned} \qquad (4.31)$$

For bosons one has to omit the trace over the Dirac-matrices. $P_{2\ell}$ denotes closed paths of lenghts 2ℓ of arbitrary orientation (we have utilized the fact that on the hypercubic lattice the path length has to be even). Note that the first non-trivial term in S_{eff} is

$$-K^4 \sum_\square \text{Tr}(U_\square + U_\square^+) \begin{cases} \times 8 & \text{for fermions} \\ \times 1 & \text{for bosons}, \end{cases} \qquad (4.32)$$

i.e. a contribution to $S^{(\text{gauge})}$. The terms contributing to S_{eff} may be understood as closed quark or boson loops; for this reason the expansion has been coined "hopping expansion" for the "hopping parameter" K. Each species of matter fields carries another hopping parameter.

A similar expansion for the propagators is obtained by

$$(Q^{-1})_{ji} = \left(\frac{1}{1-KM}\right)_{ji} = \delta_{ji} + KM_{ji} + K^2(M^2)_{ji} + \ldots \tag{4.33}$$

$$= \sum_{\substack{\text{paths} \\ P(i \to j)}} K^{\text{length}} \prod_P U \prod_P (r \pm \gamma)$$

where again the Dirac-term has to be omitted for boson propagators. The last expression shows the possibility to obtain Q^{-1} by recursion techniques (Gauss-Jacobi, Gauss-Seidel, conjugate gradient ...), e.g. like

$$(Q^{-1})_{ji}^{(n+1)} = \delta_{ji} + K M_{j\ell} (Q^{-1})_{\ell i}^{(n)} , \tag{4.34}$$

$$(Q^{-1})_{ji}^{(0)} = \delta_{ji} .$$

Expectation values for propagators of, say, mesons have now the form

$$<\bar{\psi}_x \psi_x \bar{\psi}_y \psi_y> = \frac{\int [dU] e^{-S^{(g)} - S_{\text{eff}}^{(f)}} [Q_{xx}^{-1} Q_{yy}^{-1} - Q_{xy}^{-1} Q_{yx}^{-1}]}{\int [dU] e^{-S^{(g)} - S_{\text{eff}}^{(f)}}} . \tag{4.35}$$

The hardest technical problem in the Monte Carlo-approach for this expression is to calculate the change in the effective fermion action $\Delta S_{\text{eff}}^{(f)}$ due to a tentative change of some link variable to be updated. As you have seen above, this involves all possible loops of arbitrary size running through that link. In Lang (1982a) it was tried to approximate this sum by randomly selecting a small subset of these loops. This works only for small values of K, away from the interesting critical value. Fucito (1981) suggested to use the similarity between the functional integrals for bosons and fermions (4.30) to calculate $\Delta S_{\text{eff}}^{(f)}$

in a MC-simulation for a system of pseudo-fermions. These are actually bosons with the same number of d.f. as the original fermions and interacting via

$$S^{(p.f.)} = \sum_{x,y} \phi_x^+ (Q^{(f)^+} Q^{(f)})_{xy} \phi_y . \qquad (4.36)$$

This means a whole MC-simulation for each step of the gauge field MC-run and is therefore very time consuming. Recent work (Hamber 1983) shows the applicability of the method.

In the bulk of work done up to now, however, $S_{eff}^{(f)}$ has been neglected. This amounts to neglect the vacuum polarization, the generation of quark loops from the vacuum. In this "quenched" approximation the problem reduces to find the expectation value of expressions like

$$Q_{xy}^{-1} Q_{yx}^{-1} \simeq \sum_{paths} \qquad (4.37)$$

in the presence of dynamical gauge fields. This is done in two levels. One constructs independent gauge-field configurations. For each of those one now tries to find the value of (4.37), the propagator in the presence of a frozen gauge field. Then one sums over the results for the different gauge configurations. The resulting expectation value is then analyzed to determine its exponential decay (cf. Ch. 1.2 and 3.3). One obtains dimensionless masses ma , depending on the quantum numbers of the propagating state (cf. Rebbi 1983, 1984, Creutz 1984a,b, Hasenfratz 1983,1984).

Let me briefly summarize what one has learned about the phase diagram of QCD in the $(1/g^2,K)$ plane and about the hadron spectrum. In Fig. 20, the situation for the Wilson action $(r=1)$ is sketched; note that this picture is like the one for bosons with $\lambda = 0$ in Fig. 18 (except for the precise values of K_{crit}). The critical coupling is obtained from the value of K where $am_\pi = 0$. Above this point, one would get

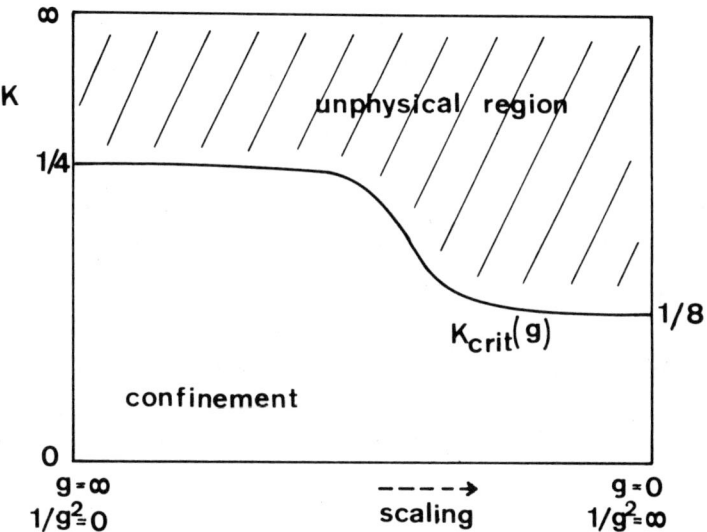

Fig. 20: The phase structure of a gauge-fermion system (fermion-action of the Wilson-type).

negative values of m_π^2. Slightly below this boundary lies the physical curve $K(1/g^2)$ where m_ρ/m_π assumes its experimental value. This function depends on the quark species and should really be called $K_u(1/g^2)$, $K_d(1/g^2)$ in this case. So, for each $1/g^2$ one needs $(n_{flavour}+1)$ experimentally determined masses. For each quark flavour one mass (or mass ratio) is needed to obtain the value of K, one extra mass is used to fix the scale.

Fig. 21 demonstrates the usual approach:

(i)   find K_u from the value m_ρ/m_π (one assumes $K_d = K_u$);
(ii) fix the scale a $(1/g^2)$ frome one of the masses.

One has traded $n_f + 1$ input masses against the unknown scale and the quark couplings; all other masses come out free of charge. Following the curves $K_{phys}(1/g^2)$ towards $1/g^2 \to \infty$ all masses should approach their physical values (cf. Ch. 3.3).

At the moment, most calculations have been performed on lattices of size $8^4 - 12^4$ for $1/g^2 \lesssim 1$. The mesons have come out roughly correct up to 10% (for fixed m_π, m_ρ, m_K), the baryons have been systematically too large by $\sim$ 30-40%. For the interpretation of these results see Fig. 22. For small $1/g^2$ the lattice graining is too coarse for a good representation of the baryons. For larger $1/g^2$ the lattice becomes fine, the results should (and do) improve -- however --, now the overall lattice size Na has become smaller, too. Since the physical object remains always the same size in physical units, this implies that boundary effects will become disturbing. The hadron has not enough space to propagate freely and to build up its wave function.

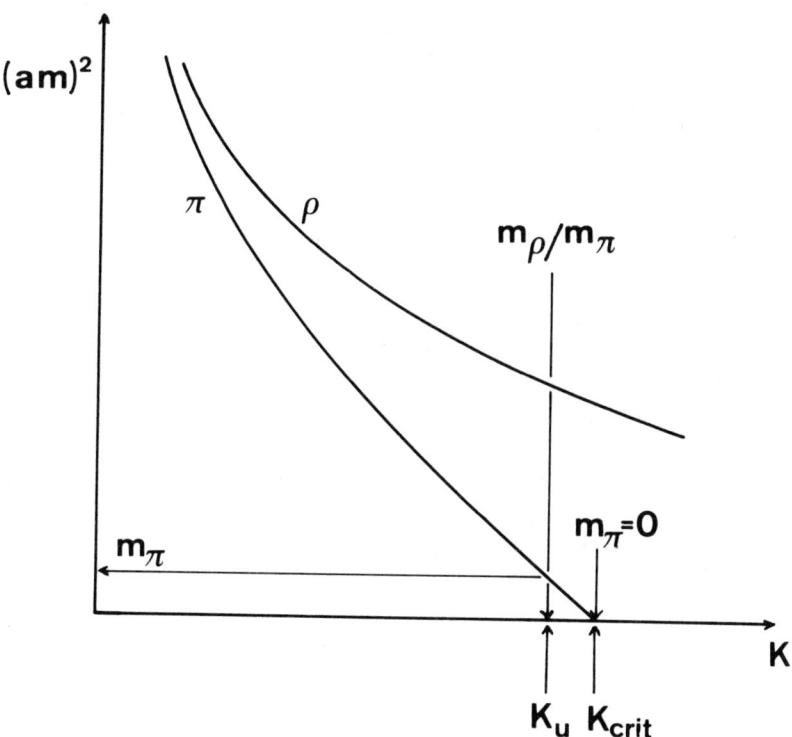

Fig. 21: Example of the method to trade the physical mass values m_π, m_ρ versus the unknown values of $a(g)$ and $K_u(g)$ or $K_{crit}(g)$. Similar curves for other values of $1/g^2$ should give $K(g)$ like in Fig. 20 and scaling $a(g)$ like in Fig. 13.

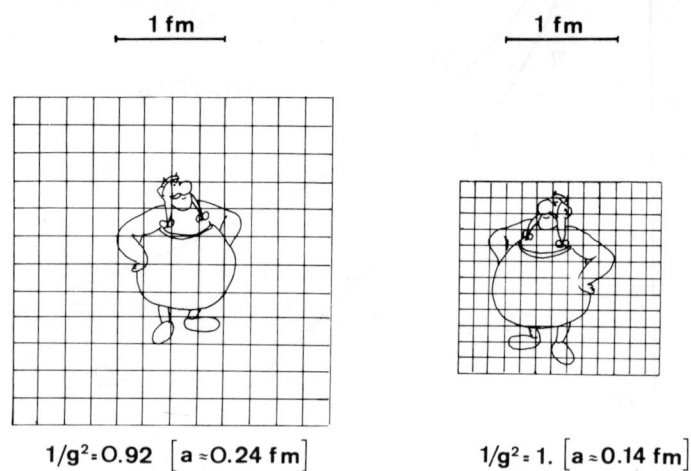

Fig. 22: The physical hadron size is fixed, the lattice spacing a decreases with increasing $1/g^2$; the lattice approximation improves while the finite size problems gain importance.

There are several possible solutions to this dilemma. One certainly should try to improve our understanding of finite size effects. Meanwhile the trend is to push the super-computer facilities to their limits.

Computers have been useful to shed light on many aspects of QFT. Only some of them have been mentioned here, mostly in relation with lattice gauge theory. In that field much progress has been made in the recent years but much more work has still to be done. One has learned a lot about the quantitative and qualitative features of confinement and other non-perturbative phenomena. The results are very promising but obviously we would like to achieve a deeper understanding of the underlying physics. Eventually we may be led to more powerful analytical and rigorous ways to attack this problems.

REFERENCES

E.S. Abers and B.W. Lee, Phys. Rep. $\underline{9C}$ (1973) 1.

S.L. Adler, Phys. Rev. $\underline{117}$ (1969) 2426.

M. Aizenmann, Phys. Rev. Lett. $\underline{47}$ (1981) 1.

M. Aizenmann, Commun. Math. Phys. $\underline{86}$ (1982) 1.

T. Banks, S. Raby, L. Susskind, J. Kogut, D.R.T. Jones, P.N. Scharbach, and D.K. Sinclair, Phys. Rev. $\underline{D15}$ (1977) 1111.

M.N. Barber, in "Phase Transitions and Critical Phenomena", Vol. 8, Ed. C. Domb and J.L. Lebowitz (Academic Press, 1983).

D. Barkai, K.J.M. Moriarty and C. Rebbi, Phys. Rev. $\underline{D30}$ (1984) 1293.

S. Belforte et al., Phys. Lett. $\underline{130B}$ (1983) 205, ibid. $\underline{131B}$ (1983) 423.

J. Bell and R. Jackiw, Nuovo Cimento $\underline{60A}$ (1969) 47.

F.A. Berezin, The Method of Second Quantization (Academic Press, 1966).

B. Berg, preprint DESY 83-120 (1983), to be published in the Proceedings of the Internat. "Internat. Symp. on the Theory of Elem. Part.", Arenshop, GDR, 1983.

B. Berg, Cargèse Lectures 1983, preprint DESY 84-012 (1984).

G. Bhanot and M. Creutz, Phys. Rev. $\underline{D24}$ (1981) 3212.

G. Bhanot, C.B. Lang and C. Rebbi, Computer Phys. Commun. $\underline{25}$ (1982) 275.

G. Bhanot, J.M. Drouffe, A. Schiller and I.O. Stamatescu, Phys. Lett. $\underline{125B}$ (1983a) 67.

G. Bhanot, M. Creutz and H. Neuberger, preprint BNL-34026 (1983b).

K. Binder (Ed.), Monte Carlo Methods (Springer, 1979).

R.C. Brower, D.A, Kessler and H. Levine, Nucl. Phys. $\underline{B205}$ (1982) 77.

D.J.E. Callaway and L. Carson, Phys. Rev. $\underline{D25}$ (1982a) 531.

D.J.E. Callaway and A. Rahman, Phys. Rev. Lett $\underline{49}$ (1982b) 613.

D.J.E. Callaway and R. Petronzio, Phys. Lett. (1984a) 189.

D.J.E. Callaway and R. Petronzio, Nucl. Phys. $\underline{B240}$ (1984b) 577.

T. Celik, J. Engles and H. Satz, Bielefeld University prepring BI-TP 83/23 (1983).

W. Celmaster, Phys. Rev. $\underline{D26}$ (1982) 2955.

N.H. Christ, R. Friedberg and T.D. Lee, Nucl. Phys. $\underline{B202}$ (1982) 198.

S. Coleman and E. Weinberg, Phys. Rev. $\underline{D7}$ (1973) 1888.

M. Creutz, Phys. Rev. Lett. $\underline{43}$ (1979a) 553.

M. Creutz, L. Jacobs and C. Rebbi, Phys. REv. $\underline{D20}$ (1979b) 1915

M. Creutz, Phys. Rev. Lett. $\underline{45}$ (1980a) 313.

M. Creutz, Phys. Rev. $\underline{D21}$ ((1980b) 1006.

M. Creutz, Quarks, Gluons and Lattices (Cambridge University Press, 1983a).

M. Creutz, L. Jacobs and C. Rebbi, Phys. Rep. $\underline{95}$ (1983b) 201.

M. Creutz, Phys. Rev. Lett. $\underline{50}$ (1983c) 1411.

G. Curci, P. Menotti and G.P. Paffuti, Phys. Lett. $\underline{130B}$ (1983) 205.

R. Dashen and D. Gross, Phys. Rev. $\underline{D23}$ (1981) 2340.

J.M. Drouffe and C. Itzykson, Phys. Rep. $\underline{38}$ (1978a) 133.

J.M. Drouffe, Phys. Rev. $\underline{D18}$ (1978b) 1174.

J.M. Drouffe, Nucl. Phys. $\underline{B170}$ [FS1] (1980) 91.

J.M. Drouffe and J.-B. Zuber, Phys. Rep. $\underline{102}$ (1983a) 1.

J.M. Drouffe and K.J.M. Moriarty, Nucl. Phys. $\underline{B220}$ [FS8] (1983b) 253.

J.M. Drouffe, J. Jurkiewicz and A. Krzywicki, preprint LPTHE Orsay 84/7 (1984).

H.G. Evertz, J. Jersak, T. Neuhaus and P. Zerwas, Nucl. Phys. $\underline{B251}$ [FS 13] (1985) 279.

E. Fradkin and S. Shenker, Phys. Rev. $\underline{D19}$ (1979) 3682.

B. Freedman, P. Smolensky and D. Weingarten, Phys. Lett. $\underline{113B}$ (1982) 481.

J. Fröhlich, Nucl. Phys. $\underline{B200}$ [FS4] (1982a) 281.

J. Fröhlich and T. Spencer, Commun. Math. Phys. $\underline{83}$ (1982b) 441.

F. Fucito, E. Marinari, G. Parisi and C. Rebbi, Nucl. Phys. $\underline{B180}$ [FS2] (1981) 369.

F. Fucito and S. Solomon, Phys. Lett. $\underline{134B}$ (1984) 230.

R.V. Gavai, F. Karsch and H. Satz, Nucl. Phys. $\underline{B220}$ [FS8] (1983) 223.

K. Gawędzki and A. Kupiainen, Phys. Rev. Lett. $\underline{54}$ (1985) 92.

V.P. Gerdt et al., JINR Dubna preprint E2-83-758 (1983).

V.P. Gerdt et al., JINR Dubna preprint E2-84-313 (1984).

P. Goddard, J. Goldstone, C. Rebbi and C.B. Thorn, Nucl. Phys. $\underline{B56}$ (1973) 109.

A. Gonzales-Arroyo and C.P. Korthals-Altes, Nucl. Phys. $\underline{B205}$ (1982) 46.

D.J. Gross and F. Wilczek, Phys. Rev. Lett. $\underline{30}$ (1973) 1343.

D.J. Gross, in "Methods in Field Theory", Proc. of the Les Houches Meeting in 1975, Eds. R. Balian and J. Zinn-Justin (North-Holland, 1976).

D.J. Gross, R.D. Pisarski and L.G. Yaffe, Rev. Mod. Phys. $\underline{53}$ (1981) 43.

S. Gupta and U.M. Heller, Phys. Lett. 138B (1984) 171.

A.H. Guth, Phys. Rev. D21 (1980) 2291.

H. Hamber, E. Marinari, G. Parisi and C. Rebbi, Phys. Lett. 124B (1983) 199.

A. Hasenfratz and P. Hasenfratz, Phys. Lett. 93B (1980) 165.

P. Hasenfratz, in "Recent Developments in High-Energy Physics", Acta Phys. Austriaca, Suppl. XXV (1983) 283 (Eds. H. Mitter and C.B. Lang; Springer, 1983).

P. Hasenfratz, preprint CERN-TH 3999/84, Erice Lectures 1984 (1984).

Homer, Odysee (-750).

K. Jansen, J. Jersak, C.B. Lang, T. Neuhaus and G. Vones, Phys. Lett. B155 (1985) 268.

K. Jansen, J. Jersak, C.B. Lang, T. Neuhaus and G. Vones, to be published in Nucl. Phys. (1985).

J. Jersak, T. Neuhaus and P. Zerwas, Phys. Lett. 133B (1983) 103.

F. Karsch, E. SEiler and I. O. Stamatescu, Phys. Lett. 131B (1983) 138.

H. Kawai, R. Nakayama and K. Seo, Nucl. Phys. B189 (1981) 40.

J. Klauder, in "Recent Developments in High-Energy Physics", Acta Phys. Austriaca, Suppl. XXV (1983) 251 (Eds. H. Mitter and C.B. Lang; Springer, 1983).

J.B. Kogut and K. Wilson, Phys. Rep. 12 (1974) 75.

J.B. Kogut and L. Susskind, Phys. Rev. D11 (1975) 395.

J.B. Kogut, Rev. Mode. Phys. 51 (1979) 659.

J.B. Kogut, Rev. Mod. Phys. 55 (1983 775.

M. Kigugawa et al., Hiroshima Univ. preprint HUPD-8208 (1982).

H. Kühnelt, C.B. Lang and G. Vones, Nucl. Phys. B230 [FS10] (1984) 16.

J. Kuti, J. Polonyi and K. Szlachanyi, Phys. Lett. 98B (1981) 199.

C.B. Lang, C. Rebbi, P. Salomonson and B.S. Skagerstam, Phys. Lett. 101B (1981a) 173.

C.B. Lang, C. Rebbi and M. Virasoro, Phys. Lett. 104B (1981b) 294.

C.B. Lang and H. Nicolai, Nucl. Phys. B200 [FS4] (1982a) 135.

C.B. Lang, C. Rebbi, P. Salomonson and B.S. Skagerstam, Phys. Rev. D26 (1982b) 2028.

C.B. Lang and C. Rebbi, Phys. Lett. 115B (1982a) 137.

C.B. Lang and M. Wiltgen, Phys. Lett. 131B (1983) 153.

C.B. Lang, Phys. Lett. 155B (1985) 399.

B. Lautrup and M. Nauenberg, Phys. Lett. 95B (1980) 63.

N.S. Manton, Phys. Lett. 96B (1980) 328.

H. Markum et al., Phys. Rev. D31 (1985) 2029.

P.T. Matthews and A. Salam, Nuovo Cim. 12 (1954) 563.

L. McLerran and B. Svetitsky, Phys. Lett. 98B (1981) 195

P. Menotti and E. Onofri, Nucl. Phys. B190 [FS3] (1981) 288.

N. Metropolis, A.W. Rosenbluth, A.H. Teller and E. Teller, J. Chem. Phys. 21 (1953) 1087.

I. Montvay, Phys. Lett. 150B (1985a) 441.

I. Montvay, preprint DESY 85-005 (1985b).

T. Munehisa and Y. Munehisa, Phys. Lett. 116B ((1983a) 508.

Y. Munehisa, Phys. Rev. D30 (1984a) 1310.

Y. Munehisa, preprint Yamanashi-84-02 (1984b).

K.H. Mütter and K. Schilling, Nucl. Phys. B200 [FS4] (1982) 362.

H.B. Nielsen and M. Ninomiya, Nucl. Phys. B185 (1981a) 20.

H.B. Nielsen and M. Ninomiya, Nucl. Phys. B193 (1981b) 173.

K. Osterwalder and R. Schrader, Commun. Math. Phys. 42 (1975) 281.

K. Osterwalder and E. Seiler, Ann. Phys. (N.Y.) 110 (1978) 440.

G. Parisi and Wu Yang-shi, Sci. Sinica 24 (1981) 483.

H.D. Politzer, Phys. Rev. Lett. 30 (1973) 1346.

C. Rebbi, Lattice Gauge Theories and Monte Carlo Simulations (World Scientific Publishing CO.: Singapore 1983).

C. Rebbi, in "Stochastic Methods and Computer Techniques in Quantum Dynamics", Acta Phys. Austriaca, Supp. XXVI (1984) 309, (Eds. H. Mitter and L. Pittner; Springer, 1984).

H. Risken, The Fokker-Planck Equation (Springer, 1984).

H. Satz, Nucl. Phys. A418 (1984a) 447c.

H. Satz, Univ. Bielefeld prepring BI-TP 84-24 (1984b).

E. Seiler, Gauge Theories as a Problem of Constructive Quantum Field Theory and Statistical Mechanics, Lecture Notes in Physics 159 (Springer, 1982).

H.S. Sharatchandra, H.J. Thun and P. Weisz, Nucl. Phys. B192 (1981) 205.

A.D. Sokal, Ann. Inst. H. Poincaré 37 (1982) 317.

J.D. Stack, Phys. Rev. D29 (1984) 1213.

I.O. Stamatescu, preprint MPI-PAE/PTh 92/83 (1983).

L. Susskind, Phys. Rev. D20 (1979) 2610.

R.H. Swendsen, Phys. Rev. Lett. 52 (1984) 1165.

K. Symanzik, in "Mathematical Problems in Theoretical Physics", Eds. R. Schrader et al. (Lecture Notes in Physics 153, Springer, 1982).

K. Symanzik, Nucl. Phys. B226 (1983a) 187.

K. Symanzik, Nucl. Phys. B226 (1983b) 205.

F.J. Wegner, J. Math. Phys. 12 (1971) 2259.

P. Weisz, Phys. Lett. 100B (1981) 331.

P. Weisz, Nucl. Phys. B221 (1983) 1.

K. Wilson, Phys. Rev. D10 (1974) 2445.

K. Wilson, in "New Phenomena in Subnuclear Physics", Ed. A. Zichichi (Plenum Press, 1977).